Respiratory Physiology of Vertebrates

How do vertebrates get the oxygen they need, or even manage without it for shorter or longer periods of time? How do they sense oxygen, how do they take it up from water or air, and how do they transport it to their tissues? Respiratory system adaptations allow numerous vertebrates to thrive in extreme environments in which oxygen availability is limited, or where there is no oxygen at all.

Written for students and researchers in comparative physiology, this authoritative summary of vertebrate respiratory physiology begins by exploring the fundamentals of oxygen sensing, uptake and transport in a textbook style. Subsequently, the reader is shown important examples of extreme respiratory performance, such as diving and high-altitude survival in mammals and birds, air breathing in fish, and those few vertebrates that can survive without any oxygen at all for several months, showing how evolution has solved the problem of life without oxygen.

GÖRAN E. NILSSON is Professor of physiology at the Department of Molecular Biosciences, University of Oslo, Norway. He has worked in the field of comparative respiratory physiology and neurobiology for more than 20 years, and has contributed to over 150 scientific papers, books, and book chapters.

Respiratory Physiology of Vertebrates

Life with and without Oxygen

GÖRAN E. NILSSON

University of Oslo

CAMBRIDGE
UNIVERSITY PRESS

CAMBRIDGE
UNIVERSITY PRESS

University Printing House, Cambridge CB2 8BS, United Kingdom

One Liberty Plaza, 20th Floor, New York, NY 10006, USA

477 Williamstown Road, Port Melbourne, VIC 3207, Australia

314-321, 3rd Floor, Plot 3, Splendor Forum, Jasola District Centre, New Delhi - 110025, India

103 Penang Road, #05-06/07, Visioncrest Commercial, Singapore 238467

Cambridge University Press is part of the University of Cambridge.

It furthers the University's mission by disseminating knowledge in the pursuit of education, learning and research at the highest international levels of excellence.

www.cambridge.org
Information on this title: www.cambridge.org/9780521703024

© Cambridge University Press 2010

First published 2010

A catalogue record for this publication is available from the British Library

ISBN 978-0-521-87854-8 Hardback
ISBN 978-0-521-70302-4 Paperback

To Peter L. Lutz

Contents

Contributors

Arnoldus Schytte Blix
Department of Arctic Biology, University of Tromsø,
NO-9037 Tromsø, Norway

Lars P. Folkow
Department of Arctic Biology, University of Tromsø,
NO-9037 Tromsø, Norway

Kathleen M. Gilmour
Department of Biology, Centre for Advanced Research in Environmental
Genomics, University of Ottawa, 30 Marie Curie, Ottawa, Ontario
K1N 6N5, Canada

Jeffrey B. Graham
Scripps Institution of Oceanography, University of California San Diego,
La Jolla, CA 92093, USA

James W. Hicks
Department of Ecology and Evolutionary Biology, University of California
at Irvine, Irvine, CA 92697, USA

Susan R. Hopkins
Departments of Medicine and Radiology, University of California San
Diego, 9500 Gilman Dr., La Jolla, CA 92093, USA

Mikko Nikinmaa
Centre of Excellence in Evolutionary Genetics and Physiology, Department
of Biology, University of Turku, FI-20014 Turku, Finland

Göran E. Nilsson
Physiology Programme, Department of Molecular Biosciences, University
of Oslo, P.O. Box 1041, NO-0316 Oslo, Norway

Steve F. Perry
Department of Biology, Centre for Advanced Research in Environmental
Genomics, University of Ottawa, 30 Marie Curie, Ottawa, Ontario
K1N 6N5, Canada

Frank L. Powell
Department of Medicine and White Mountain Research Station, University
of California San Diego, 9500 Gilman Dr., La Jolla, CA 92093, USA

David J. Randall
Department of Zoology, University of British Columbia, Vancouver, British
Columbia V6T 1Z4, Canada

Nini Skovgaard
Department of Pharmacology, Aarhus University, Wilhem Meyers Allé
4, Building 1240, DK-8000 Aarhus C, Denmark

Tobias Wang
Zoophysiology, Department of Biological Sciences, Aarhus University, C. F.
Møllers Allé, Building 1131, DK-8000
Aarhus C, Denmark

Nicholas C. Wegner
Scripps Institution of Oceanography, University of California
San Diego, La Jolla, CA 92093, USA

Preface

For good reasons, many people have a fascination with the key role that oxygen plays in the life (and death) of animals and humans. That is the theme of this book: how vertebrates get the oxygen they need, and how some even manage without it for shorter or longer periods. We therefore hope it will find a relatively wide audience. Thus, the book aims to provide a thorough introduction to the respiratory physiology of vertebrates for anyone with some basic physiological knowledge, including biologists, biomedical researchers, veterinarians, and physicians. We also hope that the book will function as a textbook for courses at the MSc and PhD student level, and we have made an effort to start treating the subject at a level intelligible for bachelor students who have had their first introductory year in biology (including some physiology). By being extensively referenced, each chapter should also function as an up-to-date review for researchers who have decided to venture into a particular area of respiratory physiology.

The first four chapters cover basic aspects of vertebrate respiration, whereas the last five chapters describe particular physiological challenges met by many vertebrates and include many examples of more-or-less extreme respiratory adaptations.

The idea for this book was born in April 2006, when I was approached by Jacqueline Garget from Cambridge University Press in connection with the Society of Experimental Biology meeting in Canterbury. At that meeting, I was organizing a session on 'Life with and without oxygen' to honor the memory of my friend Peter L. Lutz, who left us much too early, in February 2005. After some discussion, Jacqueline and I agreed that I should try to put together a comprehensive book on the subject of vertebrate respiratory physiology, rather than producing a volume of talks given by Peter's friends at this session. I knew that two journals (*The Journal of Experimental Biology* and *Comparative Biochemistry and Physiology*) were engaged in producing special issues in Peter's honor, and a book based on the session in Canterbury would inevitably be a somewhat arbitrary collection of quite specialized papers. While fearing being naïve, I aimed high and approached a number of outstanding researchers who collectively should be able to cover virtually all important aspects of vertebrate respiratory physiology. To my surprise, they all accepted the task. Indeed, they did so with enthusiasm. I am very grateful to all of them. The result is this book.

Abbreviations

ABO	air-breathing organ
ACR	air convection requirement
ADH	alcohol dehydrogenase
ADP	adenosine diphosphate
AhR	aryl hydrocarbon receptor
ALDH	aldehyde dehydrogenase
AMP	adenosine monophosphate
AMPK	AMP-activated protein kinase
AMS	acute mountain sickness
ARNT	aryl hydrocarbon receptor nuclear translocator
ASR	aquatic surface respiration
ATP	adenosine triphosphate
β_{blood}	blood capacitance coefficient
β_{gas}	air capacitance coefficient
BGB	blood–gas barrier
βO_2	O_2 capacitance coefficient
BOD	biological oxygen demand
$[Ca^{2+}]_i$	intracellular Ca^{2+} concentration
CAT	catalase
CO	carbon monoxide
CO_2	carbon dioxide
$D_L O_2$	lung diffusion capacity for oxygen
DPG	2,3-diphosphoglycerate
$D/Q\beta O_2$	equilibration coefficient
D_S	skin diffusion capacity
$D_t O_2$	tissue diffusion capacity for oxygen
f_H	heart rate

F_IO_2	fraction of oxygen in inspired air
f_R	frequency of ventilation
G	conductance
GABA	γ-amino butyric acid
$G_{diff}O_2$	transfer factor (or diffusion conductance) for O_2
GPX	glutathione peroxidase
GST	glutathione-S-transferase
HACE	high-altitude cerebral edema
HAPE	high-altitude pulmonary edema
Hb	hemoglobin
HCO_3^-	bicarbonate (hydrogencarbonate) ion
HIF	hypoxia-inducible factor
HPV	hypoxic pulmonary vasoconstriction
HRE	hypoxia response element
H_2O_2	hydrogen peroxide
H_2S	hydrogen sulfide
KO_2	Krogh's diffusion coefficient
LDH	lactate dehydrogenase
MIGET	multiple inert gas elimination technique
MSO	methionine sulfoximine
NMDA	N-methyl-D-aspartate
NMDAR	NMDA receptor
NO	nitric oxide
O_2	oxygen
O_2^-	superoxide anion
$[O_2]_a$	arterial oxygen concentration (often CaO_2)
$[O_2]_{c'}$	end capillary oxygen concentration (often $Cc'O_2$)
$[O_2]$crit	critical oxygen concentration
ODC	oxygen dissociation curve
OH•	hydroxyl radical
$[O_2]_{pv}$	oxygen concentration of pulmonary venous blood (often $C_{pv}O_2$)
$[O_2]_{sv}$	oxygen concentration of systemic venous blood (often $C_{sv}O_2$)
$[O_2]_v$	venous oxygen concentration (often C_vO_2)
P_{50}	the PO_2 at which hemoglobin is 50% saturated with O_2
P_aCO_2	partial pressure of carbon dioxide in the arteries
P_AO_2	partial pressure of oxygen in the alveoli
P_aO_2	partial pressure of oxygen in the arteries

P_A-P_a	alveolar-to-arterial PO_2 difference
PASMC	pulmonary arterial smooth muscle cell
P_B	barometric pressure
PCO_2	partial pressure of carbon dioxide
PCr	phosphocreatine
PDH	pyruvate dehydrogenase
P_EO_2	partial pressure of oxygen in exhaled air
Perf. CR	convection requirement from blood
PH_2O	partial pressure for water vapor
P_IO_2	partial pressure of oxygen in inspired air
P_LO_2	partial pressure of oxygen in the lung
P_L-P_a	PO_2 difference between lung and arteries
P_L-P_{LAt}	PO_2 difference between mixed lung gas and gas in the left atrium
$P_{mito}O_2$	mitochondrial PO_2
PO_2	partial pressure of oxygen
ΔPO_2	partial pressure difference for oxygen
PO_2crit	critical oxygen tension
P_{pv}	partial pressure in mixed pulmonary venous blood
P_vO_2	partial pressure of oxygen in the veins
$\dot{Q}_{pul}$	pulmonary blood flow
$\dot{Q}R-L$	R–L shunt flow
$\dot{Q}$	blood flow (cardiac output)
Q_{10}	metabolic rate increases for every 10°C rise in factor by which temperature
ROS	reactive oxygen species
S_aO_2	O_2 saturation in the arteries
SOD	superoxide dismutase systemic O_2 delivery
TUNEL	terminal transferase mediated dUTP nick-end labeling
UCP	uncoupling protein
$\dot{V}_A$	alveolar ventilation
$\dot{V}_b$	blood flow (often written as Q)
$\dot{V}_{CO_2}$	rate of CO_2 ventilation
$\dot{V}_D$	anatomical respiratory dead space
$\dot{V}_E$	minute ventilation
$\dot{V}_{eff}$	effective ventilation of the gas-exchange structures
Vent. CR	convection requirement from water
$\dot{V}_I$	total ventilation
$\dot{V}O_2$	rate of oxygen uptake

$\dot{V}O_{2\,max}$	maximal rate of oxygen uptake
$\dot{V}/\dot{Q}$ ratio	ventilation/perfusion ratio
V_S	stroke volume
V_T	tidal volume
Vt_b	cardiac output (often written as Q)
V_TCO_2	volume of carbon dioxide per breath
Vt_w	ventilation volume
$Vt_w/\dot{V}O_2$	volume of water flow required per unit of O_2 uptake
$\dot{V}_w$	water flow

PART I GENERAL PRINCIPLES

1

Introduction: why we need oxygen

GÖRAN E. NILSSON

The aim of this book is not only to describe the basic functions of the respiratory systems of vertebrates, and the diversity in these functions among vertebrates, but also to examine adaptations in these systems that allow numerous vertebrates to explore more or less extreme environments in which oxygen availability is limited or in which there is no oxygen at all.

For the organism to be able to respond to variable oxygen levels, it needs to be able to sense oxygen. This can be done either directly, by monitoring the level of O_2, or indirectly, by responding to changes in the energy status of tissues or cells. Even if some oxygen-sensing structures and their functions have been examined relatively thoroughly, such as the oxygen-sensing carotid bodies in mammals, it is clear that many mechanisms related to oxygen sensing are still largely unknown, particularly when it comes to the almost mysterious ability of many (perhaps most) cells to detect and respond to changing oxygen levels. Chapter 2 will describe the present state of knowledge in this very active field of research. In Chapters 3–4, we will examine the fundamental functions of the respiratory systems of air-breathing and water-breathing vertebrates, laying out the framework for the final five chapters, which deal with adaptations to particularly challenging situations for vertebrates: life at high altitude, diving, surviving in hypoxic waters, and surviving without any oxygen at all.

Oxygen is often called the molecule of life, and we almost intuitively realize the danger of being exposed to low levels of oxygen (hypoxia) or, even worse, to an environment with no oxygen at all (anoxia). We know that it is life threatening for us to have our air supply restricted (asphyxia), with the result that our blood oxygen level falls (hypoxemia), or to have a block in the blood flow to a tissue (ischemia). But why is it that hypoxia, anoxia, asphyxia, hypoxemia and ischemia are so detrimental? This is one of the most intensively studied

Respiratory Physiology of Vertebrates: Life with and without Oxygen, ed. Göran E. Nilsson. Published by Cambridge University Press. © Cambridge University Press 2010.

Table 1.1 *O₂ tension and content in air-saturated fresh water and sea water at 1 atm pressure*

Temperature		PO₂	Fresh water			Sea water (35 ppt salt)		
°C	°F	mmHg	mg/l	ml/l	mmol/l	mg/l	ml/l	mmol/l
0	32	158	14.6	10.2	0.457	11.2	7.8	0.349
5	41	158	12.8	9.1	0.399	9.9	7.0	0.308
10	50	157	11.3	8.2	0.353	8.8	6.4	0.275
15	59	156	10.1	7.5	0.315	7.9	5.9	0.248
20	68	156	9.1	6.8	0.284	7.2	5.4	0.225
25	77	154	8.3	6.3	0.258	6.6	5.0	0.206
30	86	153	7.6	5.9	0.236	6.1	4.7	0.190
35	95	151	7.0	5.5	0.218	5.6	4.5	0.176
40	104	148	6.5	5.2	0.202	5.3	4.2	0.165

Values are for 100% air saturation, and values at a lower percentage of air saturation are simply obtained by multiplying the partial pressure of O_2 (PO_2) or O_2 concentration given in the table by the percentage of air saturation (divided by 100).

questions in biomedical science. One main reason for this is that anoxia-related-diseases such as stroke and heart infarction are major killers of people in the developed world. In addition, hypoxia has much to do with the life and death of cancer cells and the complications caused by diabetes. Indeed, it is unlikely for anyone to die in a way that does not, at least finally, involve cellular anoxia.

　　Biomedical science has so far had limited success in counteracting the various detrimental effects of hypoxia-related diseases, and fresh views on these problems could be inspired by the diversity found in respiratory adaptations in vertebrates, and in the solutions that evolution has provided for survival with little or no oxygen. Indeed, hypoxia is a very common phenomenon in nature. As we shall see, it often occurs in aquatic environments (the subject of Chapter 5) and is always present at high altitude (the subject of Chapter 8). The partial pressure of oxygen at an altitude of 6000 m is less than half of that at sea level, and at the peak of Mount Everest, over which birds do fly, it has fallen to one-third. Hypoxia is common in water, because this medium holds much less oxygen than air does and is often much more stagnant than air, so oxygen can be used up readily. Even when air-equilibrated (air-saturated), one liter of water maximally holds 10.2 ml of molecular oxygen (compared with 210 ml of oxygen in 1 liter of air). Moreover, the maximal water oxygen content falls with increasing temperature and salt content (Table 1.1). For fishes, this problem comes in addition to the challenge of having to breathe in a medium that has a 50 times higher viscosity and an 800 times higher density than air, and through

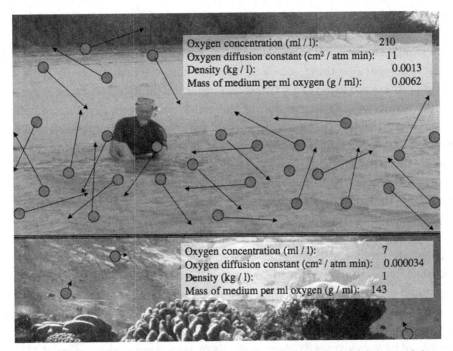

Oxygen concentration (ml / l):	210
Oxygen diffusion constant (cm² / atm min):	11
Density (kg / l):	0.0013
Mass of medium per ml oxygen (g / ml):	0.0062

Oxygen concentration (ml / l):	7
Oxygen diffusion constant (cm² / atm min):	0.000034
Density (kg / l):	1
Mass of medium per ml oxygen (g / ml):	143

Fig. 1.1 The large differences in the physicochemical properties of water and air mean that equally different demands are put on the respiratory organs of water breathers (primarily gills) and air breathers (primarily lungs). In some respects, breathing water is much more of a challenge than breathing air. The numbers show that the oxygen content of air is about 30 times higher than that of air-saturated water and that oxygen diffuses 300 000 times faster in air than in water. Moreover, the low oxygen content and the high density of water mean that a water breather will have to move about 20 000 times more mass over its respiratory surface than an air breather to get access to the same amount of oxygen. In addition, because water contains a relatively small amount of oxygen that moves very slowly, particularly in stagnant water, hypoxic conditions can readily occur in aquatic environments, presenting an additional challenge for water breathers. However, water loss by evaporation over the respiratory surface, which is a problem for air breathers, is not an issue for water breathers.

which oxygen moves through diffusion some 300 000 times more slowly than in air (Fig. 1.1). The differences in oxygen availability in water and air are further discussed at the beginning of Chapter 6.

1.1   Oxygen and cellular energy

The immediate danger of hypoxia lies in the fact that oxygen is intimately coupled to the generation of adenosine triphosphate (ATP), which drives virtually all energy-demanding processes in cells. The generation of ATP through oxidative phosphorylation in the mitochondrial respiratory chain requires molecular oxygen. Since so many key cellular functions need a constant supply of ATP, a fall in ATP levels is for most vertebrates immediately life threatening. For one thing, a lack of ATP will stop the activity of the Na^+/K^+ pump (also known as Na^+/K^+ ATPase) and other ion-pumping proteins, which rapidly results in a depolarization of the cell membrane. A depolarized cell without ATP has lost the means to control its volume, ion homeostasis, and intracellular environment, a situation that is highly detrimental and soon renders the cell necrotic.

Another problem with hypoxia is that a halt in oxidative phosphorylation means that the respiratory chain will stop pumping H^+ out of the mitochondria. Thus, not only the cells but also their mitochondria may lose the membrane potential and become depolarized. This phenomenon has been recognized as particularly problematic relatively recently, as it leads to the activation of apoptosis ('programmed cell death' or 'cell suicide'). Thus, even if oxygen supply is restored before the cells have become necrotic, they may already be doomed to die through apoptosis within hours or days (Kakkar and Singh, 2007). One way to protect the mitochondria from depolarization, which has been found to be utilized by anoxic frogs (St-Pierre et al., 2000), is to reverse the function of ATP synthase. This protein, which normally harvests the mitochondrial H^+ gradient to produce ATP, can run backwards and hydrolyze ATP while pumping H^+ out of the mitochondria. Unfortunately, this is an energy-consuming and non-sustainable mechanism that will solve the problem only temporarily and has therefore been called 'cellular treason in anoxia' (St-Pierre et al., 2000).

Apart from energy metabolism, there are several other systems in the body that demand oxygen. These include detoxification enzymes, DNA synthesis, and some steps in the synthesis and catabolism of neurotransmitters. However, the arrest of these systems during anoxia is unlikely to be immediately life threatening, and is mainly of academic importance to an animal that is unable to maintain its ATP levels.

1.2   The brain: the first organ to suffer

The brain is particularly sensitive to a reduced oxygen supply. A major reason for this is the brain's high mass-specific rate of energy use. This is primarily related to its electrical activity, which demands a high rate of ion pumping. The Na^+/K^+ pump alone may be responsible for consuming at least half of the ATP used by the brain (Hansen, 1985), and the rate of ATP turnover in the brain is about ten times faster than that of the average body tissue (Mink et al., 1981; Nilsson, 1996).

After oxygen stores have been depleted, the brain of most animals will rapidly suffer from falling ATP levels. In mammals, the oxygen present in the brain will last only a few seconds after blood flow to the brain has stopped (Hansen, 1985). The concentration of ATP in the mammalian brain is around 3 mmol kg^{-1} (Erecinska and Silver, 1994), and the brain ATP pool is turned over about once every 5–10 s (Lutz et al., 2003). Even with the additional ATP that can be generated from the 3–5 mmol kg^{-1} of phosphocreatine (PCr) that is present in the brain, ATP levels become halved in about a minute and virtually depleted within 2 min in the mammalian brain (Erecinska and Silver, 1994). The situation is not much better for cold-blooded vertebrates such as fishes, in which brain ATP levels are generally below 2 mmol kg^{-1} (DiAngelo and Heath, 1987; Van Raaij et al., 1994; DeBoeck et al., 1995; Van Ginneken et al., 1996; Ishibashi et al., 2002), and estimated rates of ATP synthesis vary between 1.3 and 5.0 mmol kg^{-1} min^{-1} at 12–26°C (Johansson et al., 1995; Nilsson, 1996). This means that the ATP pool in a fish brain is turned over about once every minute.

Thus, the brain is likely to be the first organ to lose energy charge and depolarize when an animal is exposed to severe hypoxia or anoxia. This has two consequences. First, necrotic and apoptotic processes will be initiated rapidly in the brain of an animal that has lost its oxygen supply. Secondly, the depolarized brain can no longer regulate its volume, and the brain cells will start to swell. For many vertebrates, this is a particular problem because there is simply no space in the cranial cavity to allow the brain to swell. Therefore, instead of an increase in brain volume, there will be an increase in pressure in the cranial cavity, and when this pressure rises above the blood pressure, it is no longer possible for blood to reach the brain. Even with the best health care, this is usually an irreversible situation, and consequently a lack of blood circulation in the brain is a principal legal sign of death in many countries. When it comes to brain swelling, fishes and many other cold-blooded vertebrates may be better off than mammals and birds, because they often have a brain cavity that is considerably larger than the brain, thereby allowing the brain to swell without stopping cerebral circulation: the brain of anoxia-exposed common carp (*Cyprinus carpio*) has been observed to increase in volume by 10% without impairing their subsequent recovery (Van der Linden et al., 2001).

Anoxic animals in nature have no access to emergency aid and resuscitation. For them, an energetically compromised brain will become a deadly problem before its neurons have become irreversibly damaged. This is because the brain is responsible for initiating the breathing movements necessary for moving water or air over the respiratory surfaces. In nature, having an energy-deficient brain that has stopped sending signals to the respiratory organs is a point of no return, even if ambient oxygen levels are restored.

The anoxic brain catastrophe

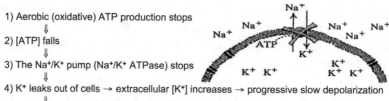

1) Aerobic (oxidative) ATP production stops
⇓
2) [ATP] falls
⇓
3) The Na⁺/K⁺ pump (Na⁺/K⁺ ATPase) stops
⇓
4) K⁺ leaks out of cells → extracellular [K⁺] increases → progressive slow depolarization
⇓
5) Rapid general depolarization of the brain caused by a massive outflow of K⁺ and inflow of Na⁺ and Ca²⁺, at least partly through voltage-sensitive channels. Lost ion gradients → reversal of transporters → release of neurotransmitters, including glutamate which activates receptor gated ion channels → outflow of K⁺ and inflow of Na⁺ and Ca²⁺. (These events occur rapidly and are likely to be mutually reinforcing.)
⇓
6) Cell swelling and lysis → increased cerebral pressure → permanent global ischemia

7) Induction of necrosis and apoptosis by various mechanisms, including:
 • Ca²⁺-activated lytic processes that breaks down proteins, lipids and DNA
 • Mitochondrial depolarization → increased mitochondrial permeability → release of apoptotic factors from the mitochondria, like cytochrome C

Fig. 1.2 Main disastrous events occurring in a mammalian brain exposed to anoxia.

In biomedical science, many efforts are being put into clarifying the details of the catastrophic events that affect an anoxic brain, with the ultimate aim of finding ways of interfering with the underlying mechanisms so that the detrimental effects of conditions such as stroke and circulatory arrest can be reduced. Thus, we have a relatively detailed knowledge about the catastrophe that occurs in the mammalian brain during anoxia (Fig. 1.2). In humans, unconsciousness occurs and electric activity is suppressed in the brain just 6–7 s after blood flow to the brain is halted (Rossen et al., 1943). This is likely to be an initial emergency response, possibly functioning to save energy by reducing ATP-consuming electrical activity. Moreover, as a standing person faints and falls, the brain moves into a lower position in relation to the heart, which increases cerebral blood pressure and thereby cerebral blood flow.

If blood or oxygen supply to the mammalian brain is not restored within seconds, progressive changes in energy status and ion homeostasis are soon apparent (Hansen, 1985; Erecinska and Silver, 1994). Virtually immediately, there is a steady, slow rise in extracellular [K⁺], probably caused by both increased K⁺ permeability of the cells and a slowdown of the Na⁺/K⁺ pump. Within a minute, the level of ATP falls to half the normoxic level, while concentration of adenosine diphosphate (ADP) is tripled and that of adenosine monophosphate (AMP) is increased by an order of magnitude. At the same time the PCr store is virtually depleted.

In rodents, the whole brain depolarizes after 1–2 min of ischemia. This is an event that is characterized by a massive outflow of K⁺, and influxes of Na⁺ and

Ca^{2+}. In larger mammals, this anoxic depolarization may take a few minutes longer due to their lower mass-specific metabolic rate. In cold-blooded vertebrates, the anoxic depolarization is further delayed. Thus, in rainbow trout (*Oncorhynchus mykiss*), the brain depolarizes after about 15–30 min of anoxia at 10–15°C (Nilsson *et al*, 1993; Hylland *et al.*, 1995). Still, the mechanisms behind the anoxic depolarization are probably similar for most vertebrates, and time differences can be fully explained by differences in metabolic rate.

Following the anoxic depolarization, there is a massive outflow of neurotransmitters from the intracellular to the extracellular compartment, where these neurotransmitters can activate their receptors. Contrary to expectations, Ca^{2+}-mediated vesicular release of neurotransmitters probably plays a minor (if any) role in this event, because vesicular transmitter release is ATP dependent, and as already discussed, very little ATP is available in this situation. Instead, the release of the neurotransmitters appears to be primarily caused by a reversal of neurotransmitter transporters. These normally harvest the ion gradients over the cell membranes to take up neurotransmitters and keep their extracellular levels low (Danbolt, 2001). However, as the ion gradients collapse, the transporters start to run backwards and release transmitters. At this point, the worst problem for the brain appears to be the release of glutamate, the major excitatory neurotransmitter in the vertebrate brain. In this uncontrolled situation, glutamate functions as an excitotoxin. In particular, the activation of two major glutamate receptor types, called NMDA and AMPA, are thought to play key roles in excitotoxic cell death in the brain. These receptors let large amounts of Ca^{2+} and Na^+ into the neurons. The resultant massive rise in intracellular $[Ca^{2+}]$ wreak havoc on the cells by, for example, activating proteolytic and lipolytic processes, as well as DNA-degrading mechanisms (see Lipton, 1999; Lutz *et al.*, 2003 for reviews). The result is either immediate necrotic cell death, which in the case of brain ischemia (stroke) will affect cells in areas completely devoid of blood flow, or slow cell death through autophagocytotic and apoptotic mechanisms. The latter two occur hours to days after blood flow has been restored and in an ischemic brain affect many cells in the so-called ischemic penumbra (the zone of suppressed blood flow surrounding the central ischemic area). There appear to be several mechanisms involved in anoxia- or ischemia-induced apoptotic cell death. One of these was recently termed 'parthanatos,' and it appears to be particularly common in ischemic/post-ischemic brain tissue. The name is derived from the death signal in this pathway, poly(ADP-ribose) ('Par') polymer, and Thanatos, the Greek personification of death and mortality. Parthanatos is biochemically and morphologically distinct from the normal (caspase-dependent) apoptosis (Harraz *et al.*, 2008).

1.3   Boosting oxygen uptake: the first option

In this book, we will not only describe how oxygen is normally handled by the organism, but also how the organism can regulate its oxygen uptake and defend itself against hypoxic conditions. The first option for an animal that experiences a fall in ambient oxygen levels is to boost the extraction of oxygen from the environment. This is primarily done by increasing the ventilation of the lungs or gills and by increasing the blood perfusion through these respiratory organs. As we shall see in Chapter 6, some animals even switch from breathing with gills to breathing with lungs. Through such adjustments, most vertebrates become what is known as 'oxygen regulators' (Prosser and Brown, 1961), i.e. they are able to regulate their oxygen extraction capacity so that oxygen uptake ($\dot{V}O_2$) is maintained at a steady level over a more-or-less wide range of ambient oxygen concentrations. There is a great species-to-species variability in how well vertebrates can do this, a variability that has its origin in how evolutionary processes have shaped the organism in response to its environment or lifestyle. For example, animals that are adapted to hypoxic habitats can typically maintain their $\dot{V}O_2$ at much lower water oxygen levels than can species that are unlikely to encounter hypoxia. The lowest level at which an animal can maintain its $\dot{V}O_2$ is denoted the critical oxygen concentration ($[O_2]$crit), or critical oxygen tension (PO_2crit) if the oxygen level is recorded as the partial pressure of oxygen (Prosser and Brown, 1961). The PO_2crit is a common measure of hypoxia tolerance in fishes (mechanisms of hypoxia tolerance in fishes are the subject of Chapter 5).

1.4   Oxygen-independent ways of making ATP

If ambient oxygen levels fall below PO_2crit, the animal has to start making ATP anaerobically. PCr can rapidly regenerate ATP from ADP, but as PCr levels are usually quite limited, ranging from about 0.5 to 5.0 mmol kg^{-1} in the brain of vertebrates (e.g. DiAngelo and Heath, 1987; Erecinska and Silver, 1994; Van Raaij et al., 1994; Van Ginneken et al., 1996), this pathway can maintain ATP levels only for one or a few minutes. To be able to maintain ATP levels longer in anoxia, anaerobic glycolysis is the only viable option. Sources of fuel other than glucose, i.e. fat and protein, are virtually useless in the absence of oxygen, because these demand a functional citric acid cycle. Without oxygen, the intimate connection between the citric acid cycle and the respiratory chain rapidly makes the citric acid cycle come to a halt (Hochachka and Somero, 2002).

Unfortunately, for most vertebrates, the anaerobic capacity of the brain is not high enough to allow it to compensate for more than a fraction of its aerobic rate of ATP production. The reason for this is that most of the chemical energy stored in glucose is left in the glycolytic end product, which in vertebrates is normally lactate. Thus, from every molecule of glucose, only two molecules of ATP are generated (an additional ATP is produced in the breakdown of glycogen to glucose). In the presence of oxygen, the complete breakdown of glucose to CO_2 and H_2O can yield up to 36 molecules of ATP, although for various reasons (including partly uncoupled mitochondria), a yield of 29 molecules of ATP is probably a more realistic figure (Brand, 2003). Thus, aerobic metabolism is able to produce about 15 times more ATP per molecule of glucose than can anaerobic glycolysis. Another serious problem with generating ATP anaerobically is that it normally leads to the production of lactate and equimolar amounts of H^+. The H^+ is actually formed during the hydrolysis of ATP rather than through glyco-lysis, but the net effect is that lactic acid is produced (see Hochachka and Somero, 2002 for review). The hydrogen ions can cause life-threatening acidosis and the lactate causes osmotic disturbances. Nevertheless, for many verte-brates, generating ATP through glycolysis during hypoxic or anoxic conditions prolongs the survival time considerably, and in some cases even allows anoxic survival for days or months (see Chapter 9).

It may be that the initial cause of anoxic death in vertebrates varies to some degree between species of vertebrate groups. Although it is clear that the rapid and severe drop in the brain ATP level initiates the anoxic catastrophe in mammals, some fishes may die from lactic acidosis rather than an inability to produce enough ATP. A study of anoxic rainbow trout and brown bullhead (*Ameiurus nebulosus*) found that ATP levels were relatively well maintained at the time when they ceased to breathe, but lactic acid levels had risen to 12–20 mmol kg^{-1}, which may have been too high for the brain to tolerate (DiAngelo and Heath, 1987; Van Raaij *et al.*, 1994). By contrast, in severely hypoxic common carp and Nile tilapia (*Oreochromis niloticus*), considerable falls in brain ATP levels have been detected (Van Raaij *et al.*, 1994; Ishibashi *et al.*, 2002), although the possibility remains that the falling ATP levels seen in some fish brains is caused by metabolic dysfunction caused by lactic acidosis or a rundown of the glycogen stores. Nevertheless, there are also striking simila-rities in the anoxic death process between mammals and fish. Measurements of extracellular K^+ and glutamate levels in the brain of anoxic rainbow trout reveal an anoxic depolarization and an outflow of K^+ and glutamate from the cells that is very similar to that observed in mammals (Nilsson *et al*, 1993; Hylland *et al.*, 1995).

In addition to boosting oxygen uptake and activating glycolytic ATP produc-tion in hypoxia or anoxia, some animals have evolved a third survival strategy: metabolic depression. Thus, during oxygen shortage they are able to lower their use of ATP so that consumption can be matched by production and ATP levels can be maintained. Moreover, there are vertebrates that totally avoid the problem of lactic acidosis by producing an alternative anaerobic end product. We will talk more about such exotic mechanisms in the final chapter of this book.

References

Brand, M. (2003). Approximate yield of ATP from glucose, designed by Donald Nicholson. *Biochem. Mol. Biol. Educ.*, **31**, 2–4.

Danbolt, N. C. (2001). Glutamate uptake. *Prog. Neurobiol.*, **65**, 1–105.

DeBoeck, G., Nilsson, G. E., Elofsson, U., Vlaeminck, A. and Blust, R. (1995). Brain monoamine levels and energy status in common carp after exposure to sublethal levels of copper. *Aquatic Toxicol.*, **33**, 265–77.

DiAngelo, C. R. and Heath, A. G. (1987). Comparison of in vivo energy metabolism in the brain of rainbow trout, Salmo gairdneri, and bullhead catfish, Ictalurus nebulosus, during anoxia. *Comp. Bioch. Physiol.*, **88B**, 297–303.

Erecinska, M. and Silver, I. A. (1994). Ions and energy in mammalian brain. *Prog. Neurobiol.*, **43**, 37–71.

Hansen, A. J. (1985). Effect of anoxia on ion distribution in the brain. *Physiol. Rev.*, **65**, 101–48.

Harraz, M. M., Dawson, T. M. and Dawson, V. L. (2008). Advances in neuronal death 2007. *Stroke*, **39**, 286–88.

Hochachka, P. W. and Somero, G. N. (2002). *Biochemical Adaptation.* New York: Oxford University Press.

Hylland, P., Nilsson, G. E. and Johansson, D. (1995). Anoxic brain failure in an ectothermic vertebrate: release of amino acids and K+ in rainbow trout thalamus. *Am. J. Physiol.*, **269**, R1077–84.

Ishibashi, Y., Ekawa, H., Hirata, H. and Kumai, H. (2002). Stress response and energy metabolism in various tissues of Nile tilapia Oreochromis niloticus exposed to hypoxic conditions. *Fish. Sci.*, **68**, 1374–83.

Johansson, D., Nilsson, G. E. and Törnblom, E. (1995). Effects of anoxia on energy metabolism in crucian carp brain slices studied with microcalorimetry. *J. Exp. Biol.*, **198**, 853–9.

Kakkar, P. and Singh, B. K. (2007). Mitochondria: a hub of redox activities and cellular distress control. *Mol. Cell. Biochem.*, **305**, 235–53.

Lipton, P. (1999). Ischemic cell death in brain neurons. *Physiol. Rev.*, **79**, 1431–568.

Lutz, P. L., Nilsson, G. E. and Prentice, H. (2003). *The Brain Without Oxygen*, 3rd edn. Dordrecht: Kluwer Academic Publishers/Springer.

Mink J. W., Blumenschine R. J. and Adams D. B. (1981). Ratio of central nervous system to body metabolism in vertebrates: its constancy and functional basis. *Am. J. Physiol.*, **241**, R203–12.

Nilsson, G. E. (1996). Brain and body oxygen requirements of Gnathonemus petersii, a fish with an exceptionally large brain. *J. Exp. Biol.*, **199**, 603–7.

Nilsson, G. E., Pérez-Pinzón, M., Dimberg, K. and Winberg, S. (1993). Brain sensitivity to anoxia in fish as reflected by changes in extracellular potassium-ion activity. *Am. J. Physiol.*, **264**, R250–3.

Prosser, C. L. and Brown, F. A. (1961). *Comparative Animal Physiology*. Philadelphia: W. B. Saunders.

Rossen, R., Kabat, H. and Andersson, J. P. (1943). Acute arrest of cerebral circulation in man. *Arch. Neurol. Psychiatry*, **50**, 510–28.

St-Pierre, J., Brand, M. D. and Boutilier, R. G. (2000). Mitochondria as ATP consumers: Cellular treason in anoxia. *Proc. Natl Acad. Sci. USA*, **97**, 8670–4.

Van der Linden, A. M. Verhoye, and Nilsson, G. E. (2001). Does anoxia induce cell swelling in carp brains? Dynamic in vivo MRI measurements in crucian carp and common carp. *J. Neurophysiol.*, **85**, 125–33.

Van Ginneken, V., Nieveen, M., VanEersel, R., Van denThillart, G. and Addink, A. (1996). Neurotransmitter levels and energy status in brain of fish species with and without the survival strategy of metabolic depression. *Comp. Biochem. Physiol.*, **114A**, 189–96.

Van Raaij, M. T. M., Bakker, E., Nieveen, M. C., Zirkzee H. and Van den Thillart, G. E. E. J. M. (1994). Energy status and free fatty acid patterns in tissues of common carp (Cyprinus carpio L.) and rainbow trout (Oncorhynchus mykiss L.) during severe oxygen restriction. *Comp. Biochem. Physiol.*, **109A**, 755–67.

2

Sensing oxygen

MIKKO NIKINMAA

2.1   Introduction

Oxygen is required as the ultimate electron acceptor in aerobic energy production. In the long run, all vertebrates need oxygen to support metabolism. In the short term, however, some animals can cope with a total lack of oxygen (anoxia), and others can tolerate reduced oxygen levels (hypoxia). Furthermore, eutrophic aquatic systems in particular are characterized by supra-atmospheric oxygen tensions (hyperoxia) during active photosynthesis of green plants. Hyperoxic conditions may also occur in the closed system of circulation, especially near the gas gland and avascular retina of fishes (Ingermann and Terwilliger, 1982; Pelster and Scheid, 1992).

With regard to oxygen requirements, there is an intricate balance between reactions that produce energy and those that consume it. It is generally agreed that energy (and oxygen) consumption is reduced when adapting to conditions of low oxygen (e.g. channel arrest) (Hochachka and Lutz, 2001). However, even in conditions in which oxygen is not limiting, adjustments of metabolic rate occur (Rissanen et al., 2006a). Because several phenomena, at both integrative and molecular levels, have turned out to be oxygen sensitive, the search for mechanisms by which oxygen is sensed has intensified in recent years.

Several questions relate to how oxygen is sensed and how oxygen-dependent responses occur. First, what is actually sensed, when apparently oxygen-dependent phenomena occur? Secondly, which molecules are utilized in sensing oxygen? Thirdly, what are the pathways used in oxygen sensing – i.e. how is the primary signal converted to be used by the effector systems in an oxygen-dependent manner? Fourthly, are the mechanisms utilized to sense oxygen the same in rapid responses, such as immediate changes in transporter activity, and

Respiratory Physiology of Vertebrates: Life with and without Oxygen, ed. Göran E. Nilsson. Published by Cambridge University Press. © Cambridge University Press 2010.

in more sustained responses involving, for example, changes in gene expression? Fifthly, how do effector systems function in different cell types and different (groups of) animals?

Some important points need to be considered when oxygen sensing is evaluated. First, oxygen sensing has mainly been studied from a biomedical viewpoint. As a consequence, the work has largely used hypoxia-intolerant mammals, such as human, rat, and mouse, as study objects. Whenever animals from other groups have been used, their biology has often not been taken into account. For example, studies on oxygen-dependent phenomena in zebrafish seldom consider that the species is a tropical cyprinid with relatively good hypoxia tolerance (Nikinmaa and Rees, 2005; Engeszer et al., 2007). Similarly, an elegant recent study showing an interaction between hypoxia-inducible-factor and heat-shock-factor function was carried out with fruit flies (Drosophila), but the study did not consider the possibility that the poikilothermic nature of the studied animal (external factors determine the body temperature of poikilothermic animals) would play a role in determining the nature of the response (Baird et al., 2006). This possibility is a distinct one, because a study with a poikilothermic animal, the nematode Caenorhabditis elegans, indicated that hypoxia-inducible factor was needed for temperature acclimation (Treinin et al., 2003), and because association between heat-shock proteins and hypoxia-inducible factor occurs during acclimation to reduced temperature in another poikilothermic animal, the teleost crucian carp (Carassius carassius) (Rissanen et al., 2006b). Secondly, although single-celled organisms such as bacteria and yeasts appear to contain a single oxygen-sensitive system regulating the expression of oxygen-sensitive genes (Bunn and Poyton, 1996), there appears to be no universal oxygen sensor in vertebrates (Lopez-Barneo et al., 2001). Thirdly, although hypoxia has been the most common stressor in oxygen-sensor studies, the physiologically effective degree of hypoxia has usually not been characterized in detail. It is usually assumed that the oxygen tension of the surrounding environment is also experienced by the cells, although studies measuring the oxygen tension of cultured cells indicate that the oxygen level experienced by the cells may deviate markedly from that in the bulk atmosphere (Pettersen et al., 2005). Furthermore, as the oxygen tension experienced by any cell in an organism depends on its location, especially its distance from an arterial blood vessel, and on its oxygen consumption, there are marked differences in cellular oxygen tension among different cell types within the body. This fact is often not taken into account when relating in vitro findings to the responses of the cells in vivo. For example, although the production of nitric oxide (NO) in macrophages responds differently to a change in oxygen tension at physiologically realistic oxygen tensions (often less than 40 mmHg in tissues) and at oxygen tensions commonly used in cell culture (often close to

Table 2.1 *A list of molecules that may be sensed*
when oxygen-dependent responses occur

Molecular oxygen	O_2
ROS (reactive oxygen species)	H_2O_2
	OH•
	O_2^-
Nitric oxide	NO
Carbon monoxide	CO
Hydrogen sulfide	H_2S
Adenosine and its phosphates	Adenosine
	AMP
	ADP
	ATP

atmospheric, i.e. approximately 150 mmHg) (Otto and Baumgardner, 2001), the relationship close to atmospheric oxygen level is usually used to describe the effect of oxygen on NO in macrophages. Fourthly, normally the distinction between oxygen-dependent phenomena and general stress responses remains unclear, because, for example, the possible difference in responses to hypoxia (i.e. oxygen being limiting but not absent) and anoxia (oxygen completely absent) is not really considered (Wenger and Gassmann, 1996).

2.2   The signal sensed

Numerous molecules might possibly function as the signal initiating oxygen-sensitive responses (Table 2.1).

2.2.1   *Molecular oxygen*

Molecular oxygen is used as a ligand or substrate by heme-containing molecules and prolyl/asparaginyl hydroxylases (Berra *et al.*, 2006; Lahiri *et al.*, 2006).

2.2.2   *Reactive oxygen species*

In many systems considered to be oxygen dependent, reactive oxygen species (ROS) may be the sensed signal that affects the activity of these systems. In this regard, it is important to remember that recent observations indicate that ROS are important in normal cellular functions. Whereas it was previously thought that ROS mainly conferred oxygen toxicity, it is now accepted that they are also important cellular-signaling molecules (Finkel, 1998; Wolin *et al.*, 2005; Halliwell and Gutteridge, 2007). The effects of ROS are usually thought to be mediated via effects on the cysteine residues of proteins (Michiels *et al.*, 2002). Two ROS, hydrogen peroxide (H_2O_2) and hydroxyl radicals (OH•), may be the most important molecules in oxygen-dependent signaling (Gloire *et al.*, 2006), for largely opposite reasons. Hydrogen peroxide is relatively stable and membrane permeant (Lesser, 2006; Halliwell and Gutteridge, 2007) and can affect the activity of tyrosine phosphatases by oxidizing cysteines in the catalytic center of the enzymes (Gloire *et al.*, 2006). Tyrosine phosphatase enzymes regulate the phosphorylation status of intracellular proteins. Because protein phosphorylation is one of the major factors affecting cellular functions, the activity of these enzymes, modulated by ROS, is of prime importance in affecting cellular functions. The cysteines of tyrosine phosphatases can also be oxidized by hydroxyl radicals. Hydroxyl radicals react with virtually all molecules they are in contact with (Halliwell and Gutteridge, 2007). The short lifetime of the hydroxyl radical (10^{-7} s) restricts its diffusion distance and effects to 4–5 nm. As a result, whenever hydroxyl radicals are involved, radical effects have a spatial dimension, even within cells (Lesser, 2006). It is very difficult to separate the effects of hydrogen peroxide and hydroxyl radicals, as hydrogen peroxide is converted to hydroxyl radicals in the Fenton reaction (Fig. 2.1), if adequate iron (or copper) ion stores are available (Halliwell

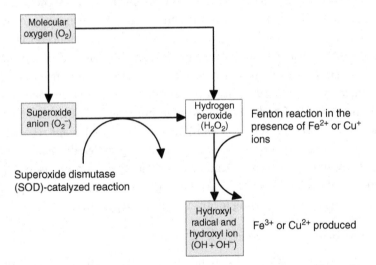

Fig. 2.1 The Fenton reaction may play a role in mediating the effects of oxygen on cellular functions. In this case molecular oxygen is converted to superoxide anion, which is dismutated to hydrogen peroxide in a reaction catalyzed by SOD (superoxide dismutase), or molecular oxygen is directly converted to hydrogen peroxide. In the presence of adequate ferrous or cuprous ion stores, highly reactive, short-lived hydroxyl radicals are produced from hydrogen peroxide in the Fenton reaction.

and Gutteridge, 1984; Bogdanova and Nikinmaa, 2001; Lesser, 2006). In addition, an important potential role has been ascribed to the superoxide anion (O_2^-), which can primarily be produced by both NADPH oxidase and mitochondria (Gonzalez et al., 2007). The mitochondrial production of superoxide anions occurs especially in complexes I–III of the respiratory chain (Gonzalez et al., 2007). Most of the available data suggest that conditions leading to a physiological hypoxia response are not severe enough to cause increased formation of superoxide anion in mitochondria, although suggestions about increased ROS production in hypoxic mitochondria have been made (Guzy and Schumacker, 2006).

2.2.3   *Nitric oxide, carbon monoxide and hydrogen sulfide*

In addition, the well-recognized gaseous signaling molecule, NO, may be utilized in oxygen-dependent signaling. Nitric oxide can react with super-oxide anion, and the peroxynitrite anion that is formed as a result is a powerful membrane-permeant oxidant (Fridovich, 1986a; Fridovich, 1986b) with a life-time near 0.1 s. Nitric oxide can also affect the oxygen affinity of mitochondrial function, for example (Koivisto et al., 1997). It has been shown that an NO synthetase isoform is induced by hypoxic conditions (Gess et al., 1997).

Carbon monoxide (CO) and hydrogen sulfide (H_2S) may also play a role in oxygen-dependent signaling. An isoform of heme oxidase is regulated by hypoxia (Lee et al., 1997). Heme oxidase is an enzyme involved in the conversion of heme to biliverdin, and has CO as one end product. Endogenously formed CO affects cellular respiration (D'Amico et al., 2006). Hypoxia-induced CO production may affect both angiogenesis and leukocyte movement across endothelium (Bussolati et al., 2004). CO may also regulate neural discharge from rat carotid body (Lahiri and Acker, 1999). It has been shown that CO, formed by heme oxidase in the carotid body, regulates potassium channel function, which is considered to be oxygen sensitive (Williams et al., 2004). Recent research indicates that H_2S may be another important signaling molecule (Wang, 2003). It appears to be involved in the oxygen-dependent regulation of vascular (Olson et al., 2006) and bladder (Dombkowski et al., 2006) muscle tone. Consequently, it may function in oxygen sensing (or transducing oxygen effects).

2.2.4   *Adenosine and its phosphates*

Changes in adenosine and adenosine phosphate concentrations are usually observed in hypoxic conditions. For example, any marked decrease in oxygen availability leads to a decrease in cellular adenosine triphosphate (ATP) concentration (Lutz and Nilsson, 2004). Similarly, ecto-5'-nucleotidase dephos-phorylates adenosine monophosphate (AMP) to adenosine in hypoxic conditions (Adair, 2005). Increased extracellular adenosine concentration is maintained, as hypoxia both increases adenosine release from the cells (Conde and Monteiro, 2004) and decreases its cellular re-uptake by decreasing the activity and production of equilibrative nucleoside transporters (Chaudary et al., 2004; Eltzschig et al., 2005). In addition, hypoxia appears to induce cellular adenosine receptors (Kong et al., 2006). As many hypoxia effects on animals have been associated with adenosine (Takagi et al., 1996; Stensløkken et al., 2004; O'Driscoll and Gorman, 2005; Kong et al., 2006; Martin et al., 2007), it appears to be one molecule used in oxygen sensing. Similarly, as hypoxia causes a decrease in cellular ATP concentration, any mechanism detecting disturbances in the energy balance with oxygen depletion would be highly useful for maintaining cellular function in hypoxia. From an energetic point of view, the ratio of adenosine diphosphate (ADP) to ATP is the primary regulated function. However, the AMP:ATP ratio varies as a square of the ADP:ATP ratio. Thus, a more sensitive regulation of the energy balance can be obtained, if AMP:ATP ratio is sensed instead of ADP:ATP ratio, and energy-producing/-consuming systems are consequently adjusted (Hardie, 2003). Adenosine monophosphate kinase senses changes in AMP and, thereby, the AMP:ATP ratio (Hardie, 2003; Hardie et al., 2006). The enzyme can thus function both in oxygen- and energy-dependent signaling.

2.3   The sensor molecules

Figure 2.2 illustrates the putative oxygen-sensing molecules and their possible locations in the cells.

2.3.1   Heme-based molecules

Two major groups of heme-containing proteins appear important in sensing oxygen in vertebrates: globins and cytochromes. In addition, some PAS domain proteins (named as such because they contain a domain similar to the one first described in the circadian protein period [Per], Ah receptor nuclear translocator protein [ARNT], and single-minded protein [Sim]) have heme as a functional group. Heme-containing proteins may have been used in O_2-, NO-, CO- and H_2S-dependent signaling in the last universal common ancestor (LUCA) of all extant organisms, and heme-based oxygen sensors are found both in prokaryotes and eukaryotes (Freitas et al., 2005; Gilles-Gonzalez and Gonzalez, 2005). Heme-based proteins bind molecular oxygen reversibly, whereupon they initiate a number of signaling cascades (Gilles-Gonzalez and Gonzalez, 2005).

2.3.1.1 The globin family of heme proteins

Hemoglobin, myoglobin, cytoglobin, and neuroglobin have all been implicated in sensing oxygen. With regard to tetrameric hemoglobin, in addition

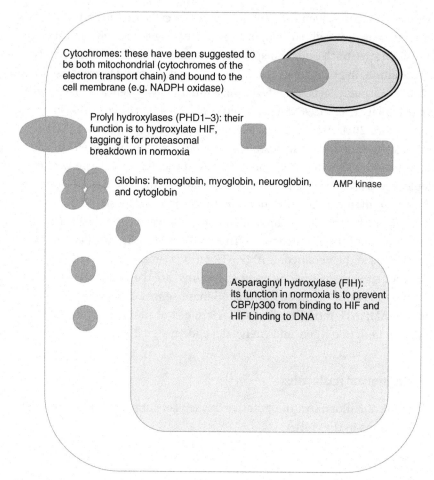

Fig. 2.2 Possible oxygen-sensing molecules and their locations in the cell. Asparaginyl hydroxylase (FIH), which catalyzes the hydroxylation of ASP803 of hypoxia-inducible factor (HIF), thereby preventing the interaction between HIF and p300/CBP, may be either nuclear or cytosolic.

to its primary function as an oxygen carrier, it is thought to be an oxygen-sensing molecule, regulating oxygen-sensitive membrane transport in erythrocytes (Gibson *et al.*, 2000). It is possible that the differential binding of oxy- and deoxy-hemoglobin to the major erythrocyte protein, band 3, is a key factor in the regulation of oxygen-dependent transport. Direct effects of hemoglobin oxygenation on sulfate transport via band 3 in human erythrocytes have been described (Galtieri *et al.*, 2002). By contrast, a direct link between hemoglobin–band 3 interaction and the oxygen dependence of transport activity of other transporters has not been found. An effect of oxygen on potassium chloride co-transport is present in pink ghosts (human erythrocyte ghosts containing band-3-bound hemoglobin), but disappears in white ghosts (ones not containing band-3-bound hemoglobin) (Khan *et al.*, 2004). Although this finding is in accordance with the idea that hemoglobin–band 3 interaction plays a role in generating the oxygen dependence of cation transport, the following observations are not. Oxygen-dependent ion transport is very pronounced in rainbow trout erythrocytes, although no oxygen-dependent interaction between the terminal (cytoplasmic) part of rainbow trout band 3 and hemoglobin has been described (Jensen *et al.*, 1998; Weber *et al.*, 2004). Furthermore, oxygen-dependent membrane transport is also present in lampreys, which lack the anion exchange pathway (Nikinmaa and Railo, 1987; Tufts and Boutilier, 1989; Virkki *et al.*, 1998).

Thus, the role and regulatory function of hemoglobin as an oxygen sensor modulating transport activity in erythrocytes remains unknown. The same is actually true for the other possible oxygen-sensing molecules of the globin family: myoglobin, cytoglobin, and neuroglobin. Myoglobin plays an important role in the storage of oxygen (Jurgens *et al.*, 2000) and the intracellular diffusion of oxygen from capillary blood to muscle mitochondria (Wittenberg and Wittenberg, 2003; Ordway and Garry, 2004). In addition, myoglobin plays a role in regulating NO function in mitochondria (Brunori, 2001a; Brunori, 2001b; Wittenberg and Wittenberg, 2003). It is possible that the myoglobin effect on NO is important in any oxygen-dependent regulation of function. In addition to muscle cells, isoforms of myoglobin are expressed in other tissues, such as liver and neural tissue (Fraser *et al.*, 2006). The transcription of the myoglobin-coding gene is often stimulated in hypoxia (van der Meer *et al.*, 2005; Fraser *et al.*, 2006), although this does not appear to be the case for neural tissue (Fraser *et al.*, 2006).

Two other globins, neuroglobin and cytoglobin, have been characterized in all vertebrates (Burmester *et al.*, 2000; Burmester *et al.*, 2002; Burmester *et al.*, 2004; Brunori and Vallone, 2007). In some vertebrates, there appear to be additional globins, such as globinX in amphibians and fish (Roesner *et al.*, 2005) and the eye-specific globin in chicken (Kugelstadt *et al.*, 2004). At present, the functions of all the 'new' globins are poorly known (Hankeln *et al.*, 2005). Whereas cytoglobin is present in most tissues, neuroglobin is primarily restricted to tissues of neural origin (Hankeln *et al.*, 2005). Phylogenetically, cytoglobin appears to belong to the same group of globins as myoglobin, whereas neuroglobin belongs to a distinct family, which is phylogenetically very ancient; neuroglobin is present both in vertebrates and in some invertebrates (Hankeln *et al.*, 2005). Neuroglobin is expressed in the carotid body, a key tissue for blood oxygen sensing, and hypoxia increases its expression there (Di Giulio *et al.*, 2006). Indeed, hypoxic conditions generally increase neuroglobin transcription, whereas both increases and lack of changes have been reported for cytoglobin (Schmidt *et al.*, 2004; Li *et al.*, 2006; Mammen *et al.*, 2006; Roesner *et al.*, 2006). It is possible that the response is tissue and species specific. Further complexities to the behavior of cytoglobin are shown by the facts that at least in teleost fish the gene coding for it has been duplicated, and that the oxygen-binding behavior of the isoforms is very different (Fuchs *et al.*, 2005). In addition to possible regulation by hypoxia, cytoglobin and neuroglobin can be regulated by redox state (Hamdane *et al.*, 2003). The possibility that cytoglobin and neuroglobin could be involved in oxygen sensing relates to the large conformational changes they show upon oxygenation. Such changes could trigger downstream regulatory cascades of oxygen-dependent functions (Pesce *et al.*, 2002).

2.3.1.2 Cytochromes

The involvement of cytochromes in oxygen sensing has been indicated in many studies (Duranteau *et al.*, 1998; Ehleben *et al.*, 1998; Porwol *et al.*, 2001; Guzy *et al.*, 2005; Guzy and Schumacker, 2006). The cytochromes involved have been suggested to be both non-mitochondrial (Porwol *et al.*, 2001) and mitochondrial (Guzy *et al.*, 2005; Guzy and Schumacker, 2006). It also appears that CO may significantly regulate the oxygen-dependent cytochrome function (Porwol *et al.*, 2001). The major cytochrome involved in oxygen sensing has been suggested to be the NADPH oxidase enzyme or a polymeric cytochrome similar to it. Traditionally NADPH oxidase has been considered to be a constituent of phago-cytic cells generating the respiratory burst (Babior, 1984; Baggiolini and Wymann, 1990; Wientjes and Segal, 1995; Dahlgren and Karlsson, 1999; DeLeo *et al.*, 1999; Decoursey and Ligeti, 2005; El Benna *et al.*, 2005). However, recent studies have indicated that it is present also in non-phagocytic cells (Infanger *et al.*, 2006; Bedard and Krause, 2007), including the carotid body (Ehleben *et al.*, 1998). NADPH oxidase consists of several subunits: G22-phox, GP91-phox, P67-phox, P47-phox, P40-phox, and Rac1/Rac2 (GTPases) (Rotrosen *et al.*, 1992; Wientjes and Segal, 1995; Bedard and Krause, 2007; Dinger *et al.*, 2007), to form a membrane-bound, multi-subunit structure (Fig. 2.3). Of the subunits, g22-phox and gp91-phox are embedded in the membrane, whereas the rest are cytosolic (Dinger *et al.*, 2007). If NADPH oxidase or a protein similar to NADPH oxidase is involved in oxygen sensing, the signaling is most likely to be mediated via oxygen-regulated ROS formation (Acker 1994a; Acker 1994b; Acker and Xue 1995; Kummer and Acker, 1995; Ehleben *et al.*, 1998; Porwol *et al.*, 2001; Acker *et al.*, 2006). Although the ROS, largely superoxide ions, produced by the NADPH oxidase of phagocytes are mainly released to the extracellular compartment, intracellular generation of ROS, compatible with them being intracellular mes-sengers, has also been demonstrated for NADPH oxidase isoforms present in non-phagocytic cells (Dinger *et al.*, 2007). Whereas clear effects of oxygenation status

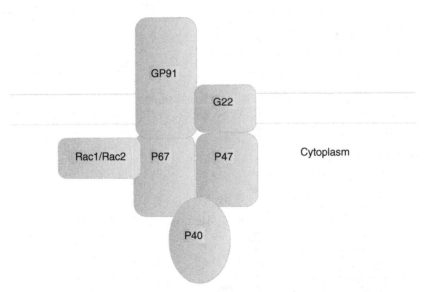

Fig. 2.3 Schematic representation of the multi-subunit protein NAD(P)H oxidase.

that are compatible with changes in, for example, oxygen-dependent ion channel function have been observed in the absorbance by cytochrome b558 (corresponding to gp91-phox subunit of NADPH oxidase) of the mammalian carotid body glomus cells (Acker and Xue, 1995), there are several studies indicating that a defective gp91-phox does not disrupt oxygen sensing in the carotid body of mice (Archer *et al.*, 1999; Roy *et al.*, 2000). Some reports suggest that p47-phox could be involved in oxygen sensing by NADPH oxidase (Sanders *et al.*, 2002). Data support the role of NADPH oxidase in oxygen sensing in pulmonary neuroepithelial bodies (Fu *et al.*, 2000; Bedard and Krause, 2007). By contrast, NADPH oxidase does not appear to play a role in oxygen sensing in the erythropoietin-producing cells of the kidney (Sanders *et al.*, 2002). The available data are restricted to a few species of mammals (humans and the laboratory rodents, rat and mouse), but already they suggest that there is pronounced cell-type-specific variation in the involvement of NADPH oxidase in oxygen sensing.

2.3.2   *Prolyl and asparaginyl hydroxylases, and the function of hypoxia-inducible factor*

Oxygen-dependent regulation of transcription generally involves a transcription factor, hypoxia-inducible factor (HIF). HIF function is described in Fig. 2.4 (see also, for example, Wenger, 2000; Bracken *et al.*, 2003). The transcriptionally active form of HIF is a dimer consisting of α- and β-subunits. The α-subunits give the oxygen sensitivity, whereas the function of the β-subunit appears to be oxygen insensitive. The β-subunit is a general dimerization partner of environmentally

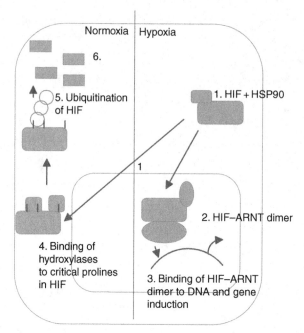

Fig. 2.4 Mechanism of HIF1 function. (1) The oxygen-dependent subunit (HIF1α) is produced continuously and interacts with HSP90 in the cytoplasm. (2) In hypoxic conditions HIF1α is stabilized and transported to the nucleus, where it forms a dimer with ARNT (HIFβ) and recruits p300/CBP. (3) The HIF1α–ARNT dimer binds to the hypoxia-response element (minimal HRE is mammals: A/GCGTG) in the promoter/enhancer area of the transcribed gene, whereby gene expression is induced. (4) In normoxia, prolyl and asparaginyl hydroxylase enzymes are active (see Fig. 2.5), and (5) HIF1α is tagged for (6) proteasomal degradation.

regulated transcription factors, and because its structure and function has been studied in most detail in connection with the aryl hydrocarbon receptor (AhR, dioxin receptor), HIF-β is often called ARNT (aryl hydrocarbon receptor nuclear translocator). There are at least three different classes of α-subunits, denoted as HIF-1α, HIF-2α (also called EPAS1), and HIF-3α (the recently characterized HIF-4α of teleost fish may be an additional subunit group [see Law *et al.*, 2006]; so far, however, HIF-3α and HIF-4α have not been found in the same species). Of the HIF-α subunit classes, HIF-1α is the one studied most in association with hypoxia.

The effect of HIF-1α on transcription is largely regulated either by affecting the stability of the protein by hydroxylation of conserved prolines (proline 402 and 564 in the human protein) and consecutive degradation of the molecule, or by affecting the interaction of the molecule with p300 and consecutive DNA binding as a result of hydroxylation of a conserved asparagine residue (ASP803) (Bracken *et al.*, 2003) (Fig. 2.5), although it now appears that especially in hypoxia-tolerant

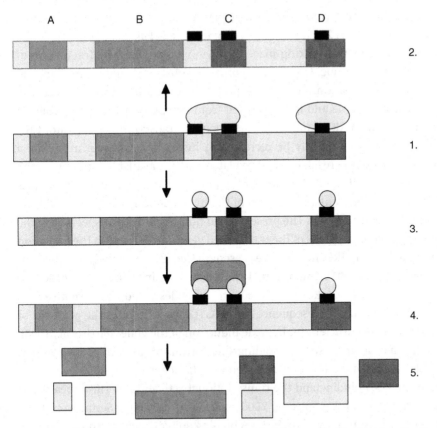

Fig. 2.5 Schematic representation of prolyl and asparaginyl hydroxylase function. Hypoxia-inducible factor 1α (HIF1α) contains: (A) the basic helix-loop-helix (BHLH) domain, which binds to DNA; (B) the PAS domain involved in dimer formation with ARNT; (C) the N-terminal transactivation domain (N-TAD), where one of the targets of prolyl hydroxylase (Proline 564, numbering on the basis of human sequence) is situated – the other target (Proline 402) is in an area preceding the N-TAD; and (D) the C-terminal transactivation domain (C-TAD), where the target for asparaginyl hydroxylase (Asparagine 803) is situated. The targets for hydroxylases are marked with black rectangles. In normoxia (1) prolyl and asparaginyl hydroxylases hydroxylate (3) their targets, whereafter (4) hydroxylated asparagine cannot interact with p300/CBP, inhibiting the DNA-binding of HIF, whereas the hydroxylated prolines interact with the VHL protein. The VHL protein is the recognition component of an E3 ubiquitin–protein ligase that targets HIF1α for proteasomal degradation (5). In hypoxia (2), the prolines and asparagine are not hydroxylated. As a consequence, the molecule is not broken down and can interact with p300/CBP. Thus, HIF can be bound to DNA and HIF-dependent induction of gene transcription can occur.

animals also the transcription of HIF may play a role in the regulation of the HIF pathway (Shams *et al.*, 2004; Law *et al.*, 2006; Rissanen *et al.*, 2006b).

Prolyl hydroxylation, taking place in normoxia, enables the interaction of HIF-α and the von Hippel-Lindau (VHL) protein, subsequent ubiquitination, and proteasomal degradation. In hypoxia, prolyl hydroxylation does not occur, so HIF-α protein is stable and is transported from cytoplasm to nucleus, where it forms a dimer with ARNT and recruits the general transcriptional activator CBP/p300 (the activator can only be recruited in hypoxia, because in normoxia a conserved asparagine residue is hydroxylated and not available for interaction). Thereafter, HIF (HIF-α + ARNT) binds to hypoxia response elements (HREs) present in the promoter/enhancer region of the hypoxia-inducible genes, and gene transcription is stimulated. The presence and number of HREs, especially in the promoter/enhancer regions of the transcribed gene, are decisive in the induction of gene expression. HREs may also be present in the introns of oxygen-dependent genes (Rees *et al.*, 2001). The minimal consensus HRE in mammals is A/GCGTG (Camenisch *et al.*, 2002). Rees *et al.* (2009) recently described an alternative HRE functioning in fish, with a sequence of GATGTG. In some cases the presence of HREs alone is not sufficient for hypoxic induction of the genes (Firth *et al.*, 1995), but additional elements such as binding sites for various molecules such as AP1, ATF1/CREB1, HNF4, or Smad3 may be required (Bracken *et al.*, 2003).

Both the DNA binding and the transcriptional activation by HIF appear to be under redox control (Lando *et al.*, 2000; Bracken *et al.*, 2003). Serine-to-cysteine mutation at a specific residue in the DNA-binding domain confers redox sensitivity of DNA binding, and the nuclear redox regulator Ref1 potentiates the hypoxic induction of a reporter gene (Lando *et al.*, 2000). There can be interaction between redox-sensitive transcription factors and HIF (Khomenko *et al.*, 2004), and cellular redox state correlates with HIF induction (Heise *et al.*, 2006a). As HIF function is affected by the redox state, it is also sensitive to ROS (Fandrey *et al.*, 2006). Also, Ca^{2+} affects the function of HIF. Thus, although the principles of how HIF regulates gene expression are clear, the fact that oxygen and other substances can affect HIF function at several different places, and the limited knowledge about oxygen levels in the cells of different organisms and different tissues, mean that there is currently no decisive information on the possible differences of oxygen affinities for oxygen-dependent gene expression between tissues and species.

In the case of both prolyl and asparaginyl hydroxylation, the hydroxylases catalyzing the respective reactions, i.e. prolyl and asparaginyl hydroxylases, use molecular oxygen as a substrate. Their function is oxygen dependent, and they thus function as oxygen sensors affecting hypoxia-inducible gene expression, as first demonstrated by Ivan *et al.* (2001) and Jaakkola *et al.* (2001). The function of prolyl hydroxylases has recently been reviewed by, for example, Hirota and

Semenza (2005), Kaelin (2005), and Fandrey *et al.* (2006). In mammals, three types of oxygen-dependent prolyl hydroxylases (PHD1–3; EGLN1–3) have been described, and the presence of a fourth (PHD4) has been deduced on the basis of genomic information (Oehme *et al.*, 2002). It appears that the PHD2 hydroxylase (EGLN1) is the most important for oxygen sensing (Berra *et al.*, 2003), although cell type-specific differences may occur (Appelhoff *et al.*, 2004). Although hydroxylation of conserved prolines (in the LXXLAP sequence) is achieved by the prolyl hydroxylases, it appears that several other residues are important for the proper functioning of the PHDs (Fandrey *et al.*, 2006). This suggests that, in addition to the properties of the enzymes themselves, the three-dimensional structure of HIF-1α affects the hydroxylation. Notably, hydrophobicity plots of HIF-1αs from various vertebrates show that the conserved proline residues are in a hydrophobic environment, suggesting that the residues are situated within the protein (Rytkonen *et al.*, 2007). Because of this, one possibility for the effect of ROS and Ca^{2+} on the HIF-1α function is that they affect some residues that are important for HIF-1α protein folding. Consequently, the three-dimensional structure of the HIF-1α protein could be changed, which would affect the accessibility of the conserved prolines toward prolyl hydroxylases. This mechanism would add another way for regulating HIF-1α function by oxygen, as ROS levels can be oxygen dependent. It would also explain why HIF-1 function has been shown to be ROS dependent in several studies (reviewed by Haddad, 2002; Kietzmann and Gorlach, 2005; Acker *et al.*, 2006), although the enzymatic hydroxylation and consecutive proteasomal breakdown of the protein do not require ROS (Fandrey *et al.*, 2006). Another possibility is that the localization of prolyl hydroxylases plays a role in the regulation of their activity, or that their function is Ca^{2+} sensitive (Fandrey *et al.*, 2006).

Oxygen-dependent regulation of the DNA binding of HIF is mediated by the function of asparaginyl hydroxylase, also known as factor-inhibiting hypoxia-inducible factor (FIH) (Kaelin, 2005). At high oxygen tension asparagine 803 (in the human protein; analogous amino acids are found in HIF1αs of all vertebrates studied to date) is hydroxylated by asparaginyl hydroxylase. As a result, the interaction of the hypoxia-inducible factor with CBP/p300 is prevented, and binding of hypoxia-inducible factor to the hypoxia response element of DNA is reduced. Because the prolyl and asparaginyl hydroxylases have different oxygen affinities – the oxygen affinity of asparaginyl hydroxylase being much higher – it is possible that the two enzymes regulate HIF function at different oxygen levels. It is also possible that different genes are regulated by the two enzymes (Dayan *et al.*, 2006), if the oxygen-dependence profile for gene induction varies between genes. The different oxygen affinities of prolyl and asparaginyl hydroxylases also give one possibility for generating differences in the oxygen-dependent regulation of gene function in different groups of animals: depending on the oxygen affinities of the hydroxylases, HIF stability and its DNA binding may have different oxygen profiles in different animal groups. However, at present this suggestion is purely hypothetical, as studies measuring the oxygen dependency of hydroxylase activities in various animal groups are not available.

The hydroxylase enzymes need α-ketoglutarate, ascorbate, and ferrous ions as co-factors (Kaelin, 2005; Halliwell and Gutteridge, 2007). The need for α-ketoglutarate as a co-factor brings together the oxygen-dependent hydroxylase function and aerobic metabolism via the citric acid cycle, as α-ketoglutarate is an intermediate of the citric acid cycle. The regulation of HIF-dependent gene expression by NO is largely due to the effects of NO on prolyl hydroxylase activity (Berchner-Pfannschmidt et al., 2007). Nitric oxide may increase HIF level in normoxia; in this case the reduced activity of prolyl hydroxylase would possibly be caused by an inhibition of interaction between oxygen and the ferrous ion in the hydroxylase.

2.3.3   AMP-activated protein kinase

The function of AMP-activated protein kinase (AMPK) has been reviewed recently (Hardie, 2003; Hardie et al., 2006; Wyatt and Evans, 2007). The enzyme can couple oxygen and energy sensing (Wyatt et al., 2007). It is composed of three subunits, the catalytic α subunit, and the regulatory β and γ subunits. The multi-subunit enzyme remains inactive even in the presence of AMP, if not phosphorylated at a crucial threonine residue (Hardie, 2003). Because the enzyme is activated by a decrease in energy charge, its function in relation to diabetes, obesity, and glucose/lipid usage in humans has been especially studied (Kim et al., 2005; Yun et al., 2005). The function of AMPK is also affected by ROS (e.g. Choi et al., 2001) and NO (e.g. Lei et al., 2005), adding to the possibilities of interaction between different regulatory pathways. As AMPK is involved in regulating cellular energy balance, its activation switches off energy-consuming, and switches on energy-producing, pathways. One of the major oxygen-consuming processes in the cells involves mRNA translation to proteins. Notably, translation is inhibited by AMPK in hypoxia, also independently from HIF regulation (Liu et al., 2006), showing the importance of energy sensing in hypoxia regulation.

2.4   Transduction systems for oxygen effects

Changes in oxygen tension can have virtually immediate effects at the level of ion transporters and longer-term effects in oxygen-dependent gene expression. Interactions between the two systems occur, and one of the important points to remember is that the long-term response to hypoxia involving gene expression may result in changes in the amounts of gene products that are involved in mediating any rapid oxygen-dependent responses.

Mitochondria, which provide the cells with energy aerobically, have been considered as one possible transducer of oxygen effects both in the short and long term. In view of a major role of prolyl hydroxylases as primary oxygen sensors involved in oxygen-dependent gene expression, explanations of mito-chondrial influence on oxygen-dependent gene expression have recently centered on how these organelles regulate prolyl hydroxylase activity (Bell et al., 2005). Although it appears that oxidative phosphorylation is not involved in mediating the effects of oxygen, ROS produced in the electron transport chain of hypoxic mitochondria may affect the activity of hydroxylases (Bell et al., 2005). Both increases and decreases in ROS with hypoxia have been observed (Chandel and Budinger, 2007). While extramitochondrial ROS production increases with increasing oxygen tension, the mitochondrial ROS production may increase in hypoxia at specific points in the electron transport system (Guzy and Schumacker, 2006). It is possible that different cell types respond differently – depending on the relative contributions of extra-mitochondrial and mitochondrial ROS generation.

Vesicles containing significant quantities of metal ions (such as, for example, perinuclear vesicles with high iron content) may take part in Fenton reactions generating hydroxyl radicals (Porwol et al., 2001; Acker et al., 2006) (Fig. 2.1), which may affect the regulation of oxygen-dependent gene expression. Alternatively, iron particles within the endoplasmic reticulum can be important in the generation of the Fenton reaction (Liu et al., 2004). Ferrous ion alters the stabilization-degradation cycle of HIF1α by affecting the prolyl hydroxylase activity: a decrease in the intracellular iron content leads to the stabilization of HIF1α (Triantafyllou et al., 2007).

2.4.1   Immediate oxygen-dependent responses

Rapid oxygen-dependent responses usually occur at the membrane level, where the ion transporter activity may be modulated. In addition, the activities of several enzymes may be modified. Events leading to the slower oxygen-dependent gene expression changes necessarily involve rapid responses in some components of the response. For example, oxygen affects the activity of the hydroxylases regulating hypoxia-inducible factor immediately, but this effect is only seen later as an effect on gene expression.

2.4.1.1 Examples of rapid oxygen-dependent responses in which oxygen sensing plays a primary role: regulation of hemoglobin–oxygen affinity

The most ancient vertebrates, hagfish and lampreys (agnathans), are not able to equilibrate anions, including bicarbonate, rapidly across the erythrocyte membrane (Ellory et al., 1987; Nikinmaa and Railo, 1987; Tufts and Boutilier,

1989). The erythrocyte membrane of all other vertebrates contains band 3 protein in sufficient quantity to make the exchange of small univalent anions rapid – equilibrium is reached maximally in a few seconds at the animal's body temperature (in contrast to equilibration times of a couple of hours in lampreys (Nikinmaa and Railo, 1987). Despite this marked difference in the rate of transport of the acid–base-relevant anion, bicarbonate, control of erythrocyte pH is possible for both agnathans and teleost fish (Nikinmaa, 1992). As intracellular pH is a major regulator of hemoglobin function, and as many ion-transport pathways affecting the steady-state H^+ distribution across the erythrocyte membrane are oxygen sensitive (Gibson et al., 2000; Nikinmaa, 2003; Drew et al., 2004), oxygen binding by hemoglobin can be rapidly regulated by oxygen-sensitive ion transport. Figure 2.6 gives an illustration of the possible mechanisms involved. The regulation of hemoglobin oxygen affinity by membrane transport has been reviewed (Nikinmaa, 1992; Nikinmaa, 2005). With regard to the hypoxia-dependent activation of the Na^+/H^+ exchange, an increase in intracellular pH occurs, increasing the hemoglobin–oxygen affinity and thereby improving oxygen loading in respiratory epithelia in times of limited environmental oxygen availability. Although there is ample documentation about the role played by erythrocytic Na^+/H^+ exchange in hypoxia acclimation (Tetens and Christensen, 1987; Fievet et al., 1988; Nikinmaa and Salama, 1998), the importance of activating K^+Cl^- co-transport at high oxygen tensions has not been shown unequivocally (Nikinmaa and Salama, 1998; Nikinmaa, 2005). The expected reduction in erythrocyte pH and volume would decrease hemoglobin–oxygen affinity. Hyperoxia may result in high oxygen tensions in all parts of the circulation. As a consequence of the high oxygen tension, only a limited amount of oxygen will be given up from hemoglobin with a normal oxygen affinity. If, however, K^+Cl^- co-transport is activated, and erythrocyte pH is decreased, thereby reducing the hemoglobin–oxygen affinity, more oxygen will be unloaded at high oxygen tensions than would be the case without the response. In addition to the effects of oxygen-sensitive ion transport in regulating hemoglobin–oxygen affinity, it may be important in controlling redox disturbances, as erythrocytes can easily be exposed to oxidative stresses because they contain large numbers of oxygen and ferrous ions and produce significant quantities of superoxide ions (Halliwell and Gutteridge, 2007). Ferrous ions are, furthermore, liberated to the cytoplasm in considerable numbers.

2.4.1.2 Examples of rapid oxygen-dependent responses in which oxygen sensing plays a primary role: regulation of ventilation I

Carotid body chemoreception is probably the most intensively studied oxygen-dependent effector system, as the function of the carotid body plays an

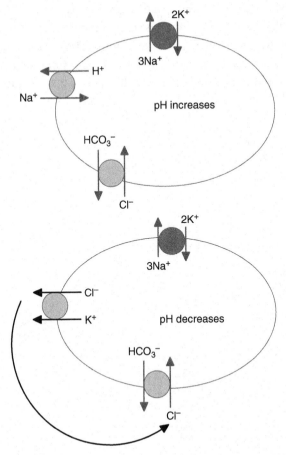

Fig. 2.6 The function of oxygen-dependent ion transport in erythrocytes. Activation of Na^+/H^+ exchange in hypoxia transports H^+ out of the cell, thereby increasing intracellular pH. As the hemoglobin–oxygen affinity increases with increasing pH, the effect improves oxygen loading at lowered oxygen availability. The degree of intracellular alkalinization depends on the relative rates of the Na^+/H^+ exchange and extracellular dehydration of bicarbonate to carbon dioxide: the greater the Na^+/H^+ exchange rate, the larger the pH increase. Activation of the K^+Cl^- co-transport (often at high oxygen tensions) causes removal of Cl^- ions from the cytoplasm. Cl^- ions are exchanged with bicarbonate via the anion exchanger to re-establish equilibrium. The removal of bicarbonate ion from the cytoplasm results in cytoplasmic acidification and decrease of hemoglobin–oxygen affinity.

important role in determining the ventilatory response to, for example, hypoxia. (In addition to the carotid body, the aortic body is also important, but much more work has been done on the oxygen sensing of carotid bodies than of aortic bodies.) The function of the carotid body has been the subject of many reviews (Gonzalez *et al.*, 1994; Gonzalez *et al.*, 1995a; Gonzalez *et al.*, 1995b; Lopez-Barneo, 2003; Lahiri *et al.*, 2006; Prabhakar, 2006; Kumar and Prabhakar, 2007). The sequence of events involved in the function of the carotid body, and seen as oxygen-dependent changes in neural activity in the respiratory center of the central nervous system, is given schematically in Fig. 2.7. Although the

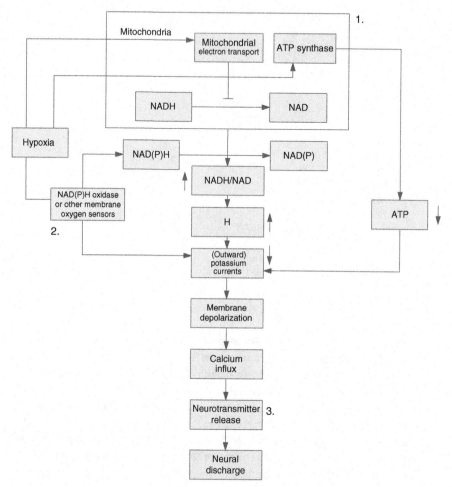

Fig. 2.7 A schematic representation of the two models for oxygen-dependent regulation of carotid body glomus cell function: (1) the mitochondrial model and (2) the membrane model. In both cases hypoxia results in membrane depolarization, often following hypoxic inhibition of outwardly directed potassium fluxes via oxygen-sensitive potassium channels. Following membrane depolarization, cytoplasmic Ca^{2+} concentration increases. Thereafter neurotransmitters are released with consequent neural discharge. The neurotransmitters released (3) are: e.g. serotonin, acetylcholine, noradrenaline, dopamine, ATP, adenosine, NO, and CO.

oxygen sensor of the carotid body has been studied intensively, its nature is still not clear. It is possible that there are several different oxygen sensors. The presence of several oxygen sensors with different affinities for oxygen makes it possible for the carotid body to respond appropriately to a wide range of oxygen tensions (Prabhakar, 2006). Surprisingly, the oxygen tensions present in the cells of the carotid body are poorly known (Gonzalez et al., 1994). In addition to sensing oxygen, the cells of the carotid body also respond to changes in, for example, carbon dioxide tension, intracellular pH, and glucose. Importantly, the ventilation rate in air-breathing vertebrates is more sensitive to changes in carbon dioxide than in oxygen. The interactions between oxygen and glucose sensing have been investigated in some detail (Zhang et al., 2007) and reviewed (Lopez-Barneo, 2003; Pardal and Lopez-Barneo, 2004), as such interaction combines two aspects of cellular metabolism: energy production and energy availability. Although it appears that HIF is not directly involved in acute oxygen sensing in the carotid body, it plays a role in the development of carotid body function, and its modulation in chronic hypoxia, for example (Kline et al., 2002; Fung and Tipoe, 2003; Roux et al., 2005). It further appears that erythropoietin (via binding to the erythropoietin receptors present in the carotid body glomus and brainstem cells) may be involved in the oxygen-dependent regulation of ventilatory activity (Soliz et al., 2005).

2.4.1.3 Examples of rapid oxygen-dependent responses in which oxygen sensing plays a primary role: regulation of ventilation II

In contrast to air-breathing vertebrates, ventilation in water breathers responds mainly to changes in ambient oxygen level (Dejours, 1975), but can also respond to changes in carbon dioxide tension and pH (Perry and Gilmour, 2002; Gilmour and Perry, 2006). Despite the fact that most fish respond to ambient hypoxia by an increase in ventilation rate and bradycardia (Gilmour and Perry, 2006), the actual oxygen-sensing cells have remained uncharacterized until recently. For a very long period of time, both external oxygen receptors (responding to changes in the oxygenation of water) and internal oxygen receptors (responding to changes in the oxygenation of blood or some other body constituent) have been considered to be present (reviewed by e.g. Perry and Gilmour, 2002). In most cases, bradycardia is caused predominantly by a decrease in ambient oxygen level, whereas ventilatory changes are associated with changes in both ambient and blood oxygenation (Milsom and Burleson, 2007). Regardless of their location, the general view has been that the sensors must be neuroendocrine cells with neural connections to the central nervous system. Microscopically, gill neuroendocrine cells were first described by Dunel-Erb et al. (1982), but the first characterization of gill neuroendocrine

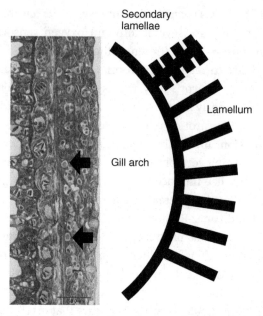

Fig. 2.8 The localization of neuroendocrine cells in fish (rainbow trout) gills. The photomicrograph was reproduced from Dunel-Erb *et al.*, 1982, with permission from the *Journal of Applied Physiology*. The neuroendocrine cells are embedded in the lamellae and are shown in the photomicrograph at the point of the arrows.

cells as oxygen-sensing cells had to wait for more than 20 years (Jonz *et al.*, 2004). In addition to zebrafish, putative oxygen-sensing cells have been characterized from the gills of the channel catfish (Burleson *et al.*, 2006). The oxygen-sensing neuroendocrine cells appear to be present in all gill arches, and characteristically contain serotonin (Milsom and Burleson, 2007). They may be homologous with the oxygen-sensing cells of the carotid body (which may have developed from the oxygen-sensing cells of the third gill arch [Milsom and Burleson, 2007]). This being the case, the oxygen-sensing and transduction mechanism to the central nervous system is probably similar to that described above for the carotid body (Fig. 2.7). The localization of neuroendocrine cells in the lamellae (Fig. 2.8) would make it possible for them to respond to changes in both the ambient and blood oxygen levels.

2.4.2   *Oxygen-dependent gene expression*

Oxygen-dependent gene expression was conclusively demonstrated in the 1990s. Early studies concentrated on red cell production, mainly the erythropoietin pathway (Fandrey, 2004; Eckardt and Kurtz, 2005; Jelkmann, 2007), but it has later become clear that many genes, possibly more than a hundred of the studied ones in mammals, may be induced by a decrease in oxygen tension (Lahiri et al., 2006). The number of genes that are inhibited by a lowered oxygen tension and the mechanisms of gene downregulation by oxygen have not been studied in detail in most cases, but altogether up to 2% of human genes may be up- or downregulated by changes in oxygen tension (Manalo et al., 2005). Hypoxic induction of genes appears to be regulated mainly transcriptionally by HIFs. A specific example of an HIF being involved in the downregulation of a gene involved in aerobic energy production (cytochrome oxidase) has also been reported recently (Fukuda et al., 2007). Regulation by HIF-1 appears to be of prime importance – HIF-1 has often been called the 'master regulator of hypoxia responses.' Notably, though, while the function of HIF-1 was originally studied and described in terms of the erythropoietin pathway, it has recently become obvious that HIF-2 may also be a very important transcriptional regulator of the erythropoietin pathway (Eckardt and Kurtz, 2005; Chavez et al., 2006; Rankin et al., 2007; Ratcliffe, 2007). The hypoxia-induced genes include those involved in oxygen transport, iron transport, angiogenesis, and energy production, affecting glucose transporters and enzymes of the glycolytic pathway etc. (Table 2.2).

Originally it was thought that the expression of HIFα proteins would be restricted to hypoxia, but recently it has become apparent that there are several instances in which HIFα is induced in normoxic conditions (Dery et al., 2005; Hirota and Semenza, 2005). The normoxic induction in mammals appears to result from increased translation involving the PI3 kinase/TOR pathway (Stiehl et al., 2002; Dery et al., 2005). Regulation of the normoxic level of HIF1α may also involve HSP90–RACK1–HIF1 α interaction: HSP90–HIF1α interaction stabilizes the protein, and RACK1–HIF1α interaction destabilizes it (Liu and Semenza, 2007; Liu et al., 2007). HSP90 and RACK 1 compete for the same binding site. Furthermore, acetylation of HSP90 decreases its binding affinity to, for example, HIF1α (Liu and Semenza, 2007), whereby HIF1α protein would be destabilized. Although HSP90 is known to participate in many cellular protein–protein interactions, the HIF1α–HSP90 interaction could bring together the temperature- and oxygen-dependent gene expression. Importantly, the few studies investigating HIF function in relation to temperature in poikilothermic animals indicate that HIF plays a role in temperature-dependent gene expression (Treinin et al., 2003; Heise et al., 2006a; Heise et al., 2006b; Rissanen et al., 2006b). The transcription factor appears to be important during long-term temperature acclimation (Rissanen et al., 2006b), and it is often present in normoxic fish (Rissanen et al., 2006b). The two regulatory pathways, one involving the regulation of the stability of the protein and the other its DNA binding, may be differently involved in temperature- and oxygen-dependent responses.

Table 2.2 *Example of genes that are under transcriptional regulation by hypoxia-inducible factor (HIF). For details see, for example: Gardner et al. (2001); Goda et al. (2003); Schnell et al. (2003); Fandrey (2004); Gorr et al. (2004); Semenza (2004); Greijer et al. (2005); Fukuda et al. (2007); and Jelkmann (2007)*

Function		Gene product
Oxygen transport		
	Iron metabolism	Ceruloplasmin
		Transferrin
		Transferrin receptor
	Red cell production	Erythropoietin
	Hemoglobin synthesis	e.g. Globin genes in *Daphnia*
	Angiogenesis	VEGF (vascular endothelial growth factor)
		VEGF-receptor 1
		Endothelin
Energy production		
	Glycolysis	Aldolase A
		Fructose-2,6-bisphosphatase 3 and 4
		Enolase
		Lactate dehydrogenase
	Substrate availability	Glucose transporter
	Mitochondrial effects	LON (mitochondrial protease involved in COX4 breakdown)
Hormonal regulation and cellular signaling		
		Leptin (could also be included in energy production; involved in lipid storage and mobilization)
		Atrial natriuretic peptide
	NO production	Nitric oxide synthase-2
	CO production	Heme oxygenase-1 (the enzyme breaks down heme into biliverdin, and CO and ferrous ions are released)
	Adrenergic signaling	α-adrenergic receptor
		Tyrosine hydroxylase (an enzyme needed in the biosynthesis of catecholamines – hydroxylating tyrosine)
Immunological responses		Thymopoietin (factor involved in T-cell development)
Cell cycle and apoptosis		p21
		p27 (both proteins are inhibitors of cyclin-dependent kinase)
		NIP3 (proapoptotic factor)
Cytoskeleton and extracellular matrix		Fibronectin
		Keratin components

2.5   Future perspectives

As pointed out in the introduction to this chapter, most of the studies on oxygen sensing have been carried out on human, mouse, or rat, which are all relatively hypoxia-intolerant mammals. Many model organisms (e.g. *Caenorhabditis elegans* and zebrafish), by contrast, are relatively hypoxia tolerant. Presently, we do not really know how this difference influences oxygen-dependent responses. Furthermore, it is not clear at all whether the responses of poikilotherms are similar to those of homeotherms. Consequently, in any future study on oxygen sensing it is important to consider the biology and evolutionary history of the studied species. Further, it is clear from this chapter that the signal sensed is often not molecular oxygen. Because of this, interactions between oxygen-dependent and other pathways (involved in redox regulation or energy metabolism, for example) may occur. Increasingly such interactions need to be taken into account. As an example, one topic in which interaction has been studied in some detail concerns the hypoxia-inducible and xenobiotically induced gene expression pathways, as both HIFα and aryl hydrocarbon receptor (AhR) use ARNT as a dimerization partner. Dimerization is required before induction of gene expression. Many, but not all, studies have observed interaction between the pathways (Chan *et al.*, 1999; Nie *et al.*, 2001; Hofer *et al.*, 2004). Also, in future studies it will be important to consider the similarities and differences between the immediate and long-term (sustained) oxygen-dependent responses. This may help in further characterizing the (probably several) oxygen sensors, their properties, and their regulation.

Acknowledgements

The author's work is supported by the Centre of Excellence grants from the Academy of Finland and the University of Turku.

References

Acker, H. (1994a). Cellular oxygen sensors. *Ann. N.Y. Acad. Sci.* **718**, 3–12.

Acker, H. (1994b). Mechanisms and meaning of cellular oxygen sensing in the organism. *Respir. Physiol.*, **95**, 1–10.

Acker, H. and Xue, D. (1995). Mechanisms of O_2 sensing in the carotid body in comparison with other O_2-sensing cells. *News Physiol. Sci.*, **10**, 211–16.

Acker, T., Fandrey, J. and Acker, H. (2006). The good, the bad and the ugly in oxygen-sensing: ROS, cytochromes and prolyl-hydroxylases. *Cardiovasc. Res.*, **71**, 195–207.

Adair, T. H. (2005). Growth regulation of the vascular system: an emerging role for adenosine. *Am. J. Physiol. Regul. Integr. Comp. Physiol.*, **289**, R283–96.

Appelhoff, R. J., Tian, Y. M., Raval, R. R., *et al.* (2004). Differential function of the prolyl hydroxylases PHD1, PHD2, and PHD3 in the regulation of hypoxia-inducible factor. *J. Biol. Chem.*, **279**, 38458–65.

Archer, S. L., Reeve, H. L., Michelakis, E., *et al.* (1999). O_2 sensing is preserved in mice lacking the gp91 phox subunit of NADPH oxidase. *Proc. Natl. Acad. Sci. USA*, 96, 7944–9.

Babior, B. M. (1984). The respiratory burst of phagocytes. *J. Clin. Invest.*, **73**, 599–601.

Baggiolini, M. and Wymann, M. P. (1990). Turning on the respiratory burst. *Trends Biochem. Sci.*, **15**, 69–72.

Baird, N. A., Turnbull, D. W. and Johnson, E. A. (2006). Induction of the heat shock pathway during hypoxia requires regulation of heat shock factor by hypoxia-inducible factor-1. *J. Biol. Chem.*, **281**, 38675–81.

Bedard, K. and Krause, K. H. (2007). The NOX family of ROS-generating NADPH oxidases: physiology and pathophysiology. *Physiol. Rev.*, **87**, 245–313.

Bell, E. L., Emerling, B. M. and Chandel, N. S. (2005). Mitochondrial regulation of oxygen sensing. *Mitochondrion*, **5**, 322–32.

Berchner-Pfannschmidt, U., Yamac, H., Trinidad, B. and Fandrey, J. (2007). Nitric oxide modulates oxygen sensing by hypoxia-inducible factor 1-dependent induction of prolyl hydroxylase 2. *J. Biol. Chem.*, **282**, 1788–96.

Berra, E., Benizri, E., Ginouves, A., Volmat, V., Roux, D. and Pouyssegur, J. (2003). HIF prolyl-hydroxylase 2 is the key oxygen sensor setting low steady-state levels of HIF-1α in normoxia. *EMBO J.*, **22**, 4082–90.

Berra, E., Ginouves, A. and Pouyssegur, J. (2006). The hypoxia-inducible-factor hydroxylases bring fresh air into hypoxia signalling. *EMBO Rep.*, **7**, 41–5.

Bogdanova, A. and Nikinmaa, M. (2001). Reactive oxygen species regulate oxygen-sensitive potassium flux in rainbow trout erythrocytes. *J. Gen. Physiol.*, **117**, 181–90.

Bracken, C. P., Whitelaw, M. L. and Peet, D. J. (2003). The hypoxia-inducible factors: key transcriptional regulators of hypoxic responses. *Cell. Mol. Life Sci.*, **60**, 1376–93.

Brunori, M. (2001a). Nitric oxide moves myoglobin centre stage. *Trends Biochem. Sci.*, **26**, 209–10.

Brunori, M. (2001b). Nitric oxide, cytochrome-c oxidase and myoglobin. *Trends Biochem. Sci.*, **26**, 21–3.

Brunori, M. and Vallone, B. (2007). Neuroglobin, seven years after. Cell. Mol. Life Sci. **64**, 1259–68.

Bunn, H. F., and Poyton, R. O. (1996). Oxygen sensing and molecular adaptation to hypoxia. *Physiol. Rev.*, **76**, 839–85.

Burleson, M. L., Mercer, S. E. and Wilk-Blaszczak, M. A. (2006). Isolation and characterization of putative O_2 chemoreceptor cells from the gills of channel catfish (*Ictalurus punctatus*). *Brain Res.*, **1092**, 100–7.

Burmester, T., Ebner, B., Weich, B. and Hankeln, T. (2002). Cytoglobin: a novel globin type ubiquitously expressed in vertebrate tissues. *Mol. Biol. Evol.*, **19**, 416–21.

Burmester, T., Haberkamp, M., Mitz, S., Roesner, A., Schmidt, M., Ebner, B., Gerlach, F., Fuchs, C., Hankeln, T. (2004). Neuroglobin and cytoglobin: genes, proteins and evolution. *IUBMB Life*, **56**, 703–7.

Burmester, T., Weich, B., Reinhardt, S. and Hankeln, T. (2000). A vertebrate globin expressed in the brain. *Nature*, **407**, 520–3.

Bussolati, B., Ahmed, A., Pemberton, H., Landis, R. C., Di Carlo, F., Haskard, D. O. and Mason, J. C. (2004). Bifunctional role for VEGF-induced heme oxygenase-1 in vivo: induction of angiogenesis and inhibition of leukocytic infiltration. *Blood*, **103**, 761–6.

Camenisch, G., Wenger, R. H. and Gassmann, M. (2002). DNA-binding activity of hypoxia-inducible factors (HIFs). *Methods Mol. Biol.*, **196**, 117–29.

Chan, W. K., Yao, G., Gu, Y. Z. and Bradfield, C. A. (1999). Cross-talk between the aryl hydrocarbon receptor and hypoxia inducible factor signaling pathways – demonstration of competition and compensation. *J. Biol. Chem.*, **274**, 12115–23.

Chandel, N. S. and Budinger, G. R. S. (2007). The cellular basis for diverse responses to oxygen. *Free Rad. Biol. Med.*, **42**, 165–74.

Chaudary, N., Naydenova, Z., Shuralyova, I. and Coe, I. R. (2004). Hypoxia regulates the adenosine transporter, mENT1, in the murine cardiomyocyte cell line, HL-1. *Cardiovasc. Res.*, **61**, 780–8.

Chavez, J. C., Baranova, O., Lin, J. and Pichiule, P. (2006). The transcriptional activator hypoxia inducible factor 2 (HIF-2/EPAS-1) regulates the oxygen-dependent expression of erythropoietin in cortical astrocytes. *J. Neurosci.*, **26**, 9471–81.

Choi, S. L., Kim, S. J., Lee, K. T., *et al.* (2001). The regulation of AMP-activated protein kinase by H_2O_2. *Biochem. Biophys. Res. Comm.*, **287**, 92–7.

Conde, S. V. and Monteiro, E. C. (2004). Hypoxia induces adenosine release from the rat carotid body. *J. Neurochem.*, **89**, 1148–56.

Dahlgren, C. and Karlsson, A. (1999). Respiratory burst in human neutrophils. *J. Immunol. Methods*, **232**, 3–14.

D'Amico, G., Lam, F., Hagen, T. and Moncada, S. (2006). Inhibition of cellular respiration by endogenously produced carbon monoxide. *J. Cell Sci.*, **119**, 2291–8.

Dayan, F., Roux, D., Brahimi-Horn, M. C., Pouyssegur, J. and Mazure, N. M. (2006). The oxygen sensor factor-inhibiting hypoxia-inducible factor-1 controls expression of distinct genes through the bifunctional transcriptional character of hypoxia-inducible factor-α. *Cancer Res.*, **66**, 3688–98.

Decoursey, T. E. and Ligeti, E. (2005). Regulation and termination of NADPH oxidase activity. *Cell. Mol. Life Sci.*, **62**, 2173–93.

Dejours, P. (1975). *Principles of Comparative Respiratory Physiology*. Amsterdam: Elsevier.

DeLeo, F. R., Allen, L. A., Apicella, M. and Nauseef, W. M. (1999). NADPH oxidase activation and assembly during phagocytosis. *J. Immunol.*, **163**, 6732–40.

Dery, M. A. C., Michaud, M. D. and Richard, D. E. (2005). Hypoxia-inducible factor 1: regulation by hypoxic and non-hypoxic activators. *Int. J. Biochem. Cell Biol.*, **37**, 535–40.

Di Giulio, C. , Bianchi, G., Cacchio, M., *et al.* (2006). Neuroglobin, a new oxygen binding protein is present in the carotid body and increases after chronic intermittent hypoxia. *Adv. Exp. Med. Biol.*, **580**, 15–19.

Dinger, B., He, L., Chen, J., *et al.* (2007). The role of NADPH oxidase in carotid body arterial chemoreceptors. *Respir. Physiol. Neurobiol.*, **157**, 45–54.

Dombkowski, R. A., Doellman, M. M., Head, S. K. and Olson, K. R. (2006). Hydrogen sulfide mediates hypoxia-induced relaxation of trout urinary bladder smooth muscle. *J. Exp. Biol.*, **209**, 3234–40.

Drew, C., Ball, V., Robinson, H., Ellory, J. C. and Gibson, J. S. (2004). Oxygen sensitivity of red cell membrane transporters revisited. *Bioelectrochemistry*, **62**, 153–8.

Dunel-Erb, S., Bailly, Y. and Laurent, P. (1982). Neuroepithelial cells in fish gill primary lamellae. *J. Appl. Physiol.*, **53**, 1342–53.

Duranteau, J., Chandel, N. S., Kulisz, A., Shao, Z. H. and Schumacker, P. T. (1998). Intracellular signaling by reactive oxygen species during hypoxia in cardiomyocytes. *J. Biol. Chem.*, **273**, 11619–24.

Eckardt, K. U. and Kurtz, A. (2005). Regulation of erythropoietin production. *Eur. J. Clin. Invest.*, **35**, 13–19.

Ehleben, W., Bolling, B., Merten, E., Porwol, T., Strohmaier, A. R. and Acker, H. (1998). Cytochromes and oxygen radicals as putative members of the oxygen sensing pathway. *Respir. Physiol.*, **114**, 25–36.

El Benna, J., Dang, P. M., Gougerot-Pocidalo, M. A. and Elbim, C. (2005). Phagocyte NADPH oxidase: a multicomponent enzyme essential for host defenses. *Arch. Immunol. Ther. Exp. (Warsz.)*, **53**, 199–206.

Ellory, J. C., Wolowyk, M. W. and Young, J. D. (1987). Hagfish (*Eptatretus stouti*). erythrocytes show minimal chloride transport activity. *J. Exp. Biol.*, **129**, 377–83.

Eltzschig, H. K., Abdulla, P., Hoffman, E., *et al.* (2005). HIF-1-dependent repression of equilibrative nucleoside transporter (ENT) in hypoxia. *J. Exp. Med.*, **202**, 1493–505.

Engeszer, R. E., Patterson, L. B., Rao, A. A. and Parichy, D. M. (2007). Zebrafish in the wild: a review of natural history and new notes from the field. *Zebrafish*, **4**, 21–40.

Fandrey, J. (2004). Oxygen-dependent and tissue-specific regulation of erythropoietin gene expression. *Am. J. Physiol. Regul. Integr. Comp. Physiol.*, **286**, R977–88.

Fandrey, J., Gorr, T. A. and Gassmann, M. (2006). Regulating cellular oxygen sensing by hydroxylation. *Cardiovasc. Res.*, **71**, 642–51.

Fievet, B., Claireaux, G., Thomas, S. and Motais, R. (1988). Adaptive respiratory responses of trout to acute hypoxia. III. Ion movements and pH changes in the red blood cell. *Respir. Physiol.*, **74**, 99–114.

Finkel, T. (1998). Oxygen radicals and signaling. *Curr. Opin. Cell Biol.*, **10**, 248–53.

Firth, J. D., Ebert, B. L. and Ratcliffe, P. J. (1995). Hypoxic regulation of lactate dehydrogenase A: interaction between hypoxia-inducible factor 1 and cAMP response elements. *J. Biol. Chem.*, **270**, 21021–7.

Fraser, J., de Mello, L. V., Ward, D., *et al.* (2006). Hypoxia-inducible myoglobin expression in nonmuscle tissues. *Proc. Natl. Acad. Sci. USA*, **103**, 2977–81.

Freitas, T. A. K., Saito, J. A., Hou, S. B. and Alam, M. (2005). Globin-coupled sensors, protoglobins, and the last universal common ancestor. *J. Inorg. Biochem.*, **99**, 23–33.

Fridovich, I. (1986a). Biological effects of the superoxide radical. *Arch. Biochem. Biophys.*, **247**, 1–11.

Fridovich, I. (1986b). Superoxide dismutases. *Adv. Enzymol. Relat. Areas Mol. Biol.*, **58**, 61–97.

Fu, X. W., Wang, D. S., Nurse, C. A., Dinauer, M. C. and Cutz, E. (2000). NADPH oxidase is an O_2 sensor in airway chemoreceptors: evidence from K^+ current modulation in wild-type and oxidase-deficient mice. *Proc. Natl. Acad. Sci. USA*, **97**, 4374–9.

Fuchs, C., Luckhardt, A., Gerlach, F., Burmester, T. and Hankeln, T. (2005). Duplicated cytoglobin genes in teleost fishes. *Biochem. Biophys. Res. Comm.*, **337**, 216–23.

Fukuda, R., Zhang, H. F., Kim, J. W., Shimoda, L., Dang, C. V. and Semenza, G. L. (2007). HIF-1 regulates cytochrome oxidase subunits to optimize efficiency of respiration in hypoxic cells. *Cell*, **129**, 111–22.

Fung, M. L. and Tipoe, G. L. (2003). Role of HIF-1 in physiological adaptation of the carotid body during chronic hypoxia. *Adv. Exp. Med. Biol.*, **536**, 593–601.

Galtieri, A., Tellone, E., Romano, L., *et al.* (2002). Band-3 protein function in human erythrocytes: effect of oxygenation-deoxygenation. *Biochim. Biophys. Acta Biomembr.*, **1564**, 214–18.

Gardner, L. B., Li, Q., Park, M. S., Flanagan, W. M., Semenza, G. L. and Dang, C. V. (2001). Hypoxia inhibits G(1)/S transition through regulation of p27 expression. *J. Biol. Chem.*, **276**, 7919–26.

Gess, B., Schricker, K., Pfeifer, M. and Kurtz, A. (1997). Acute hypoxia upregulates NOS gene expression in rats. *Am. J. Physiol. Regul. Integr. Comp. Physiol.*, **273**, R905–10.

Gibson, J. S., Cossins, A. R. and Ellory, J. C. (2000). Oxygen-sensitive membrane transporters in vertebrate red cells. *J. Exp. Biol.*, **203**, 1395–407.

Gilles-Gonzalez, M. A. and Gonzalez, G. (2005). Heme-based sensors: defining characteristics, recent developments, and regulatory hypotheses. *J. Inorg. Biochem.*, **99**, 1–22.

Gilmour, K. M. and Perry, S. F. (2006). Branchial chemoreceptor regulation of cardiorespiratory function. In *Fish Physiology vol. 25: Sensory Systems Neuroscience*, ed. T. J. Hara and B. S. Zielinski. New York: Academic Press, pp. 97–151.

Gloire, G., Legrand-Poels, S. and Piette, J. (2006). NF-B activation by reactive oxygen species: fifteen years later. *Biochem. Pharmacol.*, **72**, 1493–505.

Goda, N., Ryan, H. E., Khadivi, B., McNulty, W., Rickert, R. C. and Johnson, R. S. (2003). Hypoxia-inducible factor 1 α is essential for cell cycle arrest during hypoxia. *Mol. Cell. Biol.*, **23**, 359–69.

Gonzalez, C., Agapito, M. T., Rocher, A., *et al.* (2007). Chemoreception in the context of the general biology of ROS. *Respir. Physiol. Neurobiol.*, **157**, 30–44.

Gonzalez, C., Almaraz, L., Obeso, A. and Rigual, R. (1994). Carotid body chemoreceptors: from natural stimuli to sensory discharges. *Physiol. Rev.*, **74**, 829–98.

Gonzalez, C., Lopez-Lopez, J. R., Obeso, A., Perez-Garcia, M. T. and Rocher, A. (1995a). Cellular mechanisms of oxygen chemoreception in the carotid body. *Respir. Physiol.*, **102**, 137–47.

Gonzalez, C., Vicario, I., Almaraz, L. and Rigual, R. (1995b). Oxygen sensing in the carotid body. *Biol. Signals*, **4**, 245–56.

Gorr, T. A., Cahn, J. D., Yamagata, H. and Bunn, H. F. (2004). Hypoxia-induced synthesis of hemoglobin in the crustacean *Daphnia magna* is hypoxia-inducible factor-dependent. *J. Biol. Chem.*, **279**, 36038–47.

Greijer, A. E., van der Groep, P., Kemming, D., *et al.* (2005). Up-regulation of gene expression by hypoxia is mediated predominantly by hypoxia-inducible factor 1 (HIF-1). *J. Pathol.*, **206**, 291–304.

Guzy, R. D., Hoyos, B., Robin, E., *et al.* (2005). Mitochondrial complex III is required for hypoxia-induced ROS production and cellular oxygen sensing. *Cell Metabolism*, **1**, 401–8.

Guzy, R. D. and Schumacker, P. T. (2006). Oxygen sensing by mitochondria at complex III: the paradox of increased reactive oxygen species during hypoxia. *Exp. Physiol.*, **91**, 807–19.

Haddad, J. J. (2002). Oxygen-sensing mechanisms and the regulation of redox-responsive transcription factors in development and pathophysiology. *Respir. Res.*, **3**, article 26.

Halliwell, B. and Gutteridge, J. M. C. (1984). Oxygen toxicity, oxygen radicals, transition metals and disease. *Biochem. J.*, **219**, 1–14.

Halliwell, B. and Gutteridge, J. M. C. (2007). *Free Radicals in Biology and Medicine*, 4th edn. Oxford: Oxford University Press.

Hamdane, D., Kiger, L., Dewilde, S., *et al.* (2003). The redox state of the cell regulates the ligand binding affinity of human neuroglobin and cytoglobin. *J. Biol. Chem.*, **278**, 51713–21.

Hankeln, T., Ebner, B., Fuchs, C., *et al.* (2005). Neuroglobin and cytoglobin in search of their role in the vertebrate globin family. *J. Inorg. Biochem.*, **99**, 110–19.

Hardie, D. G. (2003). Minireview: the AMP-activated protein kinase cascade – the key sensor of cellular energy status. *Endocrinology*, **144**, 5179–83.

Hardie, D. G., Hawley, S. A. and Scott, J. (2006). AMP-activated protein kinase – development of the energy sensor concept. *J. Physiol. (London)*, **574**, 7–15.

Heise, K., Puntarulo, S., Nikinmaa, M., Abele, D. and Portner, H. O. (2006a). Oxidative stress during stressful heat exposure and recovery in the North Sea eelpout *Zoarces viviparus* L. *J. Exp. Biol.*, **209**, 353–63.

Heise, K., Puntarulo, S., Nikinmaa, M., Lucassen, M., Portner, H. O. and Abele, D. (2006b). Oxidative stress and HIF-1 DNA binding during stressful cold exposure and recovery in the North Sea eelpout (*Zoarces viviparus*). *Comp. Biochem. Physiol. A Mol. Integr. Physiol.*, **143**, 494–503.

Hirota, K. and Semenza, G. L. (2005). Regulation of hypoxia-inducible factor 1 by prolyl and asparaginyl hydroxylases. *Biochem. Biophys. Res. Comm.*, **338**, 610–16.

Hochachka, P. W. and Lutz, P. L. (2001). Mechanism, origin, and evolution of anoxia tolerance in animals. *Comp. Biochem. Physiol. B Biochem. Mol. Biol.*, **130**, 435–59.

Hofer, T., Pohjanvirta, R., Spielmann, P., *et al.* (2004). Simultaneous exposure of rats to dioxin and carbon monoxide reduces the xenobiotic but not the hypoxic response. *Biol. Chem.*, **385**, 291–4.

Infanger, D. W., Sharma, R. V. and Davisson, R. L. (2006). NADPH oxidases of the brain: distribution, regulation, and function. *Antioxid. Redox. Signal.*, **8**, 1583–96.

Ingermann, R. L. and Terwilliger, R. C. (1982). Presence and possible function of Root effect hemoglobins in fishes lacking functional swim bladders. *J. Exp. Zool.*, **220**, 171–7.

Ivan, M., Kondo, K., Yang, H. F., *et al.* (2001). HIFα targeted for VHL-mediated destruction by proline hydroxylation: implications for O_2 sensing. *Science*, **292**, 464–8.

Jaakkola, P., Mole, D. R., Tian, Y. M., *et al.* (2001). Targeting of HIFα to the von Hippel-Lindau ubiquitylation complex by O_2-regulated prolyl hydroxylation. *Science*, **292**, 468–72.

Jelkmann, W. (2007). Erythropoietin after a century of research: younger than ever. *Eur. J. Haematol.*, **78**, 183–205.

Jensen, F. B., Jakobsen, M. H. and Weber, R. E. (1998). Interaction between haemoglobin and synthetic peptides of the N-terminal cytoplasmic fragment of trout Band 3 (AE1) protein. *J. Exp. Biol.*, **201**, 2685–90.

Jonz, M. G., Fearon, I. M. and Nurse, C. A. (2004). Neuroepithelial oxygen chemoreceptors of the zebrafish gill. *J. Physiol. (London)*, **560**, 737–52.

Jurgens, K. D., Papadopoulos, S., Peters, T. and Gros, G. (2000). Myoglobin: just an oxygen store or also an oxygen transporter? *News Physiol. Sci.*, **15**, 269–74.

Kaelin, W. G. (2005). Proline hydroxylation and gene expression. *Annu. Rev. Biochem.*, **74**, 115–28.

Khan, A. I., Drew, C., Ball, S. E., Ball, V., Ellory, J. C. and Gibson, J. S. (2004). Oxygen dependence of K^+-Cl^- cotransport in human red cell ghosts and sickle cells. *Bioelectrochemistry*, **62**, 141–6.

Khomenko, T., Deng, X. M., Sandor, Z., Tarnawski, A. S. and Szabo, S. (2004). Cysteamine alters redox state, HIF-1 α transcriptional interactions and reduces duodenal mucosal oxygenation: novel insight into the mechanisms of duodenal ulceration. *Biochem. Biophys. Res. Comm.*, **317**, 121–7.

Kietzmann, T. and Gorlach, A. (2005). Reactive oxygen species in the control of hypoxia-inducible factor-mediated gene expression. *Sem. Cell Devel. Biol.*, **16**, 474–86.

Kim, K. H., Song, M. J., Chung, J., Park, H. and Kim, J. B. (2005). Hypoxia inhibits adipocyte differentiation in a HDAC-independent manner. *Biochem. Biophys. Res. Comm.*, **333**, 1178–84.

Kline, D. D., Peng, Y. J., Manalo, D. J., Semenza, G. L. and Prabhakar, N. R. (2002). Defective carotid body function and impaired ventilatory responses to chronic hypoxia in mice partially deficient for hypoxia-inducible factor 1 α. *Proc. Natl. Acad. Sci. USA*, **99**, 821–6.

Koivisto, A., Matthias, A., Bronnikov, G. and Nedergaard, J. (1997). Kinetics of the inhibition of mitochondrial respiration by NO. *FEBS Lett.*, **417**, 75–80.

Kong, T. Q., Westerman, K. A., Faigle, M., Eltzschig, H. K. and Colgan, S. P. (2006). HIF-dependent induction of adenosine A2B receptor in hypoxia. *FASEB J.*, **20**, 2242–50.

Kugelstadt, D., Haberkamp, M., Hankeln, T. and Burmester, T. (2004). Neuroglobin, cytoglobin, and a novel, eye-specific globin from chicken. *Biochem. Biophys. Res. Comm.*, **325**, 719–25.

Kumar, P. and Prabhakar, N. (2007). Sensing hypoxia: carotid body mechanisms and reflexes in health and disease. *Respir. Physiol. Neurobiol.*, **157**, 1–3.

Kummer, W. and Acker, H. (1995). Immunohistochemical demonstration of four subunits of neutrophil NAD(P)H oxidase in type I cells of carotid body. *J. Appl. Physiol.*, **78**, 1904–9.

Lahiri, S. and Acker, H. (1999). Redox-dependent binding of CO to heme protein controls PO_2-sensitive chemoreceptor discharge of the rat carotid body. *Respir. Physiol.*, **115**, 169–77.

Lahiri, S., Roy, A., Baby, S. M., Hoshi, T., Semenza, G. L. and Prabhakar, N. R. (2006). Oxygen sensing in the body. *Progr. Biophys. Mol. Biol.*, **91**, 249–86.

Lando, D., Pongratz, I., Poellinger, L. and Whitelaw, M. L. (2000). A redox mechanism controls differential DNA binding activities of hypoxia-inducible factor (HIF) 1α and the HIF-like factor. *J. Biol. Chem.*, **275**, 4618–27.

Law, S. H. W., Wu, R. S. S., Ng, P. K. S., Yu, R. M. K. and Kong, R. Y. C. (2006). Cloning and expression analysis of two distinct HIF-α isoforms – gcHIF-1α and gcHIF-4α – from the hypoxia-tolerant grass carp, *Ctenopharyngodon idellus*. *BMC Mol. Biol.*, **7** art. 15.

Lee, P. J., Jiang, B. H., Chin, B. Y., Iyer, N. V., Alam, J., Semenza, G. L. and Choi, A. M. K. (1997). Hypoxia-inducible factor-1 mediates transcriptional activation of the heme oxygenase-1 gene in response to hypoxia. *J. Biol. Chem.*, **272**, 5375–81.

Lei, B. A., Matsuo, K., Labinskyy, V., Sharma, N., Chandler, M. P., Ahn, A., *et al.* (2005). Exogenous nitric oxide reduces glucose transporters translocation and lactate production in ischemic myocardium in vivo. *Proc. Natl. Acad. Sci. USA*, **102**, 6966–71.

Lesser, M. P. (2006). Oxidative stress in marine environments: biochemistry and physiological ecology. *Annu. Rev. Physiol.*, **68**, 253–78.

Li, R. C., Lee, S. K., Pouranfar, F., *et al.* (2006). Hypoxia differentially regulates the expression of neuroglobin and cytoglobin in rat brain. *Brain Res.*, **1096**, 173–9.

Liu, L. P., Cash, T. P., Jones, R. G., Keith, B., Thompson, C. B. and Simon, M. C. (2006). Hypoxia-induced energy stress regulates mRNA translation and cell growth. *Mol. Cell*, **21**, 521–31.

Liu, Q., Berchner-Pfannschmidt, U., Moller, U., *et al.* (2004). A Fenton reaction at the endoplasmic reticulum is involved in the redox control of hypoxia-inducible gene expression. *Proc. Natl. Acad. Sci. USA*, **101**, 4302–7.

Liu, Y. V. and Semenza, G. L. (2007). RACK1 vs. HSP90 – competition for HIF-1α degradation vs. stabilization. *Cell Cycle*, **6**, 656–9.

Liu, Y. V., Baek, J. H., Zhang, H., Diez, R., Cole, R. N. and Semenza, G. L. (2007). RACK1 competes with HSP90 for binding to HIF-1 alpha and is required for O_2-independent and HSP90 inhibitor-induced degradation of HIF-1 alpha. *Mol. Cell*, **25**, 207–217.

Lopez-Barneo, J. (2003). Oxygen and glucose sensing by carotid body glomus cells. *Curr. Opin. Neurobiol.*, **13**, 493–9.

Lopez-Barneo, J., Pardal, R. and Ortega-Saenz, P. (2001). Cellular mechanism of oxygen sensing. *Annu. Rev. Physiol.*, **63**, 259–87.

Lutz, P. L., and Nilsson, G. E. (2004). Vertebrate brains at the pilot light. *Respir. Physiol. Neurobiol.* **141**, 285–96.

Mammen, P. P. A., Shelton, J. M., Ye, Q., *et al.* (2006). Cytoglobin is a stress-responsive hemoprotein expressed in the developing and adult brain. *J. Histochem. Cytochem.*, **54**, 1349–61.

Manalo, D. J., Rowan, A., Lavoie, T., *et al.* (2005). Transcriptional regulation of vascular endothelial cell responses to hypoxia by HIF-1. *Blood*, **105**, 659–69.

Martin, E. D., Fernandez, M., Perea, G., *et al.* (2007). Adenosine released by astrocytes contributes to hypoxia-induced modulation of synaptic transmission. *Glia*, **55**, 36–45.

Michiels, C., Minet, E., Mottet, D. and Raes, M. (2002). Regulation of gene expression by oxygen: NF B and HIF-1, two extremes. *Free Rad. Biol. Med.*, **33**, 1231–42.

Milsom, W. K. and Burleson, M. L. (2007). Peripheral arterial chemoreceptors and the evolution of the carotid body. *Respir. Physiol. Neurobiol.*, **157**, 4–11.

Nie, M. H., Blankenship, A. L. and Giesy, J. P. (2001). Interactions between aryl hydrocarbon receptor (AhR) and hypoxia signaling pathways. *Environ. Toxicol. Pharmacol.*, **10**, 17–27.

Nikinmaa, M. (1992). Membrane transport and the control of haemoglobin-oxygen affinity in nucleated erythrocytes. *Physiol. Rev.*, **72**, 301–21.

Nikinmaa, M. (2003). Gas transport. In *Red Cell Membrane Transport in Health and Disease*, ed. I. Bernhardt and J. C. Ellory. Berlin: Springer, pp. 489–509.

Nikinmaa, M. (2005). Gas transport. In *The Physiology of Fishes, 3rd edn*, ed. D. H. Evans and J. B. Claiborne. Boca Raton: CRC Press, pp. 153–74.

Nikinmaa, M. and Railo, E. (1987). Anion movements across lamprey (*Lampetra fluviatilis*) red cell membrane. *Biochim. Biophys. Acta*, **899**, 134–6.

Nikinmaa, M. and Rees, B. B. (2005). Oxygen-dependent gene expression in fishes. *Amer. J. Physiol. Regul. Integr. Comp. Physiol.*, **288**, R1079–90.

Nikinmaa, M. and Salama, A. (1998). Oxygen transport in fish. In *Fish Physiology vol. 17, Fish Respiration*, ed. S. F. Perry and B. L. Tufts. New York: Academic Press, pp. 141–84.

O'Driscoll, C. M. and Gorman, A. M. (2005). Hypoxia induces neurite outgrowth in PC12 cells that is mediated through adenosine A2A receptors. *Neuroscience*, **131**, 321–9.

Oehme, F., Ellinghaus, P., Kolkhof, P., *et al.* (2002). Overexpression of PH-4, a novel putative proline 4-hydroxylase, modulates activity of hypoxia-inducible transcription factors. *Biochem. Biophys. Res. Comm.*, **296**, 343–9.

Olson, K. R., Dombkowski, R. A., Russell, M. J., *et al.* (2006). Hydrogen sulfide as an oxygen sensor/transducer in vertebrate hypoxic vasoconstriction and hypoxic vasodilation. *J. Exp. Biol.*, **209**, 4011–23.

Ordway, G. A. and Garry, D. J. (2004). Myoglobin: an essential hemoprotein in striated muscle. *J. Exp. Biol.*, **207**, 3441–6.

Otto, C. M. and Baumgardner, J. E. (2001). Effect of culture PO_2 on macrophage (RAW 264.7) nitric oxide production. *Am. J. Physiol. Cell Physiol.*, **280**, C280–7.

Pardal, R. and Lopez-Barneo, J. (2004). Combined oxygen and glucose sensing in the carotid body. *Undersea Hyperbaric Med.*, **31**, 113–21.

Pelster, B. and Scheid, P. (1992). Countercurrent concentration and gas secretion in the fish swimbladder. *Physiol. Zool.*, **65**, 1–16.

Perry, S. F. and Gilmour, K. M. (2002). Sensing and transfer of respiratory gases at the fish gill. *J. Exp. Zool.*, **293**, 249–63.

Pesce, A., Bolognesi, M., Bocedi, A., *et al.* (2002). Neuroglobin and cytoglobin. Fresh blood for the vertebrate globin family. *EMBO Rep.*, **3**, 1146–51.

Pettersen, E. O., Larsen, L. H., Ramsing, N. B. and Ebbesen, P. (2005). Pericellular oxygen depletion during ordinary tissue culturing, measured with oxygen microsensors. *Cell Prolif.*, **38**, 257–67.

Porwol, T., Ehleben, W., Brand, V. and Acker, H. (2001). Tissue oxygen sensor function of NADPH oxidase isoforms, an unusual cytochrome aa3 and reactive oxygen species. *Respir. Physiol.*, **128**, 331–48.

Prabhakar, N. R. (2006). O_2 sensing at the mammalian carotid body: why multiple O_2 sensors and multiple transmitters? *Exp. Physiol.*, **91**, 17–23.

Rankin, E. B., Biju, M. P., Liu, *et al.* (2007). Hypoxia-inducible factor-2 (HIF-2) regulates hepatic erythropoietin in vivo. *J. Clin. Invest.*, **117**, 1068–77.

Ratcliffe, P. J. (2007). HIF-1 and HIF-2: working alone or together in hypoxia? *J. Clin. Invest.*, **117**, 862–5.

Rees, B. B., Bowman, J. A. and Schulte, P. M. (2001). Structure and sequence conservation of a putative hypoxia response element in the lactate dehydrogenase-B gene of *Fundulus*. *Biol. Bull.*, **200**, 247–51.

Rees, B. B., Figueroa, Y. G., Wiese, T. E., Beckman, B. S. and Schulte, P. M. (2009). A novel hypoxia response element in the lactate dehydrogenase-B gene of the killi fish *Fundulus heteroclitus*. *Comp. Biochem. Physiol. A.* **154**, 70–77.

Rissanen, E., Tranberg, H. K. and Nikinmaa, M. (2006a). Oxygen availability regulates metabolism and gene expression in trout hepatocyte cultures. *Am. J. Physiol. Regul. Integr. Comp. Physiol.*, **291**, R1507–15.

Rissanen, E., Tranberg, H. K., Sollid, J., Nilsson, G. E. and Nikinmaa, M. (2006b). Temperature regulates hypoxia-inducible factor-1 (HIF-1) in a poikilothermic vertebrate, crucian carp (*Carassius carassius*). *J. Exp. Biol.*, **209**, 994–1003.

Roesner, A., Hankeln, T. and Burmester, T. (2006). Hypoxia induces a complex response of globin expression in zebrafish (*Danio rerio*). *J. Exp. Biol.*, **209**, 2129–37.

Roesner, A., Fuchs, C., Hankeln, T. and Burmester, T. (2005). A globin gene of ancient evolutionary origin in lower vertebrates: evidence for two distinct globin families in animals. *Mol. Biol. Evol.*, **22**, 12–20.

Rotrosen, D., Yeung, C. L., Leto, T. L., Malech, H. L. and Kwong, C. H. (1992). Cytochrome b558: the flavin-binding component of the phagocyte NADPH oxidase. *Science*, **256**, 1459–62.

Roux, J. C., Brismar, H., Aperia, A. and Lagercrantz, H. (2005). Developmental changes in HIF transcription factor in carotid body: relevance for O_2 sensing by chemoreceptors. *Pediatr. Res.*, **58**, 53–7.

Roy, A., Rozanov, C., Mokashi, A., *et al.* (2000). Mice lacking in gp91 phox subunit of NAD(P)H oxidase showed glomus cell $[Ca^{2+}]_i$ and respiratory responses to hypoxia. *Brain Res.*, **872**, 188–93.

Rytkonen, K. T., Vuori, K. A. M., Primmer, C. R. and Nikinmaa, M. (2007). Comparison of hypoxia-inducible factor-1α in hypoxia-sensitive and hypoxia-tolerant fish species. *Comp. Biochem. Physiol. D: Genom. Proteom.*, **2**, 177–86.

Sanders, K. A., Sundar, K. M., He, L., Dinger, B., Fidone, S. and Hoidal, J. R. (2002). Role of components of the phagocytic NADPH oxidase in oxygen sensing. *J. Appl. Physiol.*, **93**, 1357–64.

Schmidt, M., Gerlach, F., Avivi, A., *et al.* (2004). Cytoglobin is a respiratory protein in connective tissue and neurons, which is up-regulated by hypoxia. *J. Biol. Chem.*, **279**, 8063–9.

Schnell, P. O., Ignacak, M. L., Bauer, A. L., Striet, J. B., Paulding, W. R. and Czyzyk-Krzeska, M. F. (2003). Regulation of tyrosine hydroxylase promoter activity by the von Hippel-Lindau tumor suppressor protein and hypoxia-inducible transcription factors. *J. Neurochem.*, **85**, 483–91.

Semenza, G. L. (2004). Hydroxylation of HIF-1: oxygen sensing at the molecular level. *Physiology*, **19**, 176–82.

Shams, I., Nevo, E. and Avivi, A. (2004). Ontogenetic expression of erythropoietin and hypoxia-inducible factor-1α genes in subterranean blind mole rats. *FASEB J.*, **19**, 307–9.

Soliz, J., Joseph, V., Soulage, C., *et al.* (2005). Erythropoietin regulates hypoxic ventilation in mice by interacting with brainstem and carotid bodies. *J. Physiol. (London)*, **568**, 559–71.

Stensløkken, K. O., Sundin, L., Renshaw, G. M. C. and Nilsson, G. E. (2004). Adenosinergic and cholinergic control mechanisms during hypoxia in the epaulette shark (*Hemiscyllium ocellatum*), with emphasis on branchial circulation. *J. Exp. Biol.*, **207**, 4451–61.

Stiehl, D. P., Jelkmann, W., Wenger, R. H. and Hellwig-Burgel, T. (2002). Normoxic induction of the hypoxia-inducible factor 1α by insulin and interleukin-1β involves the phosphatidylinositol 3-kinase pathway. *FEBS Lett.*, **512**, 157–62.

Takagi, H., King, G. L., Robinson, G. S., Ferrara, N. and Aiello, L. P. (1996). Adenosine mediates hypoxic induction of vascular endothelial growth factor in retinal pericytes and endothelial cells. *Invest. Ophthalmol. Visual Sci.*, **37**, 2165–76.

Tetens, V. and Christensen, N. J. (1987). Beta-adrenergic control of blood oxygen affinity in acutely hypoxia exposed rainbow trout. *J. Comp. Physiol. B*, **157**, 667–75.

Treinin, M., Shliar, J., Jiang, H. Q., Powell-Coffman, J. A., Bromberg, Z. and Horowitz, M. (2003). HIF-1 is required for heat acclimation in the nematode *Caenorhabditis elegans*. *Physiol. Genomics*, **14**, 17–24.

Triantafyllou, A., Liakos, P., Tsakalof, A., *et al.* (2007). The flavonoid quercetin induces hypoxia-inducible factor-1α (HIF-1α) and inhibits cell proliferation by depleting intracellular iron. *Free Rad. Res.*, **41**, 342–56.

Tufts, B. L. and Boutilier, R. G. (1989). The absence of rapid chloride/bicarbonate exchange in lamprey erythrocytes: implications for CO_2 transport and ion distributions between plasma and erythrocytes in the blood of *Petromyzon marinus*. *J. Exp. Biol.*, **144**, 565–76.

van der Meer, D. L. M., Van den Thillart, G. E. E. J., Witte, F., *et al.* (2005). Gene expression profiling of the long-term adaptive response to hypoxia in the gills of adult zebrafish. *Am. J. Physiol. Regul. Integr. Comp. Physiol.*, **289**, R1512–19.

Virkki, L. V., Salama, A. and Nikinmaa, M. (1998). Regulation of ion transport across lamprey (*Lampetra fluviatilis*) erythrocyte membrane by oxygen tension. *J. Exp. Biol.*, **201**, 1927–37.

Wang, R. (2003). The gasotransmitter role of hydrogen sulphide. Antioxid. *Redox Signal.*, **5**, 493–501.

Weber, R. E., Voelter, W., Fago, A., Echner, H., Campanella, E. and Low, P. S. (2004). Modulation of red cell glycolysis: interactions between vertebrate hemoglobins and cytoplasmic domains of band 3 red cell membrane proteins. *Am. J. Physiol. Regul. Integr. Comp. Physiol.*, **287**, R454–64.

Wenger, R. H. (2000). Mammalian oxygen sensing, signalling and gene regulation. *J. Exp. Biol.*, **203**, 1253–63.

Wenger, R. H. and Gassmann, M. (1996). Little difference. *Nature*, **380**, 100.

Wientjes, F. B. and Segal, A. W. (1995). NADPH oxidase and the respiratory burst. *Semin. Cell Biol.*, **6**, 357–65.

Williams, S. E. J., Wootton, P., Mason, H. S., *et al.* (2004). Hemoxygenase-2 is an oxygen sensor for a calcium-sensitive potassium channel. *Science*, **306**, 2093–7.

Wittenberg, J. B. and Wittenberg, B. A. (2003). Myoglobin function reassessed. *J. Exp. Biol.*, **206**, 2011–20.

Wolin, M. S., Ahmad, M. and Gupte, S. A. (2005). Oxidant and redox signaling in vascular oxygen sensing mechanisms: basic concepts, current controversies, and potential importance of cytosolic NADPH. *Am. J. Physiol. Lung Cell. Mol. Physiol.*, **289**, L159–73.

Wyatt, C. N., and Evans, A. M. (2007). AMP-activated protein kinase and chemotransduction in the carotid body. *Respir. Physiol. Neurobiol.*, **157**, 22–9.

Wyatt, C. N., Mustard, K. J., Pearson, S. A., *et al.* (2007). AMP-activated protein kinase mediates carotid body excitation by hypoxia. *J. Biol. Chem.*, **282**, 8092–8.

Yun, H., Lee, M., Kim, S. S. and Ha, J. (2005). Glucose deprivation increases mRNA stability of vascular endothelial growth factor through activation of AMP-activated protein kinase in DU145 prostate carcinoma. *J. Biol. Chem.*, **280**, 9963–72.

Zhang, M., Buttigieg, J. and Nurse, C. A. (2007). Neurotransmitter mechanisms mediating low-glucose signalling in cocultures and fresh tissue slices of rat carotid body. *J. Physiol. (London)*, **578**, 735–50.

3

Oxygen uptake and transport in water breathers

STEVE F. PERRY AND KATHLEEN M. GILMOUR

3.1 Introduction

Empirical studies of O_2 uptake and blood O_2 transport in fish began at least 100 years ago with the pioneering work of August Krogh (Krogh, 1904; Krogh and Leitch, 1919), who in 1941 published the seminal book on comparative respiratory physiology (*Comparative Physiology of Respiratory Mechanisms*, 1941). Catalyzed by the research of later-generation visionaries (van Dam, Scholander, Dejours, Johansen, Hughes, Shelton, Piiper, Randall, and Simpson), extensive research continues to examine the mechanisms of O_2 uptake and transport within the blood of fish. In this chapter we focus our attention on O_2 uptake and blood O_2 transport in entirely aquatic water-breathing fishes; Chapter 4 is devoted to modes of O_2 uptake in air-breathing fishes. Although some water-breathing species use skin as a supplementary route of O_2 uptake (Graham, 1997) (see Chapter 6), the gill is the predominant organ for gas transfer. Thus, in this chapter we will focus exclusively on the gill. Numerous reviews have been written previously on branchial O_2 uptake and blood O_2 transport (e.g. Jones and Randall, 1978; Randall *et al.*, 1982; Randall and Daxboeck, 1984; Malte and Weber, 1985; Butler and Metcalfe, 1988; Weber and Jensen, 1988; Cameron, 1989; Nikinmaa and Tufts, 1989; Perry and Wood, 1989; Piiper, 1989; Piiper, 1990; Randall, 1990; Thomas and Motais, 1990; Jensen, 1991; Nikinmaa, 1992; Thomas and Perry, 1992; Fritsche and Nilsson, 1993; Perry and McDonald, 1993; Nikinmaa and Boutilier, 1995; Val, 1995; Brauner and Randall, 1996; Gilmour, 1997; Nikinmaa, 1997; Val, 2000; Nikinmaa, 2001; Perry and Gilmour, 2002; Jensen, 2004; Graham, 2006; Nikinmaa, 2006). The reader is encouraged to consult these review articles, many of which are more detailed than the current overview. The intent of this review is to first cover basic concepts of O_2 transfer and transport and then to

Respiratory Physiology of Vertebrates: Life with and without Oxygen, ed. Göran E. Nilsson. Published by Cambridge University Press. © Cambridge University Press 2010.

apply these principles in understanding regulatory responses aimed at optimizing O_2 transfer and transport during stress.

3.2   Structure of the gill: integrating design and function

The functions of the gill are numerous and include respiratory gas transfer, nitrogenous waste excretion, ionic regulation, and acid–base balance. For a detailed description of fish gill structure and function, readers are encouraged to consult the comprehensive review by Evans et al. (2005). Although generally specialized for the transfer of respiratory gases between water and blood, there is tremendous diversity in gill structure and function across species, with striking differences occurring through the evolution of the primitive jawless fishes (Superclass Agnatha) to the most advanced bony fishes (Superclass Osteichthyes). In this chapter we will focus on the two groups of fish that have been most extensively studied, the cartilaginous elasmobranchs (Class Chondrichthyes, Subclass Elasmobranchii) and the ray-finned teleosts (Class Actinopterygii, Infraclass Teleostei). Detailed information concerning the structure and function of agnathan gills can be found in Strahan (1958), Bartels (1998), and Malte and Lomholt (1998).

A thorough comprehension of gill structure (external and internal) is a key first step in understanding branchial O_2 uptake. In teleosts and elasmobranchs, the gill is comprised of eight branchial arches (four on either side of the head) that are contained within two opercular cavities. In teleosts the gills are protected by an opercular flap, whereas in elasmobranchs the gills are enclosed by a layer of skin perforated with gill slits. In either case, the gills are arranged to form a sieve-like structure across which the inspired water flows (Fig. 3.1). Water flow across the gill is driven by pressure gradients between the buccal and opercular cavities that originate from oscillating buccal/spiracular and opercular/gill slit movements (see below). Protruding from the branchial arches are rows of filaments (termed primary lamellae in older literature) from which the lamellae (termed secondary lamellae in older literature) protrude (Figs 3.1 and 3.2). It is the presence of thousands of lamellae that impart the vast surface area required for high rates of O_2 uptake. Lamellar surface area (like diffusion distance; see below) is highly variable among species, being relatively low in inactive benthic species and high in active pelagic species. The parallel arrangement of the lamellae on the filaments allows water to flow unidirectionally though interlamellar water channels (Fig. 3.1). The flow of blood within the lamellae in the opposite direction allows for highly efficient counter-current gas exchange whereby arterial partial pressure of O_2 (PaO_2) can markedly exceed the PO_2 of expired (exhalent) water (see below).

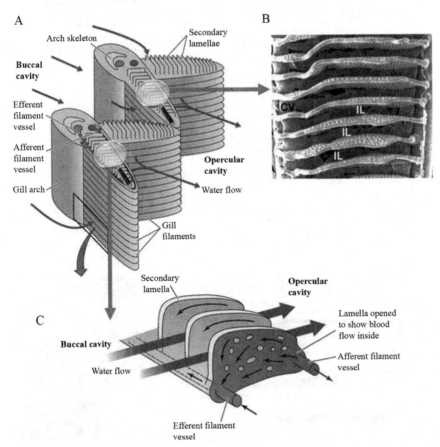

Fig. 3.1 (A) Diagrammatic representation of the gill sieve, formed by adjacent filaments, through which ventilatory water flows from the buccal to opercular cavities. Reproduced from Hill *et al.* (2004). (B) A corrosion cast of a walking catfish (*Claria batrachus*) gill showing the parallel arrangement of lamellae resulting in the formation of discrete interlamellar (IL) water channels (CV = collateral vessel); from Olson (2002) with permission. (C) Water and blood flow in opposite directions to allow counter-current gas transfer. Reproduced from Hill *et al.* (2004).

The pattern of blood flow through the gill is depicted in Fig. 3.2. Two distinct circulatory pathways (arterio-arterial and arterio-venous) are present (Olson, 2002), but only the arterio-arterial circuit contributes to O_2 transfer. Within the arterio-arterial pathway, partially deoxygenated blood enters the branchial arch via an afferent branchial artery (ABA). Afferent filament arteries (AFAs) branching from the ABA provide blood to individual filaments. In turn, afferent lamellar arterioles are derived from the AFA to enable lamellar perfusion. Oxygenated blood drains from lamellae through efferent lamellar arterioles that connect with an efferent filament artery. Blood is then delivered to an

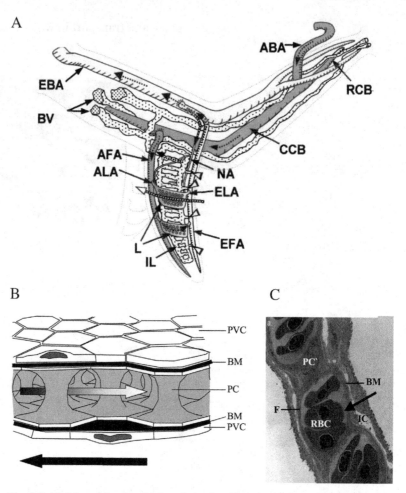

Fig. 3.2 (A) Schematic of major vessels in the gill arch and filament. The afferent branchial artery (ABA) enters the arch and bifurcates into a recurrent branch (RCB) that proceeds anterioventrally and a concurrent branch (CCB) that continues posteriodorsally. The respiratory (arterio-arterial) circulation in the filament consists of the afferent and efferent filamental arteries (AFA, EFA) and arterioles (ALA, ELA) and the lamellae (L). This is drained from the arch by the efferent branchial artery (EBA). Interlamellar vessels (IL) traverse the filament and are supplied by small feeder vessels (arrowheads) from the EFA or by nutrient vessels (NA) that arise from the basal EFA and EBA. The IL system presumably is drained from the arch by the branchial veins (BV). Thin, dotted arrows indicate direction of blood flow; large, white-on-black dotted arrow indicates path of water flow across lamellae. (B) Schematic cross-section through the lamella. Pillar cells (PC) define the blood space (indicated by the shaded arrow representing the oxygenation of blood). PCs rest on a basement membrane (BM), with the water-facing surfaces of the lamella being made up largely of pavement cells (PVC). Modified from Gilmour *et al.* (2007). (C) A transmission electron micrograph showing a gill lamella in cross-section. The water-to-blood diffusion barrier (arrow) is composed of one or more layers of lamellar epithelial cells, an interstitial compartment (IC), a basement membrane, plasma membrane of pillar cell (PC), flanges (F), and a volume of plasma.

efferent branchial artery that ultimately provides flow to the dorsal aorta and systemic circulation.

The flow of blood within individual lamellae occurs through channels formed by H-shaped pillar cells that span the width of the lamella (Fig. 3.2B); true capillaries (i.e. blood vessels formed from a single layer of endothelial cells) do not exist in the fish gill. Elongated projections (termed 'pillar cell flanges') of consecutive pillar cells are joined to delineate the perimeter of the channel. O_2 uptake occurs within the lamella across a water-to-blood diffusion pathway that is illustrated in Fig. 3.2C. The diffusion pathway consists of a barrier separating the inspired water from red blood cells (RBCs) within lamellar blood channels. The barrier is composed of one or more layers of lamellar epithelial cells, an interstitial compartment, a basement membrane, the plasma membrane of pillar cell flanges, and a volume of plasma. Diffusion distances vary markedly among species according to lifestyle and habitat. Typically, hypoxia-intolerant or active fishes possess short diffusion distances (e.g. about 0.5 μm in high-performance fishes such as tuna) (Wegner et al., 2006), whereas hypoxia-tolerant or inactive fishes exhibit longer diffusion distances (e.g. about 10 μm in bullhead) (Hughes and Morgan, 1973).

The majority of teleosts and elasmobranchs actively ventilate their gills. In teleosts, water flow is accomplished by the combined actions of cyclical buccal and opercular pumps operating out of phase to create pressure gradients that drive water flow. In the first phase of the cycle, negative pressures are created in the buccal cavity by an expansion of volume associated with a lowering of the floor of the mouth while the mouth is open. With the opercular flaps (valves) closed, water enters the mouth. In the next phase, the mouth is closed and the buccal cavity is compressed while the opercular cavity expands. This creates a pressure differential that allows water to flow from buccal to opercular cavities across the low-resistance gill sieve (note, however, that some resistance is required to establish the pressure difference across the sieve). Finally, the opercular valves open while the opercular cavity is compressed, enabling the expired water to leave. Ventilatory water flow in elasmobranchs follows similar principles except that water has two routes of entry into the buccal chamber, via the mouth and a pair of spiracles (valved openings) on top of the head. With compression of the buccal chamber, water flows across the gills and exits though external gill slits. High-performance fishes including tuna and mackerel do not actively ventilate their gills but instead swim with their mouths continually open, which forces water across the gills, in a process termed 'ram ventilation.' Ram ventilation is a metabolically efficient strategy that couples the costs of locomotion and ventilation. Fish that normally actively ventilate their gills at rest or during moderate swimming may switch to ram ventilation during high-speed swimming. Given the high costs associated with actively pumping water (Jones and Schwarzfeld, 1974), significant energetic savings can be associated with the switch to ram ventilation.

3.3   Basic principles of O_2 transfer across the gill

Blood arrives at the gill in the partially deoxygenated state, with the extent of deoxygenation at any given ambient PO_2 being largely determined by aerobic metabolic rate. Under resting normoxic conditions, mixed venous blood entering the gill is usually more than 50% saturated with O_2, thereby providing a reservoir of O_2 that can be mobilized during exercise or hypoxia. At the gill, O_2 enters the bloodstream, owing to the combined actions of diffusion and convection. Oxygen diffuses from the inspired water into the blood because of a favorable partial pressure gradient (the blood-to-water PO_2 gradient [ΔPO_2]) that is sustained by ventilation and perfusion. Thus, O_2 molecules entering the blood are carried away by perfusion, while those same O_2 molecules leaving the water are replaced by ventilation. In addition to convection, ΔPO_2 is sustained by the binding of O_2 to hemoglobin. Basically, hemoglobin acts as a sink for O_2 to minimize changes in physically dissolved (gaseous) O_2 within the plasma and thereby reduce the rate at which blood PO_2 increases (O_2 bound to hemoglobin does not exert partial pressure). For example, blood lacking hemoglobin (as in the Antarctic icefish [*Chaenocephalus aceratus*]) would rapidly reach equilibrium with inspired water oxygen levels because O_2 molecules accumulating solely in the blood (plasma) would rapidly increase PO_2. The ability of branchial blood to approach or reach equilibrium with inspired water is a measure of O_2 uptake efficiency. Clearly, it is desirable for O_2 uptake across the gill to be as efficient as possible, but it is important to realize that efficiency alone is not a reliable predictor of overall O_2 uptake. Branchial O_2 uptake in the Antarctic icefish is an excellent (albeit extreme) example of the potential disparity between uptake efficiency and molar quantities of O_2 transferred. With its low-capacitance blood, the icefish gill is likely to be highly efficient at O_2 uptake, yet the blood exiting the gill, although near equilibrium with inspired water, contains low concentrations of O_2. Thus, the molar quantities of O_2 transferred into a given volume of blood perfusing the icefish gill are very low. To achieve adequate rates of O_2 uptake to satisfy metabolism, the icefish must deliver higher volumes of blood to the gill per unit time relative to fish with greater blood O_2-carrying capacities (capacitance). Consequently, icefish have an unusually large heart, capable of sustaining exceptionally high rates of cardiac output (Holeton, 1970). Thus, in terms of optimal design, O_2 transfer at the gill should be highly efficient while also being able to achieve high overall rates of uptake without incurring unreasonable energetic costs of perfusion and ventilation.

A useful way to think about O_2 uptake is within the framework of transit time limitations. That is, the velocity of blood flow within the lamellar circulation dictates the amount of time available for O_2 diffusion. Under normal conditions, the blood may reside in the lamellar circulation for 1–3 s, and it is only during this brief period (or transit time) that diffusion may occur. Thus, conditions for O_2 diffusion must be sufficient to allow the blood to approach equilibrium with inspired water in this brief interval. For any given ΔPO_2, the rate of diffusion (referred to as diffusion conductance) is largely dictated by the water-to-blood diffusion distance and Krogh's permeation coefficient (diffusion constant · capacitance). If diffusion conductance is too low, equilibrium will not occur and, moreover, a reduction in transit time would cause a decrease in O_2 uptake efficiency and a fall in PaO_2. Gills that exhibit reduced O_2 uptake efficiency with lowered transit times are said to be diffusion limited. A characteristic feature of diffusion limitations is a fall in PaO_2 during exercise, when transit times are reduced because of increased cardiac output. Typically, fish gills are not considered to be diffusion limited with respect to O_2 transfer, although few data are available for species other than rainbow trout (Desforges et al., 2002). However, diffusion limitations on O_2 uptake may occur in the trout gill during hypoxia when ΔPO_2 is reduced (Greco et al., 1995). With its absence of diffusion limitations under normoxic conditions, the trout gill is considered to be perfusion limited for O_2 uptake. In perfusion-limited systems, O_2 uptake efficiency remains constant over the physiological range of gill transit times, and thus O_2 uptake increases as a linear function of increased perfusion (e.g. a doubling of cardiac output would result in a twofold increase in O_2 uptake). By contrast, O_2 uptake would increase less than twofold with a doubling of cardiac output in a fish experiencing gill diffusion limitations because of a drop in arterial blood PO_2 and hence O_2 concentration $[O_2]$. The severity of the reduction in $[O_2]$ would depend on the placement of the initial PaO_2 on the O_2 equilibrium curve and the extent of the PaO_2 decrease. On the one hand, at high PaO_2 the O_2 equilibrium curve is relatively shallow, and thus changes in PaO_2 in this zone will have limited impact on $[O_2]$. On the other hand, small decreases in PaO_2 on the steep portion of the O_2 equilibrium curve will promote marked reductions in $[O_2]$, and consequently significant impairment of O_2 uptake. Needless to say, there is obvious benefit to perfusion-limited branchial O_2 transfer, whereby cardiac output can be manipulated to regulate overall O_2 uptake without reducing PaO_2.

Interestingly, CO_2, a more diffusible gas than O_2 (Cameron, 1989), exhibits apparent diffusion limitations in the gill of rainbow trout (Julio et al., 2000; Desforges et al., 2002). Thus, $PaCO_2$ increases with reduced gill transit time and vice versa (Desforges et al., 2002). The underlying basis for CO_2 transfer behaving as a diffusion-limited system is depicted in Fig. 3.4. For O_2, diffusion begins as

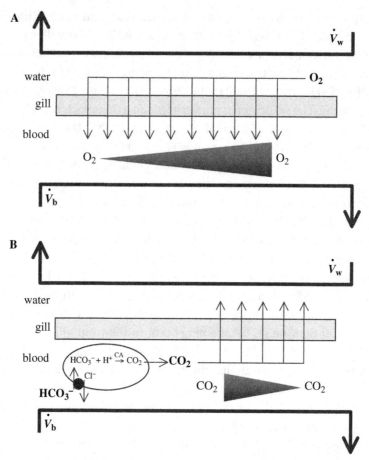

Fig. 3.3 Schematic models of O_2 (A) and CO_2 (B) transfer across the gills of typical teleost fish. In theory, any area of the gill that is both ventilated and perfused is available for O_2 diffusion, and O_2 transfer is limited primarily by perfusion. For CO_2 transfer, the conversion of plasma HCO_3^- to molecular CO_2 is limited by the slow rate of entry of HCO_3^- ions into the red blood cell. Thus, the surface area available for CO_2 diffusion out of the blood is limited (i.e. the functional surface area for CO_2 diffusion is less than the total surface area). Because of these chemical equilibrium limitations, branchial CO_2 transfer in teleost fish behaves as a diffusion-limited system. CA, carbonic anhydrase; $\dot{V}_b$, blood flow; $\dot{V}_w$, water flow.

soon as blood enters the lamellar circulation and theoretically can continue until blood exits the lamella. Thus, for O_2, diffusion can occur over the entire transit time. However, for CO_2 excretion, an initial period of transit within the lamella is required to convert plasma HCO_3^- to gaseous (diffusible) CO_2 (Perry, 1986; Tufts and Perry, 1998). Therefore, the actual transit time available for CO_2

diffusion is less than for O_2 diffusion, and this is the basis of the apparent diffusion limitations on branchial CO_2 excretion.

An obvious hallmark of optimal 'design' in a gas-exchange organ is that the rate of O_2 delivery to the respiratory surfaces should more or less equal the rate of O_2 removal by the circulation. The ratio of O_2 delivery to O_2 removal is termed the 'capacity rate ratio' (Hughes and Shelton, 1962). Because of the much lower O_2 capacitance of water relative to blood, an equivalent volume of water contains approximately 8–$30 \times$ lower total O_2 than blood. Thus, to achieve a capacity rate ratio of approximately 1, ventilatory water flow must markedly exceed blood flow. Typically, ventilation/perfusion ratios in fish range between 10 and 20 (e.g. ~10 in rainbow trout) (Cameron and Davis, 1970). The need for a high ventilation/perfusion ratio contributes to the high metabolic costs of ventilating water.

Blood and water flow in opposite directions across the respiratory surfaces of the gill. The principal advantage of this counter-current arrangement of blood and water flows is that diffusion gradients can be sustained during the entire period of gill transit. Thus, the blood entering the gill with low PO_2 exchanges gas with the inspired water containing the lowest PO_2 (i.e. the water that is about to be expired); O_2 transfer from the inspired water increases the PO_2 of the blood until the blood exiting the gill exhibits a PO_2 that may be significantly higher than that of expired water PO_2.

3.3.1   Gill remodeling and the osmorespiratory compromise

The crucial factors regulating the rate of branchial O_2 transfer have been extensively detailed in previous reviews (e.g. Randall and Daxboeck, 1984; Malte and Weber, 1985; Perry and Wood, 1989; Randall, 1990; Perry and McDonald, 1993; Gilmour, 1997; Piiper, 1998; Perry and Gilmour, 2002; Evans et al., 2005; Graham, 2006) and include diffusive conductance, convection (ventilation and perfusion), and the blood-to-water PO_2 gradient (ΔPO_2). The diffusive conductance of the fish gill is determined by functional surface area, diffusion distance and Krogh's permeation coefficient (diffusion constant · capacitance). Functional surface area and diffusion distance are labile and can be dynamically adjusted according to metabolic requirements or environmental conditions, and there are likely to be distinct advantages from such adjustments. In essence, the factors that favor high rates of branchial O_2 transfer, i.e. high surface area and a small diffusion distance, will also result in higher rates of obligatory salt and water movement across the gill. Given the relatively high costs of actively absorbing salts in fresh water and actively excreting salts in sea water, it is perhaps not surprising that diffusive conductance is matched to gas-transfer requirements, a phenomenon termed the 'osmorespiratory compromise.' Thus, diffusive conductance is kept as low as is possible without affecting O_2 delivery under resting and normoxic conditions, in order to reduce obligatory salt and water movement across the gill and minimize the energetic cost of ion pumping. Acute changes in functional surface area can be achieved by recruiting previously unperfused lamellae (lamellar recruitment) or by more uniformly perfusing individual lamellae (Booth, 1979; Farrell et al., 1980). More chronic (hours to days) and dramatic changes in functional surface area are accomplished in some species by physical covering/uncovering of lamellae (see Fig. 5.2 in Chapter 5) (Sollid et al., 2003; Brauner et al., 2004; Sollid et al., 2005; Ong et al., 2007). Crucian carp (Carassius carassius), goldfish (Carassius auratus), and mangrove killifish (Kryptolebias marmoratus) exhibit reversible gill remodeling in accordance with changes in O_2 demand or availability, whereas the arapaima (Arapaima gigas) exhibits a permanent remodeling of lamellar structure in association with a developmental transition from water to air breathing (Brauner et al. 2004). Gill remodeling is accomplished by the invasion or retraction of an interlamellar cell mass (ILCM), although the signaling mechanisms underlying proliferation of the ILCM or its removal by apoptosis are unknown (Sollid and Nilsson, 2006; Nilsson, 2007). Crucian carp and goldfish have received most attention to date. In these species, the ILCM is present in fish acclimated to cold normoxic water but is retracted in fish exposed to increasing temperature (Sollid et al. 2005) or hypoxia (Sollid et al. 2003). Thus, diffusive conductance is enhanced under conditions that require optimization of gill O_2 extraction, during periods of increased metabolism, or hypoxia. The opposite situation holds for the amphibious mangrove killifish, in which the ILCM appears when fish are exposed to aerial conditions where the gill is not functional (Ong et al. 2007).

Intuitively, ILCM appearance and the associated loss of functional surface area are predicted to be beneficial, owing to a reduction in obligatory (passive) movements of ions and water, in accordance with the concept of osmorespiratory compromise. As yet, however, experimental data in support of this notion are both scarce and indirect (plasma Cl^- levels in crucian carp with or without ILCM) (Sollid et al. 2003). A loss of ion uptake capacity in freshwater fish exhibiting an ILCM might also be predicted, because the mitochondria-rich cells (MRCs) thought to be responsible for ion uptake typically are located at the base of filaments or in the interlamellar regions, where presumably they would be covered by growth of the ILCM. However, recent research (D. Mitrovic and S. F. Perry, unpublished) indicates that the MRCs in goldfish migrate with the edge of the ILCM, thereby remaining in contact with the respiratory water (Fig. 3.5). Even so, the functional capacity of MRCs on the edge of the ILCM, where there is no obvious blood supply (G. E. Nilsson, personal communication), may be

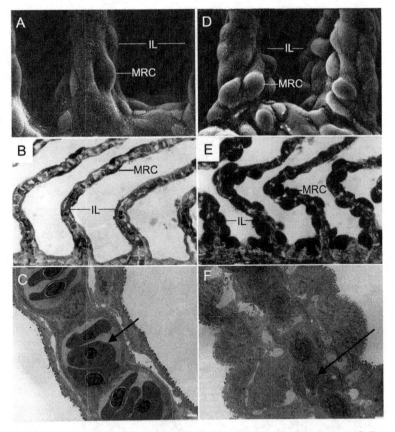

Fig. 3.4 Scanning electron (A, D), light (B, E), and transmission electron (C, F) micrographs of rainbow trout gills kept under normal conditions (A–C), or under conditions (soft water, cortisol administration, D–F) that cause a proliferation of lamellar mitochondria-rich cells (MRC). Note that the thickening of the lamellae associated with MRC proliferation reduces the thickness of the inter-lamellar (IL) water channels while increasing the water-to-blood diffusion distance (arrows).

compromised in comparison with those at the base of the filaments that are in close proximity to the filament blood vessels.

The above examples of gill remodeling illustrate the manipulation of functional surface area (and/or diffusion distance) in response to changes in O_2 availability or use. The flip side of this coin is the gill remodeling that occurs when freshwater fish are placed into ion-poor environments (see Fig. 3.4), where optimal diffusive conductance is sacrificed in an attempt to maintain ionic homeostasis. In response to reduced environmental ion availability, fish acclimated to ion-poor water experience proliferation of MRCs on the lamellae in an attempt to increase branchial capacity for ion uptake (Laurent *et al.*, 1985; Avella

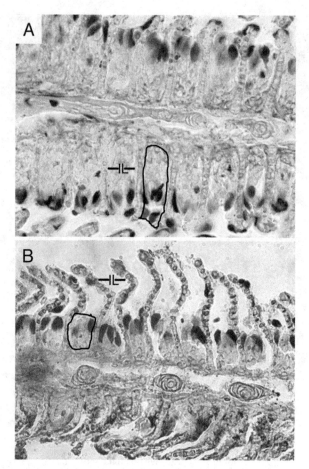

Fig. 3.5 Light micrographs depicting the external surface of the gills of goldfish maintained at 7°C under (A) normoxic or (B) hypoxic (water PO_2 = 10 mm Hg) conditions for 7 days. Under normoxic conditions, the interlamellar (IL) channels are filled by interlamellar cell masses (ILCM); a representative ILCM is outlined. Under hypoxic conditions, the volume of the ILCM is markedly diminished. Regardless of the presence or absence of an ILCM, mitochondria-rich cells (stained black) remain exposed to the water (D. Mitrovic and S. F. Perry; unpublished data).

et al., 1987; Leino et al., 1987; Perry and Laurent, 1989; Greco et al., 1996). MRC proliferation markedly increases the lamellar blood-to-water diffusion distance (Bindon et al., 1994b; Greco et al., 1996), which in turn has subtle but distinctly negative effects on gas transfer (Bindon et al., 1994a; Greco et al., 1995; Perry et al., 1996a; Perry, 1998). CO_2 movement across the gill behaves as a diffusion-limited system (reviewed by Perry and Gilmour 2002), and therefore CO_2 transfer is impaired by the increased thickness of the diffusion barrier (Greco et al.,

1995). However, O_2 transfer (a perfusion-limited system; see above) is impaired only under conditions of hypoxia (Greco et al., 1995).

3.4   Sensing of the environment and regulation of O_2 uptake at the gill

As noted above, the rate of O_2 transfer across the gill is governed by diffusive conductance, convection (ventilation and perfusion), and the blood-to-water PO_2 gradient (ΔPO_2). Well-defined cardiorespiratory reflexes, i.e. adjustments of ventilation and/or perfusion, allow fish to dynamically regulate O_2 transfer across the gill according to environmental conditions and metabolic requirements. Chemoreceptors that detect changes in external (water) and/or internal (blood) O_2 levels are the mechanism through which cardiorespiratory variables are matched to environmental O_2 availability or metabolic O_2 demand (similarly, CO_2-sensitive chemoreceptors link cardiorespiratory responses to environmental CO_2 levels). In fish, the gill is a crucial site of gas sensing, if not the most important site. Numerous detailed reviews have been written on chemoreception in fish (e.g. Shelton et al., 1986; Milsom, 1989; Smatresk, 1990; Burleson et al., 1992; Fritsche and Nilsson, 1993; Milsom, 1995a; Milsom, 1995b; Milsom et al., 1999; Gilmour, 2001; Milsom, 2002; Perry and Gilmour, 2002; Gilmour and Perry, 2007), so the aim of this section is to briefly review the cardiorespiratory reflexes mediated by branchial O_2-sensitive chemoreceptors.

Experimental evidence suggests that branchial O_2 chemoreceptors sense changes in either or both water PO_2 and blood PO_2. This could be accomplished with two distinct populations of O_2 chemoreceptors: one that is oriented to sense the external environment and another positioned to sense the internal milieu (Milsom and Brill, 1986; Burleson and Milsom, 1993), or alternatively, a single population of O_2 chemoreceptors may be strategically located within the gill epithelium to be able to sense changes in both water and blood PO_2 (Shelton et al., 1986; Milsom, 1989; Smatresk, 1990; Burleson et al., 1992; Fritsche and Nilsson, 1993; Milsom, 1995a; Milsom, 1995b;; Milsom et al., 1999; Gilmour, 2001; Milsom, 2002; Perry and Gilmour, 2002; Gilmour and Perry, 2007). Syntheses of studies on a limited number of species suggested that activation of water-sensing O_2 receptors triggers both cardiovascular and ventilatory adjustments, whereas only ventilatory responses are linked to the stimulation of blood-sensing O_2 receptors. However, close inspection of the available data for a much broader range of species (see Table 3.3 in Gilmour and Perry, 2007) indicates that this generalization oversimplifies a more complex situation in which a diversity of response patterns exists.

Chemoreception of O_2 at the gill is attributed to a specific cell type: the neuroepithelial cell (NEC). Neuroepithelial cells are not only concentrated along the leading edge of distal regions of the gill filaments, and occasionally on lamellae, ideal locations in which to monitor ventilated water, but in addition they closely resemble the O_2- and CO_2-sensing glomus (Type I) cells of the mammalian carotid body (Dunel-Erb et al., 1982; Bailly et al., 1992; Goniakowska-Witalinska et al., 1995; Zaccone et al., 1997; Sundin et al., 1998; Jonz and Nurse, 2003; Saltys et al., 2006). Like glomus cells, NECs exhibit characteristics of neurosecretory cells, including the possession of dense-cored vesicles containing synaptic vesicle protein and high levels of serotonin (Dunel-Erb et al. 1982; Bailly et al. 1992; Jonz and Nurse 2003). On the basis of such anatomical and chemical similarities between NECs and glomus cells, and their favorable location to sense water and blood gases, Dunel-Erb et al. (1982) suggested that NECs may function as O_2 chemoreceptors. In support of this function, the same group reported that NECs undergo degranulation (indicative of neurotransmitter release) in response to severe hypoxia (Bailly et al. 1992). More recently, patch clamp electrophysiology experiments have produced compelling evidence that gill NECs act as O_2 chemoreceptors. Working with cultured NECs from zebrafish (Jonz et al., 2004) or channel catfish (Ictalurus punctatus) (Burleson et al., 2006), results comparable to those obtained with mammalian glomus cells were obtained. Specifically, NECs exposed to hypoxia exhibited membrane depolarization that resulted from the inhibition of K^+ conductance. An important next step in this research area is to determine whether membrane depolarization is accompanied by neurotransmitter release. Abundant indirect evidence also supports the concept of the NEC as an O_2 sensor. In adult zebrafish, the number of NECs is increased by hypoxic exposure (Jonz et al., 2004) and decreased during hyperoxia (Vulesevic et al., 2006). In larval zebrafish, the magnitude of the hypoxic ventilatory response correlates with the maturation of the NEC, becoming maximal as the NEC becomes fully innervated (Jonz and Nurse, 2005). Moreover, numerous studies have demonstrated that branchial denervation or extirpation eliminates cardio-respiratory reflexes (see below) to changes in environmental O_2 levels (e.g. Fritsche and Nilsson, 1989; Burleson and Smatresk, 1990a; McKenzie et al., 1991; Hedrick and Jones, 1999; Sundin et al., 2000; Reid and Perry, 2003), whereas selective application to the gill of hypoxic water or pharmacological mimics evokes these reflexes (e.g. Daxboeck and Holeton, 1978; Smith and Jones, 1978; Burleson and Smatresk, 1990b; Mckenzie et al., 1995), implicating the gills as the site of O_2 sensing. Thus, it is clear that NECs are able to sense O_2, that their response resembles the well-characterized response of carotid body cells, and that cardiorespiratory reflexes to environmental O_2 originate in the

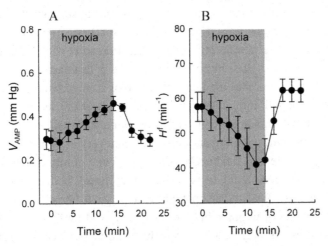

Fig. 3.6 The effects of acute ambient hypoxia (water $PO_2 = \sim 40$ mmHg) on (A) ventilation amplitude (V_{AMP}) and (B) cardiac frequency (H_f) in gulf toadfish (*Opsanus beta*). Data shown are means ± 1 SEM (S. F. Perry, K. M. Gilmour, D. McDonald and P. J. Walsh, unpublished data).

gill. What remains to be established is the direct connection between NEC stimulation and the initiation of cardiorespiratory adjustments when ambient O_2 levels are altered.

A series of well-characterized cardiorespiratory reflexes are triggered by changes in environmental O_2 level detected by the branchial O_2-sensitive chemoreceptors. Both the magnitude of the response and the magnitude of the change in ambient O_2 required to evoke the response vary from one species to another. Nevertheless, certain responses are observed very consistently. Hyperventilation in response to hypoxia (Fig. 3.6) is probably the most robust of these responses, occurring in the vast majority of species that have been examined (Gilmour and Perry, 2007). The physiological significance of hyperventilation during hypoxia is obvious, at least in those species attempting to maintain a constant metabolic rate. In addition to lamellar recruitment and gill remodeling (see above), hyperventilation is an effective (yet costly) strategy to increase the rate of branchial O_2 transfer while raising arterial PO_2. Higher arterial PO_2 is achieved because the increased water flow decreases the inspired–expired PO_2 difference, allowing the arterial blood to approach equilibrium with ventilatory water of higher mean PO_2. Bradycardia and hypertension are the most common cardiovascular responses to hypoxia (see Tables 3.1 and 3.2 in Gilmour and Perry, 2007). Bradycardia during hypoxia is induced by increased activity of cardiac parasympathetic nerves (Taylor *et al.*, 1977; Wood and Shelton, 1980). Increased blood pressure (Holeton and Randall, 1967; Wood and Shelton, 1980) reflects peripheral vasoconstriction and hence elevated systemic vascular resistance, and occurs because vascular smooth muscle α-adrenergic receptors are stimulated by sympathetic nerves or circulating catecholamines (Fritsche and Nilsson, 1990; Kinkead et al., 1991). Despite their common occurrence, the physiological benefits of bradycardia and hypertension during hypoxia remain unclear. The hypoxic bradycardia (Fig. 3.6) has been postulated to benefit gill gas-transfer efficiency (i.e. to raise PaO_2 or lower $PaCO_2$) by reducing gill transit time (if cardiac output is lowered) and/or increasing arterial pulse pressures (which may cause lamellar recruitment or increased gas permeability) (Davie and Daxboeck, 1982). However, conflicting data have been obtained from experimental tests of these hypotheses, with evidence both for (Taylor and Barrett, 1985) and against (Short et al., 1979; Perry and Desforges, 2006) a beneficial role of hypoxic bradycardia (reviewed by Farrell, 2007). An alternative hypothesis (Farrell, 2007) is that the hypoxic bradycardia enhances cardiac performance because increased diastolic residence time serves to increase O_2 delivery to the myocardium and improve cardiac contractility. Equally puzzling is the physiological benefit (if any) for gas transfer of the hypoxic hypertension (Holeton and Randall 1967; Wood and Shelton 1980). Increased blood pressure has been shown to promote lamellar recruitment (Farrell et al., 1980) and thus theoretically could enhance gas transfer, but empirical data do not support this idea (Kinkead et al., 1991; Perry and Desforges, 2006).

3.5   Blood O_2 transport

Delivery of O_2 to the tissues depends not only on O_2 transfer across the gill (see above), but also upon O_2 transport by the blood, which in turn is determined by cardiac output and arterial blood O_2 content. In the hemoglobin-lacking Antarctic icefish described above, all O_2 is carried in the blood plasma as physically dissolved O_2. More typically, however, the vast majority (~95%) of O_2 is carried in the blood chemically bound to hemoglobin within the RBCs, which increases the O_2-carrying capacity of the blood about 20-fold over that achieved with physically dissolved O_2 alone (Weber and Jensen, 1988). Arterial blood O_2 content, then, is determined by the amount of hemoglobin present and its affinity for binding O_2, as well as the PaO_2 that is set by the efficiency of O_2 transfer across the gill as described above.

Hemoglobin is a tetrameric molecule in most fish, although agnathans possess monomeric hemoglobin. O_2 binds in a reversible and cooperative fashion to the heme groups according to the prevailing partial pressure of O_2, a relationship that is described by the O_2 equilibrium curve (Fig. 3.7). The sigmoidal shape of the O_2 equilibrium curve for tetrameric hemoglobin reflects cooperativity of

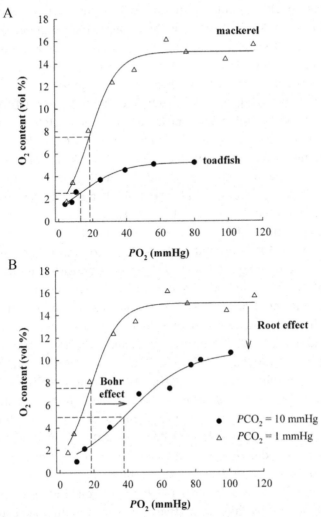

Fig. 3.7 Using the data of Root (1931), O_2 equilibrium curves are plotted (A) for the blood of mackerel (*Scomber scombrus*) and toadfish (*Opsanus tau*), and (B) for the blood of mackerel under low (1 mmHg) and high (10 mmHg) CO_2 tensions. The dotted lines indicate the estimated P_{50} value for each curve, i.e. the PO_2 at which the blood is 50% saturated, an index of hemoglobin–oxygen (Hb–O_2) binding affinity. Panel A illustrates the high blood O_2-carrying capacity and lower Hb–O_2 binding affinity that support exercise performance in the highly active mackerel, whereas the blood of the sluggish toadfish is characterized by a lower O_2-carrying capacity but relatively high Hb–O_2 binding affinity. Panel B illustrates the Bohr and Root effects in mackerel blood. Under conditions of high CO_2 and/or low pH, Hb–O_2 binding affinity is reduced (increased P_{50}) as described by the Bohr effect, while a combination of reduced affinity and reduced cooperativity results in lower blood O_2-carrying capacity, as described by the Root effect.

hemoglobin–oxygen ($Hb–O_2$) binding stemming from conformational changes with each added O_2 molecule. Deoxygenated tetrameric hemoglobin adopts a low-affinity 'tense' (T) state, whereas full oxygenation yields a high-affinity 'relaxed' (R) conformation. Binding of O_2 initiates this shift in conformation, and the progressive increase in O_2 affinity upon O_2 binding is the basis of cooperativity (Jensen, 1991; Jensen et al., 1998; Weber and Fago, 2004). A different mechanism leads to a similar outcome for agnathan hemoglobins. Although agnathan hemoglobins exist as monomers when oxygenated, they form dimers, trimers, or tetramers when deoxygenated, and this O_2-linked reversible aggregation results in apparent cooperativity of O_2 binding, with the pseudo-cooperativity being more pronounced for lamprey than for hagfish (Nikinmaa et al., 1995; Nikinmaa, 2001; Weber and Fago, 2004).

The O_2-binding affinity of hemoglobin is characterized by the P_{50}, the PO_2 at which hemoglobin is 50% saturated with O_2 (Fig. 3.7), and is influenced by temperature ($Hb–O_2$ binding affinity decreases with increasing temperature) and a suite of allosteric modulators, including H^+, CO_2, organic phosphates (in fish primarily guanosine triphosphate [GTP] and adenosine triphosphate [ATP]), and anions such as Cl; note that agnathan hemoglobins do not bind organic phosphates (Nikinmaa, 1992; Nikinmaa and Salama, 1998; Nikinmaa, 2001). Because allosteric effectors bind preferentially to the T conformation, stabilizing it, the binding of these modulators lowers $Hb–O_2$ binding affinity (Jensen, 1991; Nikinmaa, 1992; Jensen et al., 1998; Weber and Fago, 2004). Each allosteric modulator has specific binding sites on hemoglobin, leading to complex interdependencies among the various ligands (Jensen, 1991; Jensen et al., 1998). A classic example is provided by the complementary Bohr and Haldane effects, which describe the decrease in $Hb–O_2$ binding affinity that occurs with decreasing pH (Fig. 3.7) and the increase in the hemoglobin–proton binding affinity that occurs as $Hb–O_2$ saturation falls, respectively. Moreover, amino acid substitutions at these few, crucial binding sites can lead to significant variation in the sensitivity of hemoglobin to the major allosteric effectors (Jensen et al., 1998; Weber and Fago, 2004). For example, the C-terminal histidine of the β chain is a key residue implicated in the Bohr effect, but this histidine is replaced in some teleost hemoglobins (cathodic hemoglobins) by phenylalanine, leading to a loss of the pH sensitivity of $Hb–O_2$ binding (Jensen et al., 1998). Similarly, it appears that the Root effect of teleost hemoglobins can be accounted for by a handful of amino acid substitutions (Brittain, 2005). The Root effect describes the decreased affinity and cooperativity of $Hb–O_2$ binding at low pH (Fig. 3.7), an effect that can be so pronounced that full $Hb–O_2$ saturation cannot be achieved even at hyperoxic PO_2 (Brittain, 2005). Although potentially maladaptive in terms of blood O_2 transport, the Root effect is crucial to the ability of many fish to concentrate molecular O_2 in the swimbladder and eye; O_2 secretion is achieved by acidifying the blood to drive O_2 off hemoglobin despite the elevated PO_2 (Berenbrink et al., 2005; Berenbrink, 2007).

The location of hemoglobin within the RBC is significant in allowing $Hb-O_2$ binding affinity to be regulated through manipulation of the RBC intracellular environment. It is this strategy that typically is adopted within fish species to tune $Hb-O_2$ binding affinity, and hence arterial blood O_2 content, to tissue O_2 demand as environmental O_2 availability (e.g. hypoxia) and/or tissue metabolism (e.g. activity) vary. By contrast, adaptation across species to low O_2 environments and/or high activity lifestyles is apparent in the properties of hemoglobin itself. Blood O_2 content is also determined by the amount of hemoglobin present. Because the concentration of hemoglobin in RBCs is relatively constant across those species that have been examined (Perry and McDonald, 1993), hemoglobin content is varied both within individuals and across species by adjusting hematocrit (the proportion of the blood volume occupied by RBCs).

3.5.1   Regulation of blood gas transport

The three major mechanisms used by fish to adjust $Hb-O_2$ binding affinity focus on manipulating the RBC intracellular environment: pH, organic phosphate concentrations (GTP and/or ATP), and/or volume (Jensen, 1991). In a number of teleost fish, all three adjustments are achieved through an integrated suite of responses of the RBC to mobilization of the catecholamine hormones, adrenaline and noradrenaline (Randall and Perry, 1992; Thomas and Perry, 1992). Circulating catecholamine concentrations rise abruptly in teleost fish in response to a variety of acute physical or environmental stresses, such as hypoxia (e.g. Tetens and Christensen, 1987; Boutilier et al., 1988; Fievet et al., 1990; Perry and Reid, 1992a; Thomas et al., 1992; Perry and Gilmour, 1996), that require that O_2 transport be enhanced (Reid et al., 1998). Circulating catecholamines bind to RBC membrane β-adrenoreceptors, which at least in rainbow trout are of the β_{3b} type (Nickerson et al., 2003; 2004), causing cAMP-mediated activation of protein kinase A that leads to phosphorylation-induced stimulation of a unique Na^+/H^+ exchanger on the RBC membrane. Activation of βNHE, the β-adrenergic Na^+/H^+ exchanger (Borgese et al., 1992), results in the relative alkalization of the RBC owing to the extrusion of protons in exchange for plasma Na^+ (Nikinmaa, 1982; Baroin et al., 1984; Nikinmaa and Huestis, 1984; Cossins and Richardson, 1985). This process raises RBC intracellular pH under hypoxic conditions (Boutilier et al., 1988), thereby increasing $Hb-O_2$ binding affinity (lowering the P_{50} value) via the Bohr effect (Nikinmaa, 1983), a response that benefits O_2 loading into the blood from a hypoxic environment. The inward movement of Na^+ ions also promotes a reduction in the P_{50} through two mechanisms. First, the influx of Na^+ ions stimulates a compensatory activation of Na^+,K^+-ATPase, yielding a decline in cellular ATP levels that serves to enhance $Hb-O_2$ binding affinity (Ferguson et al., 1989; Val et al., 1995; reviewed by Nikinmaa and Boutilier, 1995). Secondly, accumulation of osmotically active Na^+ ions is accompanied by the entry of water, causing an increase in the RBC volume that dilutes organic phosphates, reducing hemoglobin–organic phosphate complexation and hence lowering the P_{50} (Nikinmaa and Tufts, 1989; Jensen, 1991). Cell swelling in itself can contribute to alkalization of the RBC as well, by diluting the fixed negative charges on RBC intracellular proteins in order to shift the Donnan distribution of protons across the RBC membrane (Jensen, 1991; Nikinmaa, 1992). Thus, the net result of stimulating RBC β-adrenoreceptors is increased $Hb-O_2$ binding affinity produced by the coordinated effects of RBC alkalization, decreased RBC organic phosphate levels, and RBC swelling (Fig. 3.8A).

The RBC adrenergic response is not exhibited by all teleost fish, and varies in magnitude among those species in which it does appear. Berenbrink and colleagues (Berenbrink et al., 2005) argue convincingly that the presence of the RBC adrenergic response is linked to the possession of a Root-effect hemoglobin that serves in O_2 delivery to the eye via a choroid rete. In species with Root-effect hemoglobins, systemic acidosis will jeopardize O_2 uptake. Mobilization of catecholamines (Reid et al., 1998) and subsequent activation of the RBC adrenergic response under acidotic conditions effectively uncouples RBC intracellular pH from the extracellular pH, allowing RBC pH to be maintained during extracellular acidosis (Boutilier et al., 1986; Primmett et al., 1986; Vermette and Perry, 1988a), thereby safeguarding O_2 uptake. Interestingly, in at least some of the species that fail to exhibit measurable activation of βNHE, elements of the RBC adrenergic pathway remain. For example, both brown bullhead (Ameiurus nebulosus) and American eel (Anguilla rostrata) RBCs exhibit β-adrenoreceptor stimulation linked to cAMP accumulation in the absence of measurable βNHE activation (Perry and Reid, 1992b; Szebedinszky and Gilmour, 2002).

Elevation of $Hb-O_2$ binding affinity can also be achieved independently of adrenergic phenomena, with the emphasis again being upon adjustments of RBC pH, organic phosphate concentration, and/or volume (Fig. 3.8B). These mechanisms provide the main means of adjusting $Hb-O_2$ binding affinity among the fish species that lack an RBC adrenergic response, which include an unknown number of teleost fish, as well as elasmobranchs (Tufts and Randall, 1989; Berenbrink et al., 2005) and agnathans (Nikinmaa, 1990; Tufts, 1991). In teleost (e.g. Wood and Johansen, 1973; Nikinmaa and Soivio, 1982), elasmobranch (e.g. Perry and Gilmour, 1996), and agnathan (e.g. Nikinmaa and Weber, 1984), fish acutely exposed to hypoxia, RBC pH increases. This alkalization of the RBC

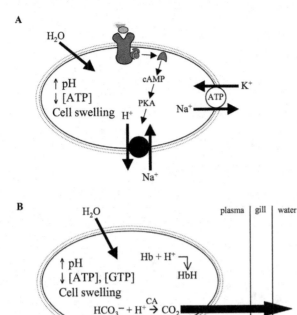

Fig. 3.8 Schematic models depict the adjustments of the red blood cell (RBC) intracellular environment that increase hemoglobin–oxygen binding affinity under hypoxic conditions. (A) The RBC β-adrenergic response. Activation of the β-adrenoreceptor by catecholamines triggers a stimulatory G protein that activates adenylate cyclase to catalyze cAMP formation. Activation of protein kinase A (PKA) by this cAMP ultimately leads to the phosphorylation of the β-adrenergic Na^+/H^+ exchanger, which extrudes protons from the RBC in exchange for Na^+ ions, thereby raising RBC pH. The resultant accumulation of Na^+ activates Na^+–K^+ exchange, which increases energy consumption and leads to a decrease in intracellular ATP levels ([ATP]). Osmotically obliged water enters the cell following the increase in intracellular Na^+, leading to cell swelling. (B) Adjustments of the RBC intracellular environment that can occur even in the absence of an RBC β-adrenergic response. Hypoxia-induced hyperventilation causes a respiratory alkalosis that lowers RBC proton levels, as will the binding of protons to deoxygenated hemoglobin (Hb) in those fish the express a Haldane effect. RBC ATP and/or GTP levels fall, although the mechanisms involved remain unclear. Cell swelling is a critical hypoxia response in agnathan RBCs. (CA Carbonic ahydrase). In either (A) or (B), the combination of alkalization, reduced organic phosphate levels and cell swelling increases hemoglobin–oxygen binding affinity.

intracellular environment evokes a Bohr effect-induced fall in P_{50}, and probably reflects a combination of factors. Most importantly, acute exposure to hypoxia stimulates hyperventilation, typically within seconds of the hypoxic water contacting the gills, resulting in a respiratory alkalosis (Fig. 3.9) (see reviews by

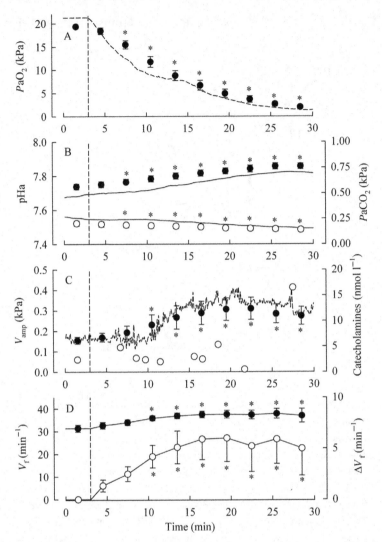

Fig. 3.9 Ventilatory variables, arterial blood gases, and pH together with circulating catecholamine concentrations are plotted as a function of time for dogfish (*Squalus acanthias*) acutely exposed to severe hypoxia. The solid lines in A–C illustrate continuously recorded arterial blood (A) PO_2, (B) pH and PCO_2, as well as (C) ventilation amplitude (V_{amp}) for a single representative fish; the plasma catecholamine data for this same fish are shown as the open symbols in C. All other symbols represent mean data (error bars are 1 standard error; $N = 12$ fish); in (B) pH data are shown as filled symbols and PCO_2 data as open symbols, whereas in (D) ventilation frequency (f_R) data are shown as filled symbols and the change in f_R (Δf_R) is plotted in open symbols. The beginning of the hypoxic period is indicated by the vertical dashed lines. An asterisk indicates a statistical difference (one-way repeated measures analysis of variance; $P < 0.05$) from the pre-hypoxic value. Hypoxic exposure elicits a hyperventilation that results in a fall in arterial PCO_2 together with an increase in arterial blood pH, i.e. a respiratory alkalosis. This respiratory alkalosis would be expected to cause alkalization of the RBC intracellular environment. Reproduced from Perry and Gilmour (1996).

Gilmour, 2001; Gilmour and Perry, 2007). Deoxygenation of hemoglobin may also contribute to RBC alkalization in those species that exhibit a Haldane effect (Jensen, 1986; Weber and Jensen, 1988; Brauner et al., 1996), i.e. not in elasmobranch fish (Lenfant and Johansen, 1966; Wood et al., 1994). Over longer periods of exposure, the respiratory alkalosis may be countered by a metabolic acidosis that arises from increased reliance of the tissues on anaerobic metabolism in the face of inadequate O_2 supply (e.g. Wood and Johansen, 1973; Butler et al., 1979). However, hypoxic exposures lasting hours to days (or longer) trigger a reduction of RBC organic phosphate pools, and this reduction constitutes the most important mechanism for increasing Hb–O_2 binding affinity during chronic hypoxia (except in agnathans) (Wood and Johansen, 1972; Wood and Johansen, 1973; Wood et al., 1975; Weber and Lykkeboe, 1978; Greaney and Powers, 1978; Soivio et al., 1980; Tetens and Lykkeboe, 1981; Rutjes et al., 2007). As discussed above, a reduction in RBC organic phosphates enhances Hb–O_2 binding affinity, both directly by relieving allosteric interactions and indirectly through its effect in raising RBC pH. The cellular mechanisms involved in the hypoxic reduction of RBC organic phosphate levels remain unclear (Nikinmaa, 2001). Interestingly, hypoxia-tolerant species tend to differ from those that rarely encounter hypoxia in both the complement of organic phosphates present in the RBC and their depletion under hypoxic conditions (Weber and Jensen, 1988; Boutilier and Ferguson, 1989). In the RBCs of hypoxia-tolerant species such as carp, tench, goldfish, and eels, GTP concentrations are relatively high, and this organic phosphate is selectively depleted during hypoxia. This strategy has two advantages: GTP is a more potent modulator of Hb–O_2 binding affinity than ATP, and selective depletion of GTP spares ATP for cellular energy metabolism. In agnathans, or more specifically lamprey, as the RBCs of hagfish appear unaffected by hypoxia (Bernier et al., 1996), RBC swelling appears to be a significant contributor to increased Hb–O_2 binding affinity during hypoxia, although the mechanism through which cell swelling is effected remains to be determined (Nikinmaa and Weber, 1984; Nikinmaa, 2001).

Increases in arterial blood O_2 content are achieved not only through the enhancement of Hb–O_2 binding affinity via the mechanisms discussed above, but also by increasing O_2-carrying capacity by raising hematocrit. Acute increases in hematocrit are achieved primarily by the recruitment of RBCs from the spleen, although in at least one fish species the liver rather than the spleen appears to function as the RBC storage site (Frangioni et al., 1997), while agnathans do not have a spleen (Fänge and Nilsson, 1985). Contraction of the smooth muscle of the spleen to release sequestered RBCs into the circulation is mediated by splenic α-adrenoreceptors that can be activated by circulating catecholamines (Perry and Vermette, 1987; Vermette and Perry, 1988b; Perry and Kinkead, 1989) or sympathetic nerves (Nilsson and Grove, 1974). In some cases, notably the Antarctic species *Pagothenia borchgrevinki*, cholinergic mechanisms have been implicated in control of the spleen and hence hematocrit (Nilsson *et al.*, 1996). Splenic contraction leading to increased blood O_2-carrying capacity occurs in response both to low environmental O_2 availability, i.e. hypoxia (Yamamoto *et al.*, 1985; Wells and Weber, 1990; Lai *et al.*, 2006), and to increased tissue O_2 demand, i.e. exercise (Yamamoto *et al.*, 1980; 1985; Yamamoto, 1988; Yamamoto and Itazawa, 1989; Wells and Weber, 1990; Pearson and Stevens, 1991; Gallaugher *et al.*, 1992).

Chronic increases in hematocrit occur during prolonged hypoxic exposure (Wood and Johansen, 1973; Lai *et al.*, 2006; Rutjes *et al.*, 2007), and probably reflect erythropoiesis. In mammals exposed to chronic hypoxia, the formation of new RBCs to increase hematocrit is stimulated by erythropoietin (EPO), with EPO induction under the transcriptional regulation of hypoxia-inducible factor (HIF). Some elements of this pathway have also been identified in fish, although much remains to be described. Both HIF (reviewed by Nikinmaa and Rees, 2005) and EPO (Chou *et al.*, 2004) genes have been characterized for several fish species. Unlike the situation in mammals, the heart rather than the kidney seems to be the main site of EPO production in fish (Chou *et al.*, 2004; Lai *et al.*, 2006), whereas the main erythropoietic tissues include the head, kidney, and/or spleen (Gallaugher and Farrell, 1998). EPO gene expression appears to be regulated by hypoxia (Chou *et al.*, 2004); levels of EPO protein detected in the kidney rise during hypoxic exposure in concert with increases in hematocrit (Lai *et al.*, 2006), and RBC formation is stimulated by EPO treatment (Taglialatela and Della Corte, 1997) – all findings that support the hypothesis of hypoxia-induced EPO production leading to erythropoiesis. However, the manner in which EPO gene expression is regulated by hypoxia and the involvement of HIF in the process are not clear, particularly as no hypoxia-responsive element was detected in the promoter to the *Fugu* EPO gene (Chou *et al.*, 2004).

To summarize, regulation of blood gas transport in response to hypoxia involves acute and/or chronic enhancement of Hb–O_2 binding affinity through modification of the RBC intracellular environment coupled with elevation of O_2-carrying capacity through increases in hematocrit. Hb–O_2 binding affinity is increased by raising RBC pH, lowering RBC organic phosphate levels, and/or increasing RBC volume. Acute increases in hematocrit are achieved via RBC recruitment from the spleen, whereas chronic elevation of hematocrit relies on the formation of new RBCs (Table 3.1).

3.5.2   *Interspecific variation in blood gas transport*

Both hematocrit and Hb–O_2 binding affinity vary widely among fish species, according to factors such as environmental O_2 availability and lifestyle (or activity level). Specific examples or case studies of the interplay among

Table 3.1 *Summary of the responses that optimize blood gas transport during hypoxia*

	Acute	Intermediate	Chronic
Hb–O_2 binding affinity	RBC alkalization	↓ RBC [ATP] and/or [GTP]	Reliance on different Hb components
	RBC swelling		
	RBC β-adrenergic response		
Hematocrit	Adrenergically mediated recruitment of RBCs from spleen		EPO-mediated RBC formation

environment, activity, and blood gas transport properties are provided below, so the objective of this section is to present the general considerations underlying trends between hematocrit or Hb–O_2 binding affinity and environmental factors or activity.

Gallaugher and Farrell (1998) summarized hematocrit values for a large number of agnathan, elasmobranch, and teleost species (see Table 2 in Gallaugher and Farrell, 1998). Scrutiny of this data set reveals hematocrits ranging from a low of ~10% to values exceeding 40–50%, with activity, environmental O_2 availability, and temperature all contributing to the variability. In general, active fish of high metabolic scope such as tuna and marlin exhibit the highest hematocrits (e.g. albacore, *Thunnus alalunga* [Cech, Jr. *et al.*, 1984]; Pacific blue marlin, *Makaira nigicans* [Dobson *et al.*, 1986]; skipjack tuna, *Katsuwonus pelamis*, and yellowfin tuna, *Thunnus albacares* [Brill and Bushnell, 1991]), whereas sedentary benthic species, particularly those from cold environments, tend to have lower hematocrits (e.g. starry flounder, *Platichthys stellatus* [Wood *et al.*, 1979a]; Antarctic species including *P. borchgrevinki*, *Trematomus bernachii*, and *T. loennbergi* [Tetens *et al.*, 1984; Wells *et al.*, 1989]). The extreme example in this regard would appear to be the Antarctic icefish (family Channichthyidae), in which hemoglobin expression and RBCs have been lost altogether (but see below). In addition, a trade-off may occur between sluggish lifestyle and environmental O_2 availability such that somewhat higher hematocrits are found in inactive species that regularly encounter hypoxia or that are considered to be hypoxia tolerant, such as tench (*Tinca tinca*) (Jensen and Weber, 1982) or brown bullhead (Gilmour and MacNeill, 2003).

Theoretical considerations suggest that hematocrit should be regulated to an optimal value (or range of values) determined by the competing effects of O_2

transport potential and viscosity (Wells and Weber, 1991; reviewed by Gallaugher and Farrell, 1998). Blood O_2 transport is proportional to blood O_2-carrying capacity such that O_2 delivery to the tissues, and hence O_2 consumption and/or exercise performance, will be limited at low hematocrit. In support of this relationship, Gallaugher et al. (1995) demonstrated reductions in both critical swimming velocity and maximal O_2 consumption in rainbow trout rendered anemic (to a low hematocrit of 8%) by blood withdrawal. Thus in fish species of naturally low hematocrit, resting cardiac output must be relatively high to maintain normal O_2 delivery, and metabolic scope is constrained by the reliance of blood O_2 transport on adjustments of cardiac output (Wood et al., 1979b). The extreme example is provided by the hemoglobinless icefish, in which resting cardiac output is several-fold higher than that of otherwise comparable, red-blooded species (Hemmingsen et al., 1972), and Egginton (1997) reported that sustained, maximal activity was difficult to elicit. Improved blood O_2-carrying capacity with increasing hematocrit is of obvious advantage, but is achieved at a cost of also increasing blood viscosity (Wells and Weber, 1991) that will, in turn, increase the cardiac work required to pump the more viscous blood. This cost is expected to be even greater at cold temperatures owing to the combined effects of low temperature and polycythemia on viscosity and may help to explain the trend for lower hematocrits in cold environments (see above). Although it is tempting to extend this argument to explain the loss of hemoglobin from icefish, the very high cardiac output required to maintain O_2 delivery in the absence of hemoglobin results in a cost of cardiac pumping that, at ~22% of resting O_2 consumption, is well above values (0.5–5.0%) for a range of temperate species, suggesting that this argument is untenable (Sidell and O'Brien, 2006). Despite these theoretical arguments for an upper limit of optimal hematocrit, experimental support is lacking. In particular, in rainbow trout subjected to blood doping, critical swimming velocity increased with hematocrit to the maximum tested (55%), while maximal oxygen consumption peaked at a hematocrit of ~42%, well above the typical value for trout. Moreover, little variation in critical swimming velocity or maximal O_2 consumption was detected over a broad range of hematocrits between the normocythemic and polycythemic states (Gallaugher et al., 1995). However, the fall in PaO_2 with exercise was hematocrit dependent (Gallaugher et al., 1995), suggesting that O_2 transfer at the gills becomes diffusion limited with polycythemia under conditions in which cardiac output is elevated and transit times for gas exchange are reduced (Gallaugher and Farrell, 1998), whereas O_2 transfer normally is perfusion limited (see above). Considerations of this nature may place an upper limit on the value of polycythemia.

Like hematocrit, Hb–O_2 binding affinity among fish species varies over a wide continuum, reflecting both environmental O_2 availability and activity (for tables of P_{50} values listing multiple species see Johansen et al., 1978a; Tetens et al., 1984; Butler and Metcalfe, 1988; Perry and McDonald, 1993; Bushnell and Jones, 1994; see also Krogh and Leitch (1919) for one of the first such comparisons). The proximate cause of interspecific variation in P_{50} is (gene-based) differences in the primary structure of globin (Jensen et al., 1998). Driving selection for such structural differences are the competing demands for high Hb–O_2 binding affinity to favor O_2 loading into the blood at the gills, particularly in environments of low O_2 availability, and low Hb–O_2 binding affinity to facilitate O_2 unloading from the blood to metabolically active tissues, particularly in highly active species. High-affinity hemoglobins are advantageous where low environmental O_2 conditions are regularly encountered, because they permit O_2 saturation of the blood at lower PaO_2. On the one hand, the ability to saturate hemoglobin at lower PO_2 may be beneficial even under normoxic conditions by allowing a lower PaO_2 to be maintained, thereby increasing the water-to-blood PO_2 gradient to maximize diffusive conductance and at the same time reducing the ventilatory requirements to maintain PaO_2, with a concomitant saving of energy. On the other hand, O_2 delivery with high-affinity hemoglobins requires that venous PO_2 be lowered to a greater extent, reducing the venous PO_2 reserve and potentially limiting metabolic scope. Consequently, the highest-affinity hemoglobins typically are found in sluggish species that inhabit environments of variable O_2 availability, whereas high-performance species usually exhibit lower-affinity hemoglobins that enable O_2 delivery with the maintenance of a considerable venous PO_2 reserve (Table 3.2).

Many teleost species are, however, able to have what is in essence the best of both worlds through the possession of hemoglobin exhibiting a pronounced Bohr effect. Addition of CO_2 and the resultant acidification of the blood at the tissues raises the P_{50} to facilitate O_2 delivery, whereas a leftward shift of the O_2 equilibrium curve occurs at the gill in accordance with CO_2 loss to favor O_2 loading. At the same time, oxygenation status-linked changes in the affinity of hemoglobin for protons benefit CO_2 excretion (Haldane effect). The increased affinity of deoxygenated hemoglobin for protons promotes CO_2 hydration in the RBC, increasing blood CO_2 content, while the release of oxylabile protons during oxygenation of the blood in the gills 'boosts' HCO_3^- dehydration and CO_2 loss (Perry and Gilmour, 1993; Perry et al., 1996b). Thus, O_2 uptake and CO_2 excretion are tightly linked in teleost fish (Fig. 3.10) (Brauner and Randall, 1996; Brauner and Randall, 1998). The possession of hemoglobin exhibiting a pronounced Bohr effect, however, goes hand-in-hand with the occurrence of the Root effect (Berenbrink et al., 2005), leaving such species vulnerable to hypoxemia (reduced blood O_2 content) during systemic acidosis. The RBC β-adrenergic response described above provides protection against such an occurrence. An

Table 3.2 P_{50} and hematocrit (hct) values for selected species of teleost fish

Common name	Species name	P_{50} (mmHg)[1]	Hct (%)	Reference(s)
High-performance species				
Skipjack tuna	*Katsuwonus pelamis*	21	41	Brill and Bushnell, 1991
Yellowfin tuna	*Thunnus albacares*	22	35	Brill and Bushnell, 1991
Kawakawa	*Euthynnus affinis*	21	34	Jones *et al.*, 1986
Moderately to highly active species				
Rainbow trout	*Oncorhynchus mykiss*	23	23	Tetens and Christensen, 1987
Brown trout	*Salmo trutta fario*	26	37	Riera *et al.*, 1993
Atlantic cod	*Gadus morhua*	25	19	Gollock *et al.*, 2006
Sea bass	*Dicentrarchus labrax*	12.8	34.4	Pichavant *et al.*, 2003
Antarctic species	*Pagothenia borchgrevinki*	21	13	Tetens *et al.*, 1984
	Dissostichus mawsoni	14.4	17.5	Tetens *et al.*, 1984
Sedentary, benthic species				
Starry flounder	*Platichthys stellatus*	8.6	14.5	Wood *et al.*, 1979a
Turbot	*Scophthalmus maximus*	12.5	16.5	Pichavant *et al.*, 2003
Antarctic species	*Rhigophila dearborni*	4.3	15	Tetens *et al.*, 1984
	Trematomus bernachii	13.5	13.5	Tetens *et al.*, 1984
	Trematomus lonnbergi	11.9	8	Tetens *et al.*, 1984
	Notothenia angustata	10.8	18.5	Tetens *et al.*, 1984
Hypoxia-tolerant species				
American eel	*Anguilla rostrata*	11.1	20	Hyde *et al.*, 1987; Perry and Reid, 1992a
Carp	*Cyprinus carpio*	7.3	20	Takeda, 1990
Crucian carp	*Carassius carassius*	0.7–1.8	35	Sollid *et al.*, 2005; G. E. Nilsson, unpublished
Goldfish	*Carassius auratus*	2.6	26	Burggren, 1982
Tench	*Tinca tinca*	6.2	23	Jensen and Weber, 1982
Amazonian species				
Pacu	*Piaractus mesopotamicus*	11.3	23	Perry *et al.*, 2004
Traira	*Hoplias malabaricus*	8.6	19	Perry *et al.*, 2004
Jeju	*Hoplerythrinus unitaeniatus*	7.7	23	Perry *et al.*, 2004
	Osteoglossum bicirrhosum	6.1	28	Johansen *et al.*, 1978b
	Erythrinus erythrinus	7.1	34	Johansen *et al.*, 1978a
	Synbranchus marmoratus	7.1	48	Johansen *et al.*, 1978a
	Hoplosternum littorale	11.1	49	Johansen *et al.*, 1978a

[1] P_{50} values measured in vivo or in vitro under conditions representative of those in vivo during normoxia.

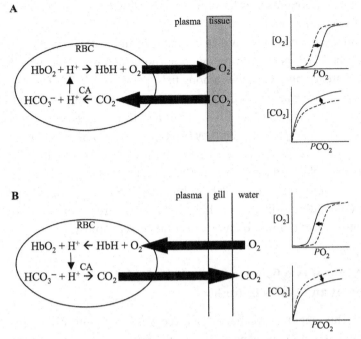

Fig. 3.10 A schematic representation of the linkage of O_2 uptake and CO_2 excretion in teleost fish is presented together with the expected impacts on the O_2 equilibrium and CO_2 combining curves. (A) The addition of CO_2 to the blood at the tissues and resultant acidification of the RBC decrease hemoglobin–oxygen (Hb–O_2) binding affinity, facilitating O_2 unloading to the tissue (Bohr effect – right shift of O_2 equilibrium curve). At the same time, deoxygenation of hemoglobin increases the affinity of hemoglobin for protons, enhancing CO_2 hydration and facilitating CO_2 loading into the blood (Haldane effect – upper line of CO_2 combining curve is for deoxygenated blood). (B) The reversal of this situation at the gills. Oxygenation of hemoglobin drives off protons (Haldane effect – lower line of CO_2 combining curve), which can be used to dehydrate HCO_3^- ions for CO_2 excretion. The loss of CO_2 also benefits O_2 uptake into the blood by reversing the Bohr effect (left shift of O_2 equilibrium curve).

additional safeguard in some species is the possession of multiple hemoglobins with different functional properties (Riggs, 1979; Weber and Jensen, 1988; Weber, 1990). The anodic hemoglobin components are found in all species and are characterized by relatively low Hb–O_2 binding affinities and pronounced pH sensitivity. In some species, such as eel, trout, and catfish, cathodic hemoglobins having high Hb–O_2 binding affinity and low pH dependence are also found and may serve to preserve blood O_2 transport during hypoxia and/or acidosis when the anodic hemoglobin is unable to load sufficient O_2 (Weber and Jensen, 1988; Jensen, 1991; Weber, 1996). In addition, the complement of hemoglobin components may be adjusted according to environmental O_2 availability (Rutjes *et al.*, 2007).

To summarize, environmental O_2 availability and activity level appear to be the two crucial factors explaining much of the tremendous interspecific variation in hematocrit and P_{50} values among fish species (Table 3.1; Fig. 3.7). Highly active species optimize O_2 delivery to the tissues by pumping blood of high hematocrit and relatively low Hb–O_2 binding affinity, reserving both cardiac scope and venous PO_2 into which they can tap at times of maximal exertion. By contrast, very sedentary species sacrifice both cardiac scope and venous PO_2 reserve in favor of optimizing cardiac efficiency, pumping low-hematocrit blood of high Hb–O_2 binding affinity. Where hypoxia is regularly encountered, both high Hb–O_2 binding affinity and high hematocrit are favored to protect O_2 delivery in the face of limited environmental O_2 availability.

3.6   A tale of two fishes: case studies of rainbow trout and starry flounder

In this chapter, we have reviewed basic principles of branchial O_2 transfer and blood O_2 transport. To conclude, we wish to comment on the marked interspecific variation in the strategies used by fish to match O_2 transfer and transport with metabolic need, habitat, and lifestyle. To do so, we will compare strategies of branchial O_2 uptake and blood O_2 transport in two species, the freshwater rainbow trout (*Oncorhynchus mykiss*) and the marine starry flounder (*Platichthys stellatus*).

3.6.1   Rainbow trout

The rainbow trout is a moderately active pelagic species capable of performing long-distance migrations between fresh water and sea water as well as short bursts of intense swimming activity. The preferred habitats of freshwater rainbow trout are well-oxygenated lakes and streams; these fish are intolerant of ambient hypoxia. The O_2 transfer and transport strategies that have evolved in rainbow trout (see Table 3.3) are consistent with their habitat and lifestyle. Trout hemoglobin exhibits a relatively low O_2-binding affinity (P_{50} = 23 mmHg) and thus a high arterial PO_2 is required to achieve complete O_2 saturation (e.g. PaO_2 = 133 mmHg) (Table 3.3). The ability of trout to achieve such an elevated PaO_2 reflects a high ventilation volume which, in turn, prevents a large reduction in PO_2 between the inspired and expired water. Consequently, the mean PO_2 of the water in contact with the respiratory epithelium (P_IO_2 + $P_EO_2/2$) is higher than it otherwise would be. The net result is that the high ventilation volume promotes an elevated PaO_2 by virtue of the high mean PO_2 of

Table 3.3 *A comparison of selected measured and calculated cardiorespiratory variables in rainbow trout* (Oncorhynchus mykiss) *and starry flounder* (Platichthys stellatus)

	Rainbow trout[1]	Starry flounder[2]
$\dot{V}O_2$ (μmol min^{-1} kg^{-1})	28.8	21.4
P_{50} (mmHg)	22.9	8.6
P_aO_2 (mmHg)	133.2	34.9
P_vO_2 (mmHg)	32	13.4
$[O_2]_a$ (mmol l^{-1})	3.42	2.05
$[O_2]_v$ (mmol l^{-1})	2.30	1.49
$[O_2]_a - [O_2]_v$ (mmol l^{-1})	1.12	0.56
Vt_w (ml min^{-1} kg^{-1})	177.5	109.2
Vt_b (ml min^{-1} kg^{-1})	18.3	39.2
Vt_w/Vt_b	10.5	2.9
Vent. CR (ml μmol^{-1})	6.2	5.1
Perf. CR (ml μmol^{-1})	0.32	1.83
P_iO_2 (mmHg)	160	139
P_EO_2 (mmHg)	86	44
ΔPO_2 (mmHg)[3]	41	67
$G_{diff}O_2$ (μmol min^{-1} kg^{-1} mmHg^{-1})[4]	0.71	0.32
O_2 diffusion efficiency (%[5])	250	32

For a definition of the terms used in this table, see the abbreviations list at the beginning of this book.

[1] Data were either copied or recalculated from Davis and Cameron (1971) or Cameron and Davis (1970).

[2] Data were either copied or recalculated from Wood *et al.* (1979a).

[3] ΔPO_2 was calculated as $(P_iO_2 + P_EO_2/2) - (P_aO_2 + P_vO_2/2)$.

[4] $G_{diff}O_2 = \dot{V}O_2/\Delta PO_2$.

[5] O_2 diffusion efficiency $= (P_aO_2 - P_vO_2)/\Delta PO_2 *100$.

the water in contact with lamellae. Although high ventilation volumes normally contribute to an increase in the water-to-blood PO_2 gradient (ΔPO_2), in trout the ΔPO_2 is relatively low (41 mmHg; Table 3.3) because of the high mean PO_2 of the blood perfusing the lamellae ($P_vO_2 + PaO_2/2$). The obligatory high ventilation volumes in trout come at considerable cost because of the significant energy required to pump water across the gills (estimates vary between 3 and 10% of overall metabolic rate). However, because of the relatively high metabolic rate of trout (28.8 μmol kg^{-1} h^{-1}; Table 3.3), the convection requirement for water (the volume of water flow required per unit of O_2 uptake; $Vt_w/\dot{V}O_2$) is comparable to other species including flounder.

The ability of the gill to transfer O_2 for any given ΔPO_2 is termed the transfer factor (Randall *et al.*, 1967) or diffusion conductance ($G_{diff}O_2$) and is defined as $\dot{V}O_2/\Delta PO_2$. Essentially, $G_{diff}O_2$ is a measure of the diffusivity of the gill, which in turn is determined by surface area, diffusion distance, and Krogh's permeation coefficient (KO_2). The relatively high $G_{diff}O_2$ in trout is consistent with their active lifestyle and the need to achieve a high PaO_2. Another index of O_2 transfer capability is the percentage efficiency of O_2 diffusion from water to blood. We believe that the arterial–venous PO_2 difference that is achieved for a given ΔPO_2 is a reasonable estimate of diffusion efficiency. Because of the counter-current arrangement of water and blood flow, percentage efficiency in fish can exceed 100; in trout the efficiency is 250%. Because percentage efficiency is strictly a measure of the ability of blood perfusing the gill to reach equilibrium with water ventilating the gill, it is related only to diffusion distance and KO_2 but, unlike $G_{diff}O_2$, it is not influenced by available surface area. An obvious negative consequence of the O_2 transfer strategy used by trout (high O_2 diffusion efficiency and $G_{diff}O_2$) is the associated high rates of water and salt movement across the gills.

3.6.2   Starry flounder

The starry flounder is a sedentary benthic marine species that is easily exhausted during exercise (Wood *et al.*, 1977; Wood and Perry, 1985). Because of its habit of burrowing into the substrate, flounders and other flatfish may experience hypoxia in the natural environment. The Hb–O_2 binding affinity is high in flounder (P_{50} = 8.6 mmHg), resulting in nearly complete Hb–O_2 saturation (Wood *et al.*, 1979a) at relatively low values of PaO_2 (34.9 mmHg; Table 3.3). There are advantages and disadvantages associated with the low PaO_2 in flounder. Advantages of the low PaO_2 include reduced ventilation volumes (and the associated energetic savings) and an increase in ΔPO_2 (67 mm Hg; Table 3.3). The high ΔPO_2 promotes the diffusion of O_2 across the gill in the face of low $G_{diff}O_2$ and percentage O_2 diffusion efficiency (0.32 and 32, respectively). Indeed, it is the ability of the flounder to saturate Hb with O_2 at low PaO_2 that allows it to exploit the strategy of reducing $G_{diff}O_2$ and percentage O_2 diffusion efficiency to minimize movements of water and salt. Although data are limited, it would appear that the anatomical gill surface area (as opposed to functional surface area) of flatfish and salmonids is similar (Hughes, 1966). Therefore, it is likely that functional surface area (extent of lamellar perfusion) is decreased while diffusion distance is increased (we are unaware of any data for flatfish) in flounder to explain the low values (in comparison with trout) of $G_{diff}O_2$ and O_2 diffusion efficiency. A disadvantage of the low PaO_2 in flounder is that it may limit the diffusion of O_2 to exercising muscle.

A striking feature of the respiratory strategy of flounder is a need for an unusually high resting cardiac output ($Vt_b = 39.2 \, ml \, min^{-1} kg^{-1}$; Table 3.3) because of their low hematocrit and resultant low arterial–venous O_2 concentration difference (approximately 50% less than in trout; Table 3.3). So, to achieve more-or-less similar rates of $\dot{V}O_2$, flounder must pump roughly twice as much blood than trout. This is the explanation for the high convection requirement for blood in flounder (approximately 6 × higher than in trout) and the unusually low ventilation–perfusion ratio (2.9; Table 3.3). It is the high resting cardiac output in flounder that probably reduces their scope for activity and underlies their poor exercise performance.

3.7   Concluding remarks

Thus, after 100 years or so of research into the modes and mechanisms of O_2 uptake and transport in fish, we have gained a reasonable appreciation of the basic principles involved. Yet despite this research effort and the fundamental importance of the process itself, many questions remain unresolved. For example, much remains to be discovered about the mechanism(s) of O_2 sensing in fish, and how the activation of these mechanisms leads to the initiation of compensatory responses. We know, for example, that the activation of branchial O_2-sensitive chemoreceptors by aquatic hypoxia triggers reflex ventilatory and cardiovascular responses, yet the cellular mechanism of O_2 sensing by the chemoreceptor, as well as the afferent pathways, central integration, and efferent pathways of this reflex are still largely unexplored. In many cases, even less is known of the sensory mechanisms and effector pathways underlying other responses to hypoxia, such as the lowering of RBC organic phosphates or gill remodeling. Finally, much of our knowledge of O_2 transfer and transport in fish rests upon relatively few species, most of which have been chosen in large part because they are amenable to laboratory study (e.g. the rainbow trout). We have yet to really come to grips with the enormous interspecific variation present among fish. Exploring and understanding this diversity remains an exciting challenge.

References

Avella, M., Masoni, A., Bornancin, M. and Mayer-Gostan, N. (1987). Gill morphology and sodium influx in the rainbow trout (*Salmo gairdneri*) acclimated to artificial freshwater environments. *J. Exp. Zool.* **241**, 159–69.

Bailly, Y., Dunel-Erb, S. and Laurent, P. (1992). The neuroepithelial cells of the fish gill filament – indolamine-immunocytochemistry and innervation. *Anat. Rec.* **233**, 143–61.

Baroin, A., Garcia-Romeu, F., Lamarre, T. and Motais, R. (1984). A transient sodium-hydrogen exchange system induced by catecholamines in erythrocytes of rainbow trout, *Salmo gairdneri*. *J. Physiol.* **356**, 21–31.

Bartels, H. (1998). The gills of hagfishes. In *The Biology of Hagfishes*, ed. J. M. Jorgensen, J. P. Lomholt, R. E. Weber, and H. Malte. London: Chapman & Hall, pp. 205–19.

Berenbrink, M. (2007). Historical reconstructions of evolving physiological complexity: O_2 secretion in the eye and swimbladder of fishes. *J. Exp. Biol.* **209**, 1641–52.

Berenbrink, M., Koldkjær, P., Kepp, O. and Cossins, A. R. (2005). Evolution of oxygen secretion in fishes and the emergence of a complex physiological system. *Science* **307**, 1752–7.

Bernier, N. J., Fuentes, J. and Randall, D. J. (1996). Adenosine receptor blockade and hypoxia-tolerance in rainbow trout and Pacific hagfish II. Effects on plasma catecholamines and erythrocytes. *J. Exp. Biol.* **199**, 497–507.

Bindon, S. F., Fenwick, J. C. and Perry, S. F. (1994b). Branchial chloride cell proliferation in the rainbow trout, Oncorhynchus mykiss: implications for gas transfer. *Can. J. Zool.* **72**, 1395–402.

Bindon, S. D., Gilmour, K. M., Fenwick, J. C. and Perry, S. F. (1994a). The effect of branchial chloride cell proliferation on respiratory function in the rainbow trout, *Oncorhynchus mykiss. J. Exp. Biol.* **197**, 47–63.

Booth, J. H. (1979). Circulation in trout gills: the relationship between branchial perfusion and the width of the lamellar blood space. *Can. J. Zool.* **57**, 2183–5.

Borgese, F., Sardet, C., Cappadoro, M., Pouyssegur, J. and Motais, R. (1992). Cloning and expression of a cAMP-activated Na^+/H^+ exchanger – evidence that the cytoplasmic domain mediates hormonal regulation. *Proc. Natl. Acad. Sci. USA* **89**, 6765–9.

Boutilier, R. G. and Ferguson, R. A. (1989). Nucleated red cell function: metabolism and pH regulation. *Can. J. Zool.* **67**, 2986–93.

Boutilier, R. G., Iwama, G. K. and Randall, D. J. (1986). The promotion of catecholamine release in rainbow trout, *Salmo gairdneri*, by acute acidosis: Interactions between red cell pH and haemoglobin oxygen-carrying capacity. *J. Exp. Biol.* **123**, 145–57.

Boutilier, R. G., Dobson, G., Hoeger, U. and Randall, D. J. (1988). Acute exposure to graded levels of hypoxia in rainbow trout (*Salmo gairdneri*): metabolic and respiratory adaptations. *Respir. Physiol.* **71**, 69–82.

Brauner, C. J. and Randall, D. J. (1996). The interaction between oxygen and carbon dioxide movements in fishes. *Comp. Biochem. Physiol. A* **113**, 83–90.

Brauner, C. J. and Randall, D. J. (1998). The linkage between oxygen and carbon dioxide transport. In *Fish Physiology vol. 17 Fish Respiration*, ed. S. F. Perry, and B. L. Tufts. San Diego: Academic Press, pp. 283–319.

Brauner, C. J., Gilmour, K. M. and Perry, S. F. (1996). Effect of haemoglobin oxygenation on Bohr proton release and CO_2 excretion in the rainbow trout. *Respir. Physiol.* **106**, 65–70.

Brauner, C. J., Matey, V., Wilson, J. M., Bernier, N. J. and Val, A. L. (2004). Transition in organ function during the evolution of air-breathing; insights from Arapaima gigas, an obligate air-breathing teleost from the Amazon. *J. Exp. Biol.* **207**, 1433–8.

Brill, R. W. and Bushnell, P. G. (1991). Effects of open- and closed-system temperature changes on blood oxygen dissociation curves of skipjack tuna, Katsuwonus pelamis, and yellowfin tuna, Thunnus albacares. *Can. J. Zool.* **69**, 1814–21.

Brittain, T. (2005). Root effect hemoglobins. *J. Inorg. Biochem.* **99**, 120–9.

Burggren, W. W. (1982). 'Air gulping' improves blood oxygen transport during aquatic hypoxia in the goldfish *Carassius auratus*. *Physiol. Zool.* **55**, 327–34.

Burleson, M. L. and Milsom, W. K. (1993). Sensory receptors in the 1st gill arch of rainbow trout. *Resp. Physiol.* **93**, 97–110.

Burleson, M. L. and Smatresk, N. J. (1990a). Effects of sectioning cranial nerves IX and X on cardiovascular and ventilatory reflex responses to hypoxia and NaCN in channel catfish. *J. Exp. Biol.* **154**, 407–20.

Burleson, M. L. and Smatresk, N. J. (1990b). Evidence for two oxygen-sensitive chemoreceptor loci in channel catfish, *Ictalurus punctatus*. *Physiol. Zool.* **63**, 208–21.

Burleson, M. L. Mercer, S. E. and Wilk-Blaszczak, M. A. (2006). Isolation and characterization of putative O_2 chemoreceptor cells from the gills of channel catfish (*Ictalurus punctatus*). *Brain Res.* **1092**, 100–7.

Burleson, M. L., Smatresk, N. J. and Milsom, W. K. (1992). Afferent inputs associated with cardioventilatory control in fish. In *Fish Physiology vol. XIIB The Cardiovascular System*, ed. W. S. Hoar, D. J. Randall and A. P. Farrell. San Diego: Academic Press, pp. 389–423.

Bushnell, P. G. and Jones, D. R. (1994). Cardiovascular and respiratory physiology of tuna: adaptations for support of exceptionally high metabolic rates. *Env. Biol. Fish.* **40**, 303–18.

Butler, P. J. and Metcalfe, J. D. (1988). Cardiovascular and respiratory systems. In *Physiology of Elasmobranch Fishes*, ed. T. J. Shuttleworth. Berlin: Springer-Verlag, pp. 1–47.

Butler, P. J., Taylor, E. W. and Davison, W. (1979). The effect of long term, moderate hypoxia on acid-base balance, plasma catecholamines and possible anaerobic end products in the unrestrained dogfish *Scyliorhinus canicula*. *J. Comp. Physiol.* **132**, 297–303.

Cameron, J. N. (1989). *The Respiratory Physiology of Animals*. New York: Oxford University Press, pp. 1–353.

Cameron, J. N. and Davis, J. C. (1970). Gas exchange in rainbow trout (*Salmo gairdneri*) with varying blood oxygen capacity. *J. Fish Res. Bd. Can.* **27**, 1069–85.

Cech, J. J., Jr., Laurs, R. M. and Graham, J. B. (1984). Temperature-induced changes in blood gas equilibria in the albacore, *Thunnus alalunga*, a warm-bodied tuna. *J. Exp. Biol.* **109**, 21–34.

Chou, C.-F., Tohari, S., Brenner, S. and Venkatesh, B. (2004). Erythropoietin gene from a teleost fish, Fugu rubripes. *Blood* **104**, 1498–503.

Cossins, A. R. and Richardson, P. A. (1985). Adrenalin-induced Na^+/H^+ exchange in trout erythrocytes and its effects upon oxygen-carrying capacity. *J. Exp. Biol.* **118**, 229–46.

Davie, P. S. and Daxboeck, C. (1982). Effect of pulse pressure on fluid exchange between blood and tissues in trout gills. *Can. J. Zool.* **60**, 1000–6.

Davis, J. C. and Cameron, J. N. (1971). Water flow and gas exchange at the gills of rainbow trout, *Salmo gairdneri. J. Exp. Biol.* **54**, 1–18.

Daxboeck, C. and Holeton, G. F. (1978). Oxygen receptors in the rainbow trout, *Salmo gairdneri. Can. J. Zool.* **56**, 1254–9.

Desforges, P. R., Harman, S. S., Gilmour, K. M. and Perry, S. F. (2002). The sensitivity of CO_2 excretion to changes in blood flow in rainbow trout is determined by carbonic anhydrase availability. *Am. J. Physiol.* **282**, R501–8.

Dobson, G., Wood, S. C., Daxboeck, C. and Perry, S. F. (1986). Intracellular buffering and oxygen transport in the Pacific blue marlin (*Makaira nigricans*): adaptations to high-speed swimming. *Physiol. Zool.* **59**, 150–6.

Dunel-Erb, S., Bailly, Y. and Laurent, P. (1982). Neuroepithelial cells in fish gill primary lamellae. *J. Appl. Physiol.* **53**, 1342–53.

Egginton, S. (1997). A comparison of the response to induced exercise in red- and white-blooded Antarctic fishes. *J. Comp. Physiol. B* **167**, 129–34.

Evans, D. H., Piermarini, P. M. and Choe, K. P. (2005). The multifunctional fish gill: dominant site of gas exchange, osmoregulation, acid-base regulation, and excretion of nitrogenous waste. *Physiol. Rev.* **85**, 97–177.

Fänge, R. and Nilsson, S. (1985). The fish spleen: structure and function. *Experientia* **41**, 152–7.

Farrell, A. P. (2007). Tribute to P. L. Lutz: a message from the heart – why hypoxic bradycardia in fishes? *J. Exp. Biol.* **210**, 1715–25.

Farrell, A. P., Sobin, S. S., Randall, D. J. and Crosby, S. (1980). Intralamellar blood flow patterns in fish gills. *Am. J. Physiol.* **239**, R428–36.

Ferguson, R. A., Tufts, B. L. and Boutilier, R. G. (1989). Energy metabolism in trout red cells: consequences of adrenergic stimulation *in vivo* and in vitro. *J. Exp. Biol.* **143**, 133–47.

Fievet, B., Caroff, J. and Motais, R. (1990). Catecholamine release controlled by blood oxygen tension during deep hypoxia in trout: effect on red blood cell Na/H exchanger activity. *Respir. Physiol.* **79**, 81–90.

Frangioni, G., Berti, R. and Borgioli, G. (1997). Hepatic respiratory compensation and haematological changes in the cave cyprinid, *Phreatichthys andruzzii. J. Comp. Physiol. B* **167**, 461–7.

Fritsche, R. and Nilsson, S. (1989). Cardiovascular responses to hypoxia in the Atlantic cod, *Gadus morhua. Exp. Biol.* **48**, 153–60.

Fritsche, R. and Nilsson, S. (1990). Autonomic nervous control of blood pressure and heart rate during hypoxia in the cod, *Gadus morhua. J. Comp. Physiol. B* **160**, 287–92.

Fritsche, R. and Nilsson, S. (1993). Cardiovascular and ventilatory control during hypoxia. In *Fish Ecophysiology*, ed. J. C. Rankin and F. B. Jensen. London: Chapman & Hall, pp. 180–206.

Gallaugher, P. and Farrell, A. P. (1998). Hematocrit and blood oxygen-carrying capacity. In *Fish Physiology vol. 17 Fish Respiration*, ed. S. F. Perry and B. L. Tufts. San Diego: Academic Press, pp. 185–227.

Gallaugher, P., Axelsson, M. and Farrell, A. P. (1992). Swimming performance and haematological variables in splenectomized rainbow trout, *Oncorhynchus mykiss*. *J. Exp. Biol.* **171**, 301–14.

Gallaugher, P., Thorarensen, H. and Farrell, A. P. (1995). Hematocrit in oxygen transport and swimming in rainbow trout (*Oncorhynchus mykiss*). *Respir. Physiol.* **102**, 279–92.

Gilmour, K. M. (1997). Gas exchange. In *The Physiology of Fishes*, ed. D. H. Evans. Boca Raton: CRC Press, pp. 101–27.

Gilmour, K. M. (2001). The CO_2/pH ventilatory drive in fish. *Comp. Biochem. Physiol. A* **130**, 219–40.

Gilmour, K. M. and MacNeill, G. K. (2003). Apparent diffusion limitations on branchial CO_2 transfer are revealed by severe experimental anaemia in brown bullhead (*Ameiurus nebulosus*). *Comp. Biochem. Physiol. A* **135**, 165–75.

Gilmour, K. M. and Perry, S. F. (2007). Branchial chemoreceptor regulation of cardiorespiratory function. In *Fish Physiology Vol. 25 Sensory Systems Neuroscience*, ed. B. Zielinski and T. J. Hara. San Diego: Academic Press, pp. 97–151.

Gilmour, K. M., Bayaa, M., Kenney, L., McNeill, B. and Perry, S. F. (2007). Type IV carbonic anhydrase is present in the gills of spiny dogfish (*Squalus acanthias*). *Am. J. Physiol.* **292**, R556–67.

Gollock, M. J., Currie, S., Petersen, L. H. and Gamperl, A. K. (2006). Cardiovascular and haematological responses of Atlantic cod (Gadus morhua) to acute temperature increase. *J. Exp. Biol.* **209**, 2961–70.

Goniakowska-Witalinska, L., Zaccone, G., Fasulo, S., Mauceri, A., Licata, A. and Youson, J. (1995). Neuroendocrine cells in the gills of the bowfin *Amia calva*. An ultrastructural and immunocytochemical study. *Folia Histochem. Cytobiol.* **33**, 171–7.

Graham, J. B. (1997). *Air-breathing Fishes*. San Diego: Academic Press.

Graham, J. B. (2006). Aquatic and aerial respiration. In *The Physiology of Fishes*, ed. D. H. Evans and J. B. Claiborne. Boca Raton: CRC Press, pp. 85–152.

Greaney, G. S. and Powers, D. A. (1978). Allosteric modifiers of fish hemoglobins: in vitro and in vivo studies of the effect of ambient oxygen and pH on erythrocyte ATP concentrations. *J. Exp. Zool.* **203**, 339–50.

Greco, A. M., Fenwick, J. C. and Perry, S. F. (1996). The effects of softwater acclimation on gill morphology in the rainbow trout, *Oncorhynchus mykiss*. *Cell. Tiss. Res.* **285**, 75–82.

Greco, A. M., Gilmour, K. M., Fenwick, J. C. and Perry, S. F. (1995). The effects of soft-water acclimation on respiratory gas transfer in the rainbow trout, *Oncorhynchus mykiss*. *J. Exp. Biol.* **198**, 2557–67.

Hedrick, M. S. and Jones, D. R. (1999). Control of gill ventilation and air breathing in the bowfin *Amia Calva*. *J. Exp. Biol.* **202**, 87–94.

Hemmingsen, E. A., Douglas, E. L., Johansen, K. and Millard, R. W. (1972). Aortic blood flow and cardiac output in the hemoglobin-free fish *Chaenocephalus aceratus*. *Comp. Biochem. Physiol.* **43A**, 1045–51.

Hill, R. W., Wyse, G. A. and Anderson, M. (2004). *Animal Physiology.* Sunderland, Massachusetts: Sinauer Associates.

Holeton, G. F. (1970). Oxygen uptake and circulation by a hemoglobinless Antarctic fish (*Chaenocephalus aceratus* Lonnberg) compared with three red-blooded Antarctic fish. *Comp. Biochem. Physiol.* **34**, 457–71.

Holeton, G. F. and Randall, D. J. (1967). Changes in blood pressure in the rainbow trout during hypoxia. *J. Exp. Biol.* **46**, 297–305.

Hughes, G. M. (1966). The dimensions of fish gills in relation to their function. *J. Exp. Biol.* **45**, 177–95.

Hughes, G. M. and Morgan, M. (1973). The structure of fish gills in relation to their respiratory function. *Biol. Rev.* **48**, 419–75.

Hughes, G. M. and Shelton, G. (1962). Respiratory mechanisms and their nervous control in fish. *Adv. Comp. Physiol. Biochem.* **1**, 275–364.

Hyde, D. A., Moon, T. W. and Perry, S. F. (1987). Physiological consequences of prolonged aerial exposure in the Americal eel, *Anguilla rostrata*: blood respiratory and acid-base status. *J. Comp. Physiol. B* **157**, 635–42.

Jensen, F. B. (1986). Pronounced influence of Hb–O_2 saturation on red cell pH in tench blood in vivo and in vitro. *J. Exp. Zool.* **238**, 119–24.

Jensen, F. B. (1991). Multiple strategies in oxygen and carbon dioxide transport by haemoglobin. In *Physiological Strategies for Gas Exchange and Metabolism*, ed. A. J. Woakes, M. K. Grieshaber and C. R. Bridges. Society for Experimental Biology Seminar Series. Cambridge: Cambridge University Press, pp. 55–78.

Jensen, F. B. (2004). Red blood cell pH, the Bohr effect, and other oxygenation-linked phenomena in blood O_2 and CO_2 transport. *Acta Physiol Scand.* **182**, 215–27.

Jensen, F. B. and Weber, R. E. (1982). Respiratory properties of tench blood and hemoglobin adaptation to hypoxic-hypercapnic water. *Mol. Physiol.* **2**, 235–50.

Jensen, F. B., Fago, A. and Weber, R. E. (1998). Hemoglobin structure and function. In *Fish Physiology Vol. 17 Fish Respiration*, ed. S. F. Perry and B. L. Tufts. San Diego: Academic Press, pp. 1–40.

Johansen, K., Mangum, C. P. and Lykkeboe, G. (1978a). Respiratory properties of the blood of Amazon fishes. *Can. J. Zool.* **56**, 898–906.

Johansen, K., Mangum, C. P. and Weber, R. E. (1978b). Reduced blood O_2 affinity associated with air breathing in osteoglossid fishes. *Can. J. Zool.* **56**, 891–7.

Jones, D. R. and Randall, D. J. (1978). The respiratory and circulatory sytems during exercise. In *Fish Physiology Vol. VII Locomotion.* ed. W. S. Hoar and D. J. Randall. San Diego: Academic Press, pp. 425–500.

Jones, D. R. and Schwarzfeld, T. (1974). The oxygen cost to the metabolism and efficiency of breathing in trout (*Salmo gairdneri*). *Respir. Physiol.* **21**, 241–53.

Jones, D. R., Brill, R. W. and Mense, D. C. (1986). The influence of blood gas properties on gas tensions and pH of ventral and dorsal aortic blood in free-swimming tuna, *Euthynnus affinis. J. Exp. Biol.* **120**, 201–13.

Jonz, M. G. and Nurse, C. A. (2003). Neuroepithelial cells and associated innervation of the zebrafish gill: a confocal immunofluorescence study. *J. Comp. Neurol.* **461**, 1–17.

Jonz, M. G. and Nurse, C. A. (2005). Development of oxygen sensing in the gills of zebrafish. *J. Exp. Biol.* **208**, 1537–1549.

Jonz, M. G., Fearon, I. M. and Nurse, C. A. (2004). Neuroepithelial oxygen chemoreceptors of the zebrafish gill. *J. Physiol.* **560**, 737–52.

Julio, A. E., Desforges, P. and Perry, S. F. (2000). Apparent diffusion limitations for carbon dioxide excretion in rainbow trout (*Oncorhynchus mykiss*) are relieved by intravascular injections of carbonic anhydrase. *Respir. Physiol.* **121**, 53–64.

Kinkead, R., Fritsche, R., Perry, S. F. and Nilsson, S. (1991). The role of circulating catecholamines in the ventilatory and hypertensive responses to hypoxia in the Atlantic cod (*Gadus morhua*). *Physiol. Zool.* **64**, 1087–109.

Krogh, A. (1904). Some experiments on the cutaneous respiration of vertebrate animals. *Skand. Arch. Physiol.* **16**, 348–67.

Krogh, A. (1941). *Comparative Physiology of Respiratory Mechanisms.* Philadelphia: University of Pennsylvania Press, pp. 1–172.

Krogh, A. and Leitch, I. (1919). The respiratory function of the blood in fishes. *J. Physiol.* **52**, 288–300.

Lai, J. C. C., Kakuta, I., Mok, H. O. L., Rummer, J. L. and Randall, D. J. (2006). Effects of moderate and substantial hypoxia on erythropoietin levels in rainbow trout kidney and spleen. *J. Exp. Biol.* **209**, 2734–8.

Laurent, P., Hobe, H. and Dunel-Erb, S. (1985). The role of environmental sodium chloride relative to calcium in gill morphology of freshwater salmonid fish. *Cell. Tiss. Res.* **240**, 675–92.

Leino, R. L., McCormick, J. H. and Jensen, K. M. (1987). Changes in gill histology of fathead minnows and yellow perch transferred to soft water or acidified soft water with particular reference to chloride cells. *Cell. Tiss. Res.* **250**, 389–99.

Lenfant, C. and Johansen, K. (1966). Respiratory function in the elasmobranch *Squalus suckleyi* G. *Respir. Physiol.* **1**, 13–29.

Malte, H. and Lomholt, J. P. (1998). Ventilation and gas exchange. In *The Biology of Hagfishes*, ed. J. M. Jorgensen, J. P. Lomholt, R. E. Weber and H. Malte. London: Chapman & Hall, pp. 223–34.

Malte, H. and Weber, R. E. (1985). A mathematical model for gas exchange in the fish gill based on non-linear blood gas equilibrium curves. *Respir. Physiol.* **62**, 359–74.

McKenzie, D. J., Burleson, M. L. and Randall, D. J. (1991). The effects of branchial denervation and pseudobranch ablation on cardioventilatory control in an air-breathing fish. *J. Exp. Biol.* **161**, 347–65.

McKenzie, D. J., Taylor, E. W., Bronzi, P. and Bolis, C. L. (1995). Aspects of cardioventilatory control in the adriatic sturgeon (*Acipenser naccarii*). *Respir. Physiol.* **100**, 45–53.

Milsom, W. K. (1989). Mechanisms of ventilation in lower vertebrates: adaptations to respiratory and nonrespiratory constraints. *Can. J. Zool.* **67**, 2943–55.

Milsom, W. K. (1995a). Regulation of respiration in lower vertebrates: role of CO_2/pH chemoreceptors. *In Advances in Comparative and Environmental Physiology Vol. 21 Mechanisms of Systemic Regulation: Acid-base Regulation, Ion Transfer and Metabolism*, ed. N. Heisler. Berlin: Spinger-Verlag, pp. 62–104.

Milsom, W. K. (1995b). The role of CO_2/pH chemoreceptors in ventilatory control. *Braz. J. Med. Biol. Res.* **28**, 1147–60.

Milsom, W. K. (2002). Phylogeny of CO_2/H^+ chemoreception in vertebrates. *Respir. Physiol. Neurobiol.* **131**, 29–41.

Milsom, W. K. and Brill, R. W. (1986). Oxygen sensitive afferent information arising from the first gill arch of yellowfin tuna. *Respir. Physiol.* **66**, 193–203.

Milsom, W. K., Sundin, L., Reid, S., Kalinin, A. and Rantin, F. T. (1999). Chemoreceptor control of cardiovascular reflexes. In *Biology of Tropical Fishes*, ed. A. L. Val and V. M. F. Almeida-Val. Manaus: INPA, pp. 363–74.

Nickerson, J. G., Drouin, G., Perry, S. F. and Moon, T. W. (2004). *In vitro* regulation of β-adrenoreceptor signaling in the rainbow trout, *Oncorhynchus mykiss. Fish Physiol. Biochem.* **27**, 157–71.

Nickerson, J. G., Dugan, S. G., Drouin, G., Perry, S. F. and Moon, T. W. (2003). Activity of the unique β-adrenergic Na^+/H^+ exchanger in trout erythrocytes is controlled by a novel $β_3$-AR subtype. *Am. J. Physiol.* **285**, R526–35.

Nikinmaa, M. (1982). Effects of adrenaline on red cell volume and concentration gradient of protons across the red cell membrane in the rainbow trout, *Salmo gairdneri. Mol. Physiol.* **2**, 287–97.

Nikinmaa, M. (1983). Adrenergic regulation of haemoglobin oxygen affinity in rainbow trout red cells. *J. Comp. Physiol. B* **152**, 67–72.

Nikinmaa, M. (1990). *Vertebrate Red Blood Cells.* Berlin: Springer-Verlag.

Nikinmaa, M. (1992). Membrane transport and control of hemoglobin-oxygen affinity in nucleated erythrocytes. *Physiol. Rev.* **72**, 301–21.

Nikinmaa, M. (1997). Oxygen and carbon dioxide transport in vertebrate erythrocytes: an evolutionary change in the role of membrane transport. *J. Exp. Biol.* **200**, 369–80.

Nikinmaa, M. (2001). Haemoglobin function in vertebrates: evolutionary changes in cellular regulation in hypoxia. *Respir. Physiol.* **128**, 317–29.

Nikinmaa, M. (2006). Gas transport. In *The Physiology of Fishes*, ed. D. H. Evans and J. B. Claiborne. Boca Raton: CRC Press, pp. 153–74.

Nikinmaa, M. and Boutilier, R. G. (1995). Adrenergic control of red cell pH, organic phosphate concentrations and haemoglobin function in teleost fish. In *Advances in Comparative and Environmental Physiology Vol. 21 Mechanisms of Systemic Regulation: Respiration and Circulation*, ed. N. Heisler. Berlin: Springer-Verlag, pp. 107–33.

Nikinmaa, M. and Huestis, W. H. (1984). Adrenergic swelling of nucleated erythrocytes: cellular mechanisms in a bird, domestic goose, and two teleosts, striped bass and rainbow trout. *J. Exp. Biol.* **113**, 215–24.

Nikinmaa, M. and Rees, B. B. (2005). Oxygen-dependent gene expression in fishes. *Am. J. Physiol.* **288**, R1079–90.

Nikinmaa, M. and Salama, A. (1998). Oxygen transport in fish. In *Fish Physiology Vol. 17 Fish Respiration*, ed. S. F. Perry and B. L. Tufts. San Diego: Academic Press, pp. 141–84.

Nikinmaa, M. and Soivio, A. (1982). Blood oxygen transport of hypoxic *Salmo gairdneri. J. Exp. Zool.* **219**, 173–8.

Nikinmaa, M. and Tufts, B. L. (1989). Regulation of acid and ion transfer across the membrane of nucleated erythrocytes. *Can. J. Zool.* **67**, 3039–45.

Nikinmaa, M. and Weber, R. E. (1984). Hypoxic acclimation in the lamprey, Lampetra fluviatilis: organismic and erythrocytic responses. *J. Exp. Biol.* **109**, 109–19.

Nikinmaa, M., Airaksinen, S. and Virkki, L. V. (1995) Haemoglobin function in intact lamprey erythrocytes: interactions with membrane function in the regulation of gas transport and acid-base balance. *J. Exp. Biol.* **198**, 2423–30.

Nilsson, G. E. (2007). Gill remodeling in fish–a new fashion or an ancient secret? *J. Exp. Biol.* **210**, 2403–9.

Nilsson, S. and Grove, D. J. (1974). Adrenergic and cholinergic innervation of the spleen of the cod: Gadus morhua. *Eur. J. Pharmacol.* **28**, 135–43.

Nilsson, S., Forster, N. E., Davison, W. and Axelsson, M. (1996). Nervous control of the spleen in the red-blooded Antarctic fish, *Pagothenia borchgrevinki*. *Am. J. Physiol.* **39**, R599–604.

Olson, K. R. (2002). Vascular anatomy of the fish gill. *J. Exp. Zool.* **293**, 214–31.

Ong, K. J., Stevens, E. D. and Wright, P. A. (2007). Gill morphology of the mangrove killifish (Kryptolebias marmoratus) is plastic and changes in response to terrestrial air exposure. *J. Exp. Biol.* **210**, 1109–15.

Pearson, M. P. and Stevens, E. D. (1991). Size and hematological impact of the splenic erythrocyte reservoir in rainbow trout, *Oncorhynchus mykiss*. *Fish Physiol. Biochem.* **9**, 39–50.

Perry, S. F. (1986). Carbon dioxide excretion in fish. *Can. J. Zool.* **64**, 565–72.

Perry, S. F. (1998). Relationships between branchial chloride cells and gas transfer in freshwater fish. *Comp. Biochem. Physiol. A* **119**, 9–16.

Perry, S. F. and Desforges, P. R. (2006). Does bradycardia or hypertension enhance gas transfer in rainbow trout (*Oncorhynchus mykiss*) exposed to hypoxia or hypercarbia? *Comp. Biochem. Physiol. A* **144**, 163–72.

Perry, S. F. and Gilmour, K. M. (1993). An evaluation of factors limiting carbon dioxide excretion by trout red blood cells in vitro. *J. Exp. Biol.* **180**, 39–54.

Perry, S. F. and Gilmour, K. M. (1996). Consequences of catecholamine release on ventilation and blood oxygen transport during hypoxia and hypercapnia in an elasmobranch (*Squalus acanthias*) and a teleost (*Oncorhynchus mykiss*). *J. Exp. Biol.* **199**, 2105–18.

Perry, S. F. and Gilmour, K. M. (2002). Sensing and transfer of respiratory gases at the fish gill. *J. Exp. Zool.* **293**, 249–63.

Perry, S. F. and Kinkead, R. (1989). The role of catecholamines in regulating arterial oxygen content during acute hypercapnic acidosis in rainbow trout (*Salmo gairdneri*). *Respir. Physiol.* **77**, 365–78.

Perry, S. F. and Laurent, P. (1989). Adaptational responses of rainbow trout to lowered external NaCl concentration: contribution of the branchial chloride cell. *J. Exp. Biol.* **147**, 147–68.

Perry, S. F. and McDonald, D. G. (1993). Gas exchange. In *The Physiology of Fishes*, ed. D. H. Evans. Boca Raton: CRC Press, pp. 251–78.

Perry, S. F. and Reid, S. D. (1992a). Relationship between blood O_2 content and catecholamine levels during hypoxia in rainbow trout and American eel. *Am. J. Physiol.* **263**, R240–9.

Perry, S. F. and Reid, S. D. (1992b). The relationship between β-adrenoceptors and adrenergic responsiveness in trout (*Oncorhynchus mykiss*) and eel (*Anguilla rostrata*) erythrocytes. *J. Exp. Biol.* **167**, 235–50.

Perry, S. F. and Vermette, M. G. (1987). The effects of prolonged epinephrine infusion on the physiology of the rainbow trout, *Salmo gairdneri* I. Blood respiratory, acid-base and ionic states. *J. Exp. Biol.* **128**, 235–53.

Perry, S. F. and Wood, C. M. (1989). Control and coordination of gas transfer in fishes. *Can. J. Zool.* **67**, 2961–70.

Perry, S. F., Reid, S. G., Gilmour, K. M., *et al.* (2004). A comparison of adrenergic stress responses in three tropical teleosts exposed to acute hypoxia. *Am. J. Physiol.* **287**, R188–97.

Perry, S. F., Wood, C. M., Walsh, P. J. and Thomas, S. (1996b). Fish red blood cell carbon dioxide transport *in vitro*: a comparative study. *Comp. Biochem. Physiol.* **113A**, 121–30.

Perry, S. F., Reid, S. G., Wankiewicz, E., Iyer, V. and Gilmour, K. M. (1996a). Physiological responses of rainbow trout (Oncorhynchus mykiss) to prolonged exposure to softwater. *Physiol. Zool.* **69**, 1419–41.

Pichavant, K., Maxime, V., Soulier, P., Boeuf, G. and Nonnotte, G. (2003). A comparative study of blood oxygen transport in turbot and sea bass: effect of chronic hypoxia. *J. Fish Biol.* **62**, 928–37.

Piiper, J. (1989). Factors affecting gas transfer in respiratory organs of vertebrates. *Can. J. Zool.* **67**, 2956–60.

Piiper, J. (1990). Modeling of gas exchange in lungs, gills and skin. In *Advances in Comparative and Environmental Physiology*, ed. R. G. Boutilier. Berlin: Springer-Verlag, pp. 15–44.

Piiper, J. (1998). Branchial gas transfer models. *Comp. Biochem. Physiol.* **119A**, 125–30.

Primmett, D. R. N., Randall, D. J., Mazeaud, M. M. and Boutilier, R. G. (1986). The role of catecholamines in erythrocyte pH regulation and oxygen transport in rainbow trout (*Salmo gairdneri*) during exercise. *J. Exp. Biol.* **122**, 139–48.

Randall, D. J. (1990). Control and co-ordination of gas exchange in water breathers. In *Advances in Comparative and Environmental Physiology*, ed. R. G. Boutilier. Berlin: Springer-Verlag, pp. 253–78.

Randall, D. J. and Daxboeck, C. (1984). Oxygen and carbon dioxide transfer across fish gills. In *Fish Physiology Vol. XA*, ed. W. S. Hoar and D. J. Randall. New York: Academic Press, pp. 263–314.

Randall, D. J. and Perry, S. F. (1992). Catecholamines. In *Fish Physiology Vol. XIIB The Cardiovascular System*, ed. W. S. Hoar, D. J. Randall and A. P. Farrell. San Diego: Academic Press, pp. 255–300.

Randall, D. J., Holeton, G. F. and Stevens, E. D. (1967). The exchange of oxygen and carbon dioxide across the gills of rainbow trout. *J. Exp. Biol.* **46**, 339–48.

Randall, D. J., Perry, S. F. and Heming, T. A. (1982). Gas transfer and acid-base regulation in salmonids. *Comp. Biochem. Physiol. B* **73**, 93–103.

Reid, S. G. and Perry, S. F. (2003). Peripheral O_2 chemoreceptors mediate humoral catecholamine secretion from fish chromaffin cells. *Am. J. Physiol.* **284**, R990–9.

Reid, S. G., Bernier, N. J. and Perry, S. F. (1998). The adrenergic stress response in fish: control of catecholamine storage and release. *Comp. Biochem. Physiol.* **C120**, 1–27.

Riera, M., Prats, M. T., Palacios, L., Pesquero, J. and Planas, J. (1993). Seasonal adaptations in oxygen transport in brown trout *Salmo trutta fario. Comp. Biochem. Physiol.* **106A**, 695–700.

Riggs, A. F. (1979). Studies of the hemoglobins of Amazonian fishes: an overview. *Comp. Biochem. Physiol.* **62A**, 257–71.

Root, R. W. (1931). The respiratory functions of the blood of marine fishes. *Biol. Bull.* **61**, 427–56.

Rutjes, H. A., Nieveen, M. C., Weber, R. E., Witte, F. and van den Thillart, G. E. E. J. M. (2007). Multiple strategies of Lake Victoria cichlids to cope with lifelong hypoxia include hemoglobin switching. *Am. J. Physiol.* **293**, R1376–83.

Saltys, H. A., Jonz, M. G. and Nurse, C. A. (2006). Comparative study of gill neuroepithelial cells and their innervation in teleosts and *Xenopus* tadpoles. *Cell Tissue Res.* **323**, 1–10.

Shelton, G., Jones, D. R. and Milsom, W. K. (1986). Control of breathing in ectothermic vertebrates. In *Handbook of Physiology, Section 3. The Respiratory System, Vol. 2, Control of Breathing*, eds. N. S. Cherniak and J. G. Widdicombe. Bethesda: American Physiological Society, pp. 857–909.

Short, S., Taylor, E. W. and Butler, P. J. (1979). The effectiveness of oxygen transfer during normoxia and hypoxia in the dogfish (*Scyliohinus canicula* L.) before and after cardiac vagotomy. *J. Comp. Physiol. B* **132**, 289–95.

Sidell, B. D. and O'Brien, K. M. (2006). When bad things happen to good fish: the loss of hemoglobin and myoglobin expression in Antarctic icefishes. *J. Exp. Biol.* **209**, 1791–802.

Smatresk, N. J. (1990). Chemoreceptor modulation of endogenous respiratory rhythms in vertebrates. *Am. J. Physiol.* **259**, R887–97.

Smith, F. M. and Jones, D. R. (1978). Localization of receptors causing hypoxic bradycardia in trout (*Salmo gairdneri*). *Can. J. Zool.* **56**, 1260–5.

Soivio, A., Nikinmaa, M. and Westman, K. (1980). The blood oxygen binding properties of hypoxic *Salmo gairdneri. J. Comp. Physiol.* **136**, 83–7.

Sollid, J. and Nilsson, G. E. (2006). Plasticity of respiratory structures–adaptive remodeling of fish gills induced by ambient oxygen and temperature. *Respir. Physiol. Neurobiol.* **154**, 241–51.

Sollid, J., Weber, R. E. and Nilsson, G. E. (2005). Temperature alters the respiratory surface area of crucian carp Carassius carassius and goldfish Carassius auratus. *J. Exp. Biol.* **208**, 1109–16.

Sollid, J., De Angelis, P., Gundersen, K. and Nilsson, G. E. (2003). Hypoxia induces adaptive and reversible gross morphological changes in crucian carp gills. *J. Exp. Biol.* **206**, 3667–73.

Strahan, R. (1958). The velum and the respiratory current of Myxine. *Acta. Zool.* **39**, 227–40.

Sundin, L., Holmgren, S. and Nilsson, S. (1998). The oxygen receptor of the teleost gill? *Acta. Zool.* **79**, 207–14.

Sundin, L., Reid, S. G., Rantin, F. T. and Milsom, W. K. (2000). Branchial receptors and cardiorespiratory reflexes in the neotropical fish, Tambaqui (*Colossoma macropomum*). *J. Exp. Biol.* **203**, 1225–39.

Szebedinszky, C. and Gilmour, K. M. (2002). High plasma buffering and the absence of a red blood cell β-NHE response in brown bullhead (*Ameiurus nebulosus*). *Comp. Biochem. Physiol. A* **133**, 399–409.

Taglialatela, R. and Della Corte, F. (1997). Human and recombinant erythropoietin stimulate erythropoiesis in the goldfish *Carassius auratus*. *Eur. J. Histochem.* **41**, 301–4.

Takeda, T. (1990). Ventilation, cardiac output and blood respiratory parameters in the carp, *Cyprinus carpio*, during hyperoxia. *Respir. Physiol.* **81**, 227–40.

Taylor, E. W. and Barrett, D. J. (1985). Evidence of a respiratory role for the hypoxic bradycardia in the dogfish *Scyliohinus canicula* L. *Comp. Biochem. Physiol. A* **80**, 99–102.

Taylor, E. W., Short, S. and Butler, P. J. (1977). The role of the cardiac vagus in the response of the dogfish *Scyliorhinus canicula* to hypoxia. *J. Exp. Biol.* **70**, 57–75.

Tetens, V. and Christensen, N. J. (1987). Beta-adrenergic control of blood oxygen affinity in acutely hypoxia exposed rainbow trout. *J. Comp. Physiol. B* **157**, 667–75.

Tetens, V. and Lykkeboe, G. (1981). Blood respiratory properties of rainbow trout, *Salmo gairdneri*: responses to hypoxia acclimation and anoxic incubation of blood in vitro. *J. Comp. Physiol.* **145**, 117–25.

Tetens, V., Wells, R. M. G. and DeVries, A. L. (1984). Antarctic fish blood: respiratory properties and the effects of thermal acclimation. *J. Exp. Biol.* **109**, 265–79.

Thomas, S. and Motais, R. (1990). Acid-base balance and oxygen transport during acute hypoxia in fish. *Comp. Physiol.* **6**, 76–91.

Thomas, S. and Perry, S. F. (1992). Control and consequences of adrenergic activation of red blood cell Na$^+$/H$^+$ exchange on blood oxygen and carbon dioxide transport. *J. Exp. Zool.* **263**, 160–75.

Thomas, S., Perry, S. F., Pennec, Y. and Maxime, V. (1992). Metabolic alkalosis and the response of the trout, *Salmo fario*, to acute severe hypoxia. *Respir. Physiol.* **87**, 91–104.

Tufts, B. L. (1991). Acid-base regulation and blood gas transport following exhaustive exercise in an agnathan, the sea lamprey *Petromyzon marinus*. *J. Exp. Biol.* **159**, 371–85.

Tufts, B. L. and Perry, S. F. (1998). Carbon dioxide transport and excretion. In *Fish Physiology Vol. 17 Fish Respiration*, ed. S. F. Perry and B. L. Tufts. New York: Academic Press, pp. 229–81.

Tufts, B. L. and Randall, D. J. (1989). The functional significance of adrenergic pH regulation in fish erythrocytes. *Can. J. Zool.* **67**, 235–8.

Val, A. L. (1995). Oxygen transfer in fish: morphological and molecular adjustments. *Braz. J. Med. Biol. Res.* **28**, 1119–27.

Val, A. L. (2000). Organic phosphates in the red blood cells of fish. *Comp. Biochem. Physiol. A Mol. Integr. Physiol.* **125**, 417–35.

Val, A. L., Lessard, J. and Randall, D. J. (1995). Effects of hypoxia on rainbow trout (*Oncorhynchus mykiss*): intraerythrocytic phosphates. *J. Exp. Biol.* **198**, 305–10.

Vermette, M. G. and Perry, S. F. (1988a). Adrenergic involvement in blood oxygen transport and acid-base balance during hypercapnic acidosis in the rainbow trout, *Salmo gairdneri*. *J. Comp. Physiol. B* **158**, 107–15.

Vermette, M. G. and Perry, S. F. (1988b). Effects of prolonged epinephrine infusion on blood respiratory and acid-base states in the rainbow trout: alpha and beta effects. *Fish Physiol. Biochem.* **4**, 189–202.

Vulesevic, B., McNeill, B. and Perry, S. F. (2006). Chemoreceptor plasticity and respiratory acclimation in the zebrafish, Danio rerio. *J. Exp. Biol.* **209**, 1261–73.

Weber, R. E. (1990). Functional significance and structural basis of multiple hemoglobins with special reference to ectothermic vertebrates. In *Animal Nutrition and Transport Processes. 2. Transport, Respiration and Excretion: Comparative and Environmental Aspects*, ed. J.-P. Truchot and B. Lahlou. Basel: S. Karger, pp. 58–75.

Weber, R. E. (1996). Hemoglobin adaptations in Amazonian and temperate fish with special reference to hypoxia, allosteric effectors and functional heterogeneity. In *Physiology and Biochemistry of the Fishes of the Amazon*, ed. A. L. Val, V. M. F. Almeida-Val and D. J. Randall. Manaus, Brazil: INPA, pp. 75–90.

Weber, R. E. and Fago, A. (2004). Functional adaptation and its molecular basis in vertebrate hemoglobins, neuroglobins and cytoglobins. *Respir. Physiol. Neurobiol.* **144**, 141–59.

Weber, R. E. and Jensen, F. B. (1988). Functional adaptations in hemoglobins from ectothermic vertebrates. *Annu. Rev. Physiol.* **50**, 161–79.

Weber, R. E. and Lykkeboe, G. (1978). Respiratory adaptations in carp blood. Influences of hypoxia, red cell organic phosphates, divalent cations and CO_2 on hemoglobin-oxygen affinity. *J. Comp. Physiol.* **128**, 127–37.

Wegner, N. C., Sepulveda, C. A. and Graham, J. B. (2006). Gill specializations in high-performance pelagic teleosts with reference to striped marlin (*Tetrapturus audax*) and wahoo (*Acanthocybium solandri*). *Bull. Mar. Sci.* **79**, 747–59.

Wells, R. M. G. and Weber, R. E. (1990). The spleen in hypoxic and exercised rainbow trout. *J. Exp. Biol.* **150**, 461–6.

Wells, R. M. G. and Weber, R. E. (1991). Is there an optimal haematocrit for rainbow trout, *Oncorhynchus mykiss* (Walbaum)? An interpretation of recent data based on blood viscosity measurements. *J. Fish Biol.* **38**, 53–65.

Wells, R. M. G., Grigg, G. C., Beard, L. A. and Summers, G. (1989). Hypoxic responses in a fish from a stable environment: blood oxygen transport in the Antarctic fish *Pagothenia borchgrevinki*. *J. Exp. Biol.* **141**, 97–111.

Wood, S. C. and Johansen, K. (1972). Adaptation to hypoxia by increased HbO_2 affinity and decreased red cell ATP concentration. *Nature* **237**, 278–9.

Wood, S. C. and Johansen, K. (1973). Blood oxygen transport and acid-base balance in eels during hypoxia. *Am. J. Physiol.* **225**, 849–51.

Wood, C. M. and Perry, S. F. (1985). Respiratory, circulatory, and metabolic adjustments to exercise in fish. In *Circulation, Respiration and Metabolism*, ed. R. Gilles. Berlin: Springer-Verlag, pp. 2–22.

Wood, C. M. and Shelton, G. (1980). The reflex control of heart rate and cardiac output in the rainbow trout: interactive influences of hypoxia, haemorrhage, and systemic vasomotor tone. *J. Exp. Biol.* **87**, 271–84.

Wood, S. C., Johansen, K. and Weber, R. E. (1975). Effects of ambient PO_2 on hemoglobin-oxygen affinity and red cell ATP concentrations in a benthic fish, *Pleuronectes platessa. Respir. Physiol.* **25**, 259–67.

Wood, C. M., McMahon, B. R. and McDonald, D. G. (1977). An analysis of changes in blood pH following exhausting activity in the starry flounder, *Platichthys stellatus. J. Exp. Biol.* **69**, 173–85.

Wood, C. M., McMahon, B. R. and McDonald, D. G. (1979a). Respiratory gas exchange in the resting starry flounder, *Platichthys stellatus*: a comparison with other teleosts. *J. Exp. Biol.* **78**, 167–79.

Wood, C. M., McMahon, B. R. and McDonald, D. G. (1979b). Respiratory, ventilatory, and cardiovascular responses to experimental anaemia in the starry flounder, *Platichthys stellatus. J. Exp. Biol.* **82**, 139–62.

Wood, C. M., Perry, S. F., Walsh, P. J. and Thomas, S. (1994). HCO_3^- dehydration by the blood of an elasmobranch in the absence of a Haldane effect. *Respir. Physiol.* **98**, 319–37.

Yamamoto, K. (1988). Contraction of spleen in exercised freshwater teleost. *Comp. Biochem. Physiol.* **89A**, 65–6.

Yamamoto, K. and Itazawa, Y. (1989). Erythrocyte supply from the spleen of exercised carp. *Comp. Biochem. Physiol.* **92A**, 139–44.

Yamamoto, K., Itazawa, Y. and Kobayashi, H. (1980). Supply of erythrocytes into the circulating blood from the spleen of exercised fish. *Comp. Biochem. Physiol.* **65A**, 5–11.

Yamamoto, K., Itazawa, Y. and Kobayashi, H. (1985). Direct observation of fish spleen by an abdominal window method and its application to exercised and hypoxic yellowtail. *Japan. J. Ichthyol.* **31**, 427–33.

Zaccone, G., Fasulo, S., Ainis, L. and Licata, A. (1997). Paraneurons in the gills and airways of fishes. *Microsc. Res. Tech.* **37**, 4–12.

4

Oxygen uptake and transport in air breathers

NINI SKOVGAARD, JAMES W. HICKS, AND TOBIAS WANG

4.1   Introduction

Air-breathing vertebrates constitute a large group of diverse animals belonging to different taxonomic classes. Air breathing evolved independently in different groups of fish and early tetrapods, and extant species employ an array of different air-breathing organs that are derived from various existing structures, such as the gastrointestinal tract or the buccopharyngeal cavity (see Chapter 6). True lungs in terrestrial vertebrates develop embryologically as a ventral outpocketing of the posterior pharynx into a paired structure that extends into the peritoneal cavity. The entrance to the lung through the pharynx is guarded by the glottis, and the lungs are perfused by a pulmonary artery that carries oxygen-poor blood to the respiratory surfaces in the lungs, while a pulmonary vein returns oxygen-rich blood to the heart. Although the lungs of extant air-breathing vertebrates share a common embryological development and overall arrangement, there are large structural differences, from the simple sac-like lungs of amphibians to the complex structure of the alveolar lungs of mammals and the parabronchial lungs of birds. Regardless of the structural variation, in all air-breathing vertebrates the gas-exchange organs provide adequate exchange of O_2 and CO_2 to meet the variable metabolic needs of the animal.

Vertebrates supply the majority of their energetic requirements through aerobic metabolism. As the product of aerobic metabolism, adenosine triphospate (ATP), cannot be effectively stored, the oxygen-transport process represents a continual balance between delivery of oxygen (supply) and the use of ATP (demand). This balance is achieved primarily through cardiovascular and ventilatory adjustments that, in concert, help maintain an adequate delivery of

Respiratory Physiology of Vertebrates: Life with and without Oxygen, ed. Göran E. Nilsson. Published by Cambridge University Press. © Cambridge University Press 2010.

oxygen to the metabolizing tissues. This is particularly evident during periods of increased demand, e.g. increased activity such as walking, running, flying, or swimming. In these instances the vast majority of vertebrates exhibit very rapid responses in convective oxygen transport (ventilation, blood flow) to match the increased oxygen demands in the tissues. By contrast, during periods of reduced supply, e.g. hypoxemia resulting from cardiac shunt, a variety of cellular mechanisms can reduce the O_2 demands of the tissues. This hypoxia-induced hypometabolism (the hypoxic metabolic response) (Hochachka et al., 1996; Hochachka and Lutz, 2001) appears to be a common metabolic strategy in both endothermic and ectothermic animals (Hicks and Wang, 2004). In addition, during periods of reduced oxygen supply, many vertebrates can augment these cellular/biochemical adjustments with behavioral changes that contribute to an overall reduction in energy requirements, e.g. reductions in body temperature. The hypoxia-induced hypothermia (anapyrexia) results from a reduction in thermoregulatory set points, and is not the result of impaired thermoregulation (Wood, 1991). The Q_{10} (difference in respiration rate for every 10°C rise in temperature) for metabolism is approximately 2 for most vertebrates. Thus, the metabolic rate will be halved for each 10°C fall in body temperature, and for each 1°C reduction there is an 11% energetic saving.

In vertebrates, the oxygen-transport system exhibits variability in structure at each transport step; regardless, the delivery of oxygen to the tissues represents a highly integrated, complex system that adjusts delivery in response to changes in O_2 supply and tissue energetic demands. These adjustments are well coordinated and can occur rapidly (seconds), over periods of days and weeks (phenotypic plasticity; acclimatization) or generations (genetic changes; adaptations). The goals of this chapter are to review the quantitative models that describe gas exchange and transport in vertebrates. In addition, specific examples of the integrative responses of extant vertebrates to changes in oxygen supply and demand will be reviewed. Finally, the oxygen-transport system of vertebrates will be discussed within the context of vertebrate evolution, specifically focusing on the possible mechanisms and conditions that influenced the structural and functional properties of the cardiopulmonary system and the relationship of these changes to the levels in atmospheric oxygen over the past 550 million years (Phanerozoic Eon).

4.2   General models for O_2 uptake and transport

In all air-breathing vertebrates, oxygen diffuses from the air in the gas-exchange organ and is transported convectively by the blood to the tissues where it diffuses from the capillary and is used for respiration in the mitochondria. Lungs are the most common gas-exchange organs in air-breathing vertebrates, but many air-breathing fishes use other structures (see Chapter 6), and amphibians rely extensively on O_2 uptake and CO_2 excretion over the skin. This cutaneous gas exchange does not appear to be regulated to the same extent as pulmonary gas exchange and will, therefore, be treated separately below.

The convective and diffusive transfer processes of the respiratory gases can be accurately described by relatively simple equations (the Fick principle and Fick's law of diffusion, respectively). These fundamental transport equations, along with the classical equations that describe the behavior of gases (ideal gas law, Dalton's law of partial pressures, Henry's law) provide the foundation for mathematical models that can accurately describe and predict the behavior of complex and diverse vertebrate gas-transport systems. For example, within the comparative literature, mathematical analyses of gas exchange have been particularly useful for the understanding of functional differences among various types of gas-exchange organs (e.g. septated and alveolar lungs of squamates and mammals vs. the parabronchial lungs of birds) (Piiper and Scheid, 1975) and in macroevolutionary analyses such as the concept of symmorphosis (Taylor and Weibel, 1981).

4.2.1   The oxygen-transport cascade

The transport of oxygen from the surrounding air to the mitochondria in the cells is illustrated schematically in Fig. 4.1, and results from the interaction of four transfer steps that operate in series: ventilation; diffusion of O_2 from the air to the blood; circulation; and diffusion of O_2 into the cells. The transfer rate of O_2 ($\dot{V}O_2$) at each step is quantified by the Fick principle (convective steps) and Fick's law (diffusive steps). The overall diffusive driving force for this process is the difference between the ambient partial pressure of oxygen (PO_2) and the PO_2 in the mitochondria. At each step, there can be reductions in PO_2. This reduction results from physical factors (e.g. humidification of inspired air that causes a rise in water vapor pressure and, therefore, a fall in PO_2) and physiological functions (e.g. ventilation/perfusion heterogeneity in the gas-exchange organ or shunt). The overall decrease in PO_2 as oxygen moves from the ambient respiratory medium to the mitochondria is referred to as the 'the oxygen-transport cascade'.

The mass transfer of oxygen, i.e. the rate of oxygen uptake ($\dot{V}O_2$), at each of these four steps in the oxygen-transport cascade can be quantified as a function of the partial pressure difference (ΔPO_2) and the conductance (G) (Fig. 4.1). Thus, G_{diff} is the diffusing capacity of the lung and the tissues (D_LO_2 and D_tO_2, respectively), whereas G_{vent} and G_{perf} are the products of convective flow rates

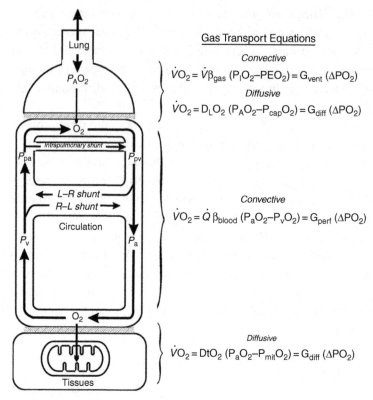

Gas Transport Equations

Convective

$$\dot{V}O_2 = \dot{V}\beta_{gas}\ (P_IO_2 - PEO_2) = G_{vent}\ (\Delta PO_2)$$

Diffusive

$$\dot{V}O_2 = D_LO_2\ (P_AO_2 - P_{cap}O_2) = G_{diff}\ (\Delta PO_2)$$

Convective

$$\dot{V}O_2 = \dot{Q}\ \beta_{blood}\ (P_aO_2 - P_vO_2) = G_{perf}\ (\Delta PO_2)$$

Diffusive

$$\dot{V}O_2 = DtO_2\ (P_aO_2 - P_{mit}O_2) = G_{diff}\ (\Delta PO_2)$$

Fig. 4.1 A schematic diagram of the oxygen cascade, expanded to include intrapulmonary and cardiac shunts. Oxygen transport from the air to the metabolizing tissue consists of four steps that function in series: ventilation, diffusion of O_2 from the air to the blood, circulation and diffusion of O_2 into the cells. The transfer rate of O_2 ($\dot{V}O_2$) at each step is represented by either convective or diffusive equations that can be simplified as the product between the PO_2 difference and a conductance (G). P_v, systemic venous PO_2; P_{pa}, pulmonary arterial PO_2; P_{pv}, $P_{cap}O_2$, pulmonary capillary PO_2; pulmonary venous PO_2; P_a, systemic arterial PO_2. See text for description of gas transport equations (From Wang and Hicks, 2002).

(ventilation [V] or cardiac output [Q]) and their respective capacitance coefficients (β_{gas} or β_{blood}). These capacitance coefficients represent the relationship between oxygen concentration and partial pressure. For oxygen, as for all other ideal gases, there is a linear relationship between concentration and partial pressure in water and air, and the capacitance coefficient. In other words, the solubility is independent of PO_2 and, in the case of water, merely reflects the Bunsen solubility. In blood, however, where oxygen binds cooperatively to hemoglobin, the relationship between PO_2 and O_2 concentration is characterized by the sigmoidal oxygen dissociation curve, and the capacitance coefficient, therefore, varies considerably and will often attain maximal values around P_{50} (PO_2 at 50% saturation of hemoglobin). The capacitance coefficient, at a given PO_2, will, furthermore, be directly proportional to the hemoglobin concentration. The oxygen-binding characteristics, i.e. both oxygen affinity and the capacity for oxygen binding, therefore, form essential components in the description of gas exchange. These parameters vary considerably among different taxonomic groups, and both affinity and capacity are amendable to environmental conditions, such as hypoxia or temperature. Furthermore, because of the Bohr effect, which describes the reduction in blood oxygen affinity during acidification, the transport of CO_2 and O_2 are intimately linked.

At rest, $\dot{V}O_2$ may be regarded as independent of oxygen delivery. In this case, the venous partial pressure can be viewed as dependent variables determined by metabolism and the arterial blood gases. By contrast, whenever the demands for oxygen are increased (e.g. during physical activity or digestion), $\dot{V}O_2$ becomes a dependent variable of the oxygen transport, and the maximal rate of oxygen consumption ($\dot{V}O_{2max}$) is limited by oxygen delivery and is, therefore, determined by the structural and functional constraints of the oxygen-transport system (Wagner, 1996). These constraints include the diffusive limitations in the tissues and in the lungs, as well as the capacities of the lungs and the cardiovascular systems to convey the mass transport of gas and blood.

4.3   Pulmonary gas exchange

4.3.1   Overview of lung ventilation in vertebrates

The lungs of vertebrates differ enormously among the various taxonomic groups in their morphological features, from simple sac-like structures to systems with complex branching patterns. Regardless, in all animals lung ventilation is accomplished by anatomical structures that generate pressure differentials between the ambient air and the gas-exchange organ. This serves the same purpose, bringing the ambient air into close contact with blood and facilitating gas exchange. In amphibians and most air-breathing fish, a positive pressure pump ventilates the lungs, where contraction of the buccal cavity forces air into the lungs through the open glottis. Exhalation is normally caused by passive recoil of the lungs. In many anurans, such as toads (*Bufo*), the breathing pattern is rather complex and can alternate between tidal ventilation of the lungs and so-called lung inflation events, where the lungs are progressively filled by many subsequent inhalations followed by a single, large exhalation. In reptiles and mammals, the lungs are ventilated by the creation of a sub-atmospheric pressure that draws air into the lungs. The sub-atmospheric pressure is formed by contraction of the intercostal muscles and, in mammals, the thoracic diaphragm, which enlarge the thoracic cavity. Some lizards appear to supplement costal ventilation with buccal pumping, and the contribution of the ventilatory mechanism may be particularly important during locomotion, when the costal muscle also serves an important role in stabilizing the trunk. For example, in the Savannah monitor lizard (*Varanus exanthematicus*), buccal ventilation during activity augments hypaxial ventilation (Owerkowicz *et al.*, 1999). In this example, preventing buccal ventilation reduces maximal oxygen transport and negatively impacts aerobic performance. In non-avian reptiles, crocodilians have a very unique system for ventilating the lung. In addition to the hypaxial musculature involvement in changing the volume of the thoracic cavity, crocodilians also use the movement of the liver to further expand the thoracic volume. In these animals, muscles (diaphragmaticus) are attached to the liver, and as these muscles contract, the liver is pulled towards the pelvis. Because the base of the lungs is attached to the dome of the liver, the displacement of the liver increases the lung volume.

The respiratory system of birds is unique among air-breathing vertebrates and is the only system that is not ventilated bidirectionally, with the air inhaled and exhaled through the same airways. It consists of two parallel lungs (parabronchi) and a number of air sacs. These air sacs act as a 'bellows' system that during both inspiration and expiration creates a unidirectional flow of air through the parabronchi, where gas exchange takes place.

4.3.2   *Variation in lung structure in vertebrates*

The structure of the lungs varies enormously among the different air breathers. In mammals, the airways form a highly complex fractal branching structure, the bronchial tree, which divides into hundreds of millions of alveoli, the functional unit of gas exchange (Weibel and Gomez, 1962). The parabronchial avian lung consists of parallel parabronchi, a rigid structure with limited compliance, in which gas exchange takes place in the periparabronchial tissue, where air and blood capillaries are interwoven (Duncker, 1972; Maina, 2006). Lungs of amphibians and reptiles are structurally simpler than mammalian and avian lungs. They contain a simple airway conduction system (trachea that may divide into primary bronchi). This airway conduction system brings air into a simple lung, which is shaped like a sac and has a single chamber, with a central air-filled cavity that opens radially into the parenchyma. In some species, the lungs have more chambers and are divided by one or several septae. Gas exchange takes place in honeycombed faveola in the lung wall between the air and the pulmonary capillaries (Perry, 1989). However, the complexity of the reptilian lung structure does not appear to be correlated with overall oxygen-transport capacity. For example, some tegu lizards and varanid lizards have very similar maximal rates of oxygen consumption, but very different lung structures. In the tegu lizard, the lung is sac-like with few subdivisions, whereas the varanid lizard possesses a very complex lung structure, with many subdivisions. In the tegu, overall surface area for gas exchange is smaller than that in the varanid, but the diffusional barrier is thinner.

In amphibians, gas exchange over the skin is an important mode of respiration, and in some species, where the lungs are greatly reduced or entirely absent, the skin is the only method of respiration. Respiratory gases are exchanged between the surrounding air and blood in the cutaneous capillary network (Burggren and West, 1982; Feder and Burggren, 1985).

4.3.3   Ventilation and the composition of lung gases

Given the tidal ventilation of the lungs, it is possible to describe the mass transfer of the ventilated air (minute ventilation, $\dot{V}_E$) as a multiple of the frequency of ventilation (f_R) and tidal volume (V_T), which represents the volume of air in each breath:

$$\dot{V}_E = f_R \times V_T \tag{4.1}$$

The trachea and most of the branches of the airways are thick, rigid and solid structures without perfusion and do not take part in gas exchange. The volume of these structures is, therefore, called the anatomical respiratory dead space (V_D), and the effective ventilation of the gas-exchange structures ($\dot{V}_{eff}$) can accordingly be described as:

$$\dot{V}_{eff} = f_R \times (V_T - V_D) \tag{4.2}$$

In mammals, $\dot{V}_{eff}$ is the same as alveolar ventilation ($\dot{V}_A$), but given that the lungs of other vertebrates do not have alveoli, $\dot{V}_{eff}$ is a better term to describe how much air is being delivered to the sites of gas exchange. There is variation in V_T among species, but it often represents somewhere between 10 and 30% of normal lung volume. Because V_D is an anatomically determined and fixed volume, it is more effective for gas exchange to increase V_T than f_R. Nevertheless, V_T is constrained by the vital capacity of lungs, and it is generally more expensive to raise V_T than to augment f_R.

The oxygen partial pressure of the lung gas within the lung (P_LO_2; normally referred to as alveolar PO_2 (P_AO_2) in mammals) is determined by $\dot{V}_{eff}$ relative to $\dot{V}O_2$. Thus, the dependence of P_LO_2 on V_{eff} and $\dot{V}O_2$ can be described as:

$$P_LO_2 = P_IO_2 - (\dot{V}O_2/\dot{V}_{eff})(P_B - PH_2O) \tag{4.3}$$

where P_IO_2 is the partial pressure of oxygen in the inspired air, P_B is the barometric pressure, and PH_2O is the water vapor pressure at body temperature. This equation provides the intuitive prediction that P_LO_2 increases as $\dot{V}_{eff}$ is elevated and that $\dot{V}_{eff}$ must increase proportionally to $\dot{V}O_2$ to maintain P_LO_2 when metabolic demands are elevated. The ratio between $\dot{V}_{eff}$ and $\dot{V}O_2$ (air convection requirements, ACR) is useful in this context because it provides a robust prediction of the respective effects of ventilation and metabolism on lung gases.

Both P_LO_2 and $\dot{V}_{eff}$ are difficult to measure, but because CO_2 equilibrates readily in the lungs and because the partial pressure of CO_2 in the arterial blood (P_aCO_2) is relatively unaffected by inhomogeneities and shunts, it is often convenient to calculate P_LO_2 from the alveolar gas equation:

$$P_LO_2 = P_IO_2 - P_aCO_2[F_IO_2 + (1 - F_IO_2)/RER] \tag{4.4}$$

where F_IO_2 is the fraction of O_2 in the inspired air and RER is the respiratory gas-exchange ratio ($\dot{V}CO_2/\dot{V}O_2$). It is necessary to take RER into account, because the volume of oxygen extracted from the lung gas does not equal the volume of CO_2 added when RER differs from unity. Normally, RER attains values between 0.7 and 1.0, depending on the metabolic fuel, but in many reptiles, and amphibians the metabolic acidosis that results from lactic acid production during exercise may cause RER to increase above 2.0, and failure to take such deviations from the normal condition into account could seriously affect the estimated P_LO_2.

The alveolar gas equation is often simplified as:

$$P_LO_2 = P_IO_2 - P_aCO_2/RER \tag{4.5}$$

which suffices in most cases and is conceptually more straightforward. As discussed in more detail below, it is important to consider that large cardiac shunts, which are common in most amphibians and reptiles, increase P_aCO_2 above PCO_2 of the blood leaving the lungs, which would lead to an underestimation of P_LO_2 if ignored. A theoretical model, however, shows that these effects are relatively minor (T. Wang, unpublished).

4.3.4   Pulmonary oxygen uptake through diffusion

The exchange of gases between lung gas and blood within the lungs takes place across the blood–gas barrier (BGB), which separates the lung gas from the capillary blood. The transfer of oxygen across this barrier takes place by passive diffusion and, therefore, is dictated by the partial pressure difference. During steady-state conditions, this can be described by Fick's first law of diffusion, where oxygen uptake ($\dot{V}O_2$) is proportional to the difference in PO_2 (ΔPO_2) between lung gas and blood, the driving force for diffusion, as well as the pulmonary diffusing capacity for oxygen (D_LO_2):

$$\dot{V}O_2 = D_LO_2 \times \Delta PO_2 \tag{4.6}$$

The diffusing capacity is a product of Krogh's diffusion coefficient (KO_2), which accounts for the physicochemical properties of the gas and the BGB, and the 'anatomical diffusion factor,' which is the respiratory surface area (A) relative to the thickness (l) of the BGB:

$$\dot{V}O_2 = KO_2 \times A/1 \times \Delta PO_2 \tag{4.7}$$

Thus, diffusion of oxygen through the BGB is directly proportional to the respiratory surface area and inversely proportional to the thickness of the BGB. Therefore, a larger respiratory surface area and a thinner BGB will increase the pulmonary diffusing capacity for oxygen.

The transition from ectothermy in amphibians and reptiles to endothermy in birds and mammals was associated with an approximate ten fold rise in resting and maximal rates of oxygen consumption (e.g. Bennett and Ruben, 1979). This increased need for oxygen required that all steps of the oxygen cascade be improved. In the lungs, enlarged structural complexity allowed for smaller gas-exchange units and increased surface area, which, in concert with a thinner BGB, increased the pulmonary diffusing capacity for oxygen (Perry, 1989; West, 2003). In air-breathing vertebrates, the degree of subdivision of the lung increases through amphibian, reptilian, mammalian, and avian lungs (Maina, 1998). Interestingly, the independent evolution of endothermy in the synapsids and archosaurs resulted is very different lung structures, each capable of maintaining high oxygen flux rates: the mammalian alveolar lung and the avian parabronchial lung. The extensive subdivision of the avian lung results in the largest respiratory surface density (surface area per unit volume of the gas-exchange tissue [$mm^2 \ g^{-1}$]) found in air-breathing vertebrates. However, birds have small lungs relative to body weight compared with other air-breathing vertebrates (Maina 1998). Thus, the largest mass-specific respiratory surface area in birds, reported in a hummingbird called the sparkling violet-ear (*Colibri coruscans*), is only 87 $cm^2 \ g^{-1}$ (Dubach, 1981). This can be compared with the 138 $cm^2 \ g^{-1}$ record in mammals, reported in Wahlberg's epauletted fruit bat (*Epomophorus wahlbergi*) (Maina et al., 1982). In general, however, the respiratory surface area is within the same range in mammals and birds (Perry, 1989).

The BGB is composed of three layers: the capillary endothelium, the interstitial layer, and an epithelial layer (West, 2003). The basic tripartite structure of the BGB has been highly conserved through evolution from the first air-breathing vertebrates to endothermic birds and mammals (West, 2003). The thickness of the BGB decreases from amphibians, reptiles, mammals, and birds (Maina and West, 2005), and the sparkling violet-ear humming bird has a BGB that is only 0.1 μm thick (Dubach, 1981). A thin BGB will increase the diffusing capacity for oxygen; however, at the same time the BGB needs to keep the strength to resist the high pulmonary capillary pressure developed during exercise. The evolution of the very thin BGB in endothermic vertebrates required complete cardiac division and separation of the pulmonary and systemic circulations to allow for low pulmonary blood pressure to protect the respiratory epithelium from a high vascular pressure that could cause pulmonary edema (West, 2003). In birds, the large respiratory surface area, in combination with the extremely thin BGB, results in the largest diffusing capacity for oxygen among vertebrates (Perry, 1989) and may have evolved to supply the oxygen consumption during the energetically expensive flapping flight. This is further supported by the fact that among mammals, flying bats have evolved the thinnest lungs and largest respiratory surface area (Perry, 1989).

The functional pulmonary diffusing capacity for oxygen (D_LO_2) can be determined experimentally using the CO rebreathing technique (Krogh and Krogh, 1910). Comparing D_LO_2 between different groups of air-breathing vertebrates reveals that amphibians and reptiles have a D_LO_2 that is an order of magnitude lower than the D_LO_2 in birds and mammals (Glass, 1991), which correlates with the approximately ten fold lower oxygen consumption in amphibians and reptiles compared with birds and mammals.

4.3.5   Gas-exchange efficiency

In a 'perfect' or 'ideal' lung, there are no diffusive limitations to gas exchange, and blood equilibrates swiftly with PO_2 in the lungs. In many animals, however, the PO_2 of blood that returns to the heart from the lungs (pulmonary venous return) is lower than lung PO_2, which indicates that equilibration has not occurred. In mammals, gas-exchange efficiency is conventionally evaluated from the alveolar–arterial PO_2 difference (P_A–P_a difference), which describes how well air and blood equilibrate. Because many air-breathing vertebrates have non-alveolar lungs, as well as intra cardiac shunts that lower arterial PO_2 (P_aO_2) through venous admixture, it is more convenient to evaluate gas-exchange efficiencies from the PO_2 difference between mixed lung gas and the left atrium (P_L–P_{LAt} difference).

Figure 4.2 shows the different types of gas-exchange organs in air breathers. In the avian lung, blood flows perpendicular to the unidirectional air flow through the parabronchi. This arrangement of the respiratory medium relative to blood flow ('cross-current' gas-exchange model) results in a highly efficient gas exchanger: *theoretically* the most efficient among air-breathing vertebrates, where PO_2 in pulmonary venous blood flowing from the lung can exceed PO_2 in expired air. Thus, the cross-current gas-exchange model can reach a negative P_L–P_{LAt} difference (Piiper and Scheid, 1972). In amphibian, reptilian, and mammalian

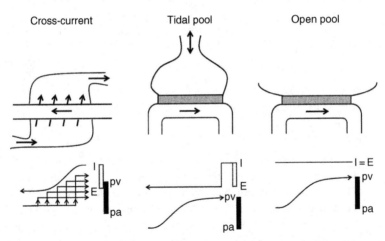

Fig. 4.2 Different gas-exchange systems in air breathers. In the cross-current model (bird lungs), blood flows perpendicular to the unidirectional air flow through the parabronchi, which forms a highly efficient gas exchanger where pulmonary venous blood flowing from the lung can exceed the PO_2 in expired air. The tidal pool model (lungs of mammals, reptiles, and amphibians) is an efficient gas exchanger with an arterial PO_2 that is close to but does not exceed alveolar PO_2. The open pool model (amphibian skin) is the least efficient gas exchanger because of diffusion limitation, with arterial PO_2 being lower than ambient PO_2. The lower figures represent inspired (I), expired (E), pulmonary arterial (pa), and pulmonary venous (pv) PO_2 values (based on Piiper and Scheid, 1977).

lungs ('tidal pool' gas-exchange model), or amphibian skin ('open pool' gas-exchange model), *in theory* blood equilibrates with air and the P_L–P_{LAt} difference reaches zero in ideal situations (Piiper and Scheid, 1972). Although theoretical models predict differences in the gas-exchange efficiency of the various gas-exchange organs (with cross-current > ventilated > open system), in vivo, these differences are often minimized. Three physiological mechanisms account for gas exchange to deviate from theoretically ideal levels with P_L–P_{LAt} differences above zero. These mechanisms include intrapulmonary shunts, diffusion limitations, and ventilation to perfusion ($\dot{V}/\dot{Q}$ heterogeneity) (Fig. 4.3) (Scheid and Piiper, 1997).

4.3.6   Gas-exchange inefficiency (deviation from ideal models)

Under resting normoxic conditions, the P_L–P_{LAt} difference averages between 4 and 10 mmHg in mammals and birds (Piiper, 1990), whereas it often exceeds 20 mmHg in reptiles and amphibians (Burggren and Shelton, 1979; Glass, 1991; Hicks and White, 1992). The contribution of intrapulmonary shunts and $\dot{V}/\dot{Q}$ heterogeneity to the measured P_L–P_{LAt} difference can be

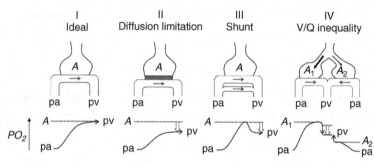

Fig. 4.3 Physiological mechanisms that can cause difference in the partial pressure of lung gas compared with that of pulmonary venous blood (P_AO_2–$P_{pv}O_2$) difference. The ideal lung lacks a P_AO_2–$P_{pv}O_2$ difference. In the case of diffusion limitation, the P_AO_2–$P_{pv}O_2$ difference is due to incomplete diffusive equilibration. In the presence of intrapulmonary shunts, admixture of pulmonary arterial blood causes a drop in end-capillary oxygen content. In the case of ventilation-perfusion inequality, the P_AO_2–$P_{pv}O_2$ difference is caused by mixing of blood from the two compartments with different $\dot{V}/\dot{Q}$ ratios (based on Piiper, 1993).

quantified using the multiple inert gas elimination technique (MIGET), and any remaining P_L–P_{LAt} difference is ascribed to diffusion limitations (Wagner *et al.*, 1974a; Wagner *et al.*, 1974b). The relative contributions to the gas-exchange inefficiencies in amphibians remain to be quantified, and the following discussion includes only reptiles, mammals, and birds. In birds and mammals, each gas-exchange unit may not receive the same level of ventilation and blood flow. Consequently, some units will have a $\dot{V}/\dot{Q} > 1$ and others will have a $\dot{V}/\dot{Q} < 1$. The resulting $\dot{V}/\dot{Q}$ heterogeneity appears to be the prevailing cause for gas-exchange inefficiency under resting normoxic conditions (Hopkins *et al.*, 1999; Schmitt *et al.*, 2002; Powell and Hopkins, 2004). The P_L–P_{LAt} may also deviate from zero due to intrapulmonary shunts, which is a portion of pulmonary blood flow that bypasses the respiratory medium and, therefore, does not partake in gas exchange. Intrapulmonary shunts result in oxygen-poor blood mixing with oxygen-rich blood, with the resulting admixture having a reduced oxygen content and PO_2. Intrapulmonary shunts are virtually absent in mammals and birds (Hopkins *et al.*, 1999; Schmitt *et al.*, 2002; Powell and Hopkins, 2004). In reptiles, $\dot{V}/\dot{Q}$ heterogeneity in addition to large intrapulmonary shunts (which can be 5% or more of total pulmonary blood flow) contribute to the gas-exchange inefficiency (Powell and Gray, 1989; Hopkins *et al.*, 1995; Hopkins *et al.*, 1996). In addition, intrapulmonary gas diffusion limitations ('stratification') and the lower and inhomogenously distributed oxygen diffusing capacity of the lung in reptiles may contribute to the larger P_L–P_{LAt} difference in reptiles (for a review see Wang *et al.*, 1998b).

4.3.7   Diffusion limitation

If there is no diffusion limitation, blood flowing through a capillary will quickly attain gas partial pressures equivalent to the respiratory medium ('I, Ideal' in Fig. 4.3). Conversely, if diffusion limits gas exchange, respiratory gases will not equilibrate when blood transverses the capillary ('II, diffusion limitation' in Fig. 4.3). The degree of diffusion limitation is determined by the diffusing capacity (D) relative to the perfusive conductance ($\dot{Q}\beta O_2$), where $\dot{Q}$ is blood flow and βO_2 is the capacitance coefficient for blood (the change in blood oxygen content for a given change in PO_2). This ratio, $D/\dot{Q}\beta O_2$, the equilibration coefficient, allows predictions of limitations of gas exchange (e.g. Piiper, 1961; Scheid and Piiper, 1997). At high $D/\dot{Q}\beta O_2$, full equilibration is attained faster than at low values and the gas exchange is primarily perfusion limited, whereas at low $D/\dot{Q}\beta O_2$ gas exchange is primarily or completely diffusion limited. A diffusion limitation results from either a decreased D and/or an increase in $\dot{Q}$ or β.

The skin is an important site for respiratory gas exchange in amphibians. Amphibian skin has the simplest structure of all gas-exchange organs, and the respiratory surface is in direct contact with the respiratory environment – hence the term 'open pool' gas exchange (Scheid and Piiper, 1997). However, the stagnant layer of air surrounding the 'non-ventilated' skin increases the distance of diffusion, lowers the diffusing capacity for skin (D_S), and therefore increases $D_S/\dot{Q}\beta O_2$ (Feder and Pinder, 1988; Malvin, 1988). Therefore, cutaneous respiration is primarily diffusion limited. The diffusion limitation can be reduced by movement (skin ventilation), reducing the stagnant layer around the respiratory tissue. Skin ventilation occurs more frequently during hypoxia and after exercise, when oxygen demand is increased (Feder and Pinder, 1988). Although the amphibian skin gas exchange is mainly diffusion limited, some salamanders are cable of increasing their oxygen uptake substantially during exercise by increasing D_S, which is likely to occur through capillary recruitment (Burggren and Moalli, 1984).

4.3.8   Ventilation/perfusion heterogeneity

Ventilation/perfusion ($\dot{V}/\dot{Q}$) heterogeneity reflects that different parts of the lung have different $\dot{V}/\dot{Q}$ ratios. Thus, it refers to a spatial rather than a temporal heterogeneity. This lowers gas-exchange efficiency and increases the $P_L–P_{LAt}$ difference, whereas a decrease in the overall $\dot{V}/\dot{Q}$ ratio, merely calculated as overall ventilation relative to pulmonary blood flow, would lower P_aO_2 without increasing the $P_L–P_{LAt}$ difference. Figure 4.4 illustrates a classic three-compartment lung model showing how $\dot{V}/\dot{Q}$ differences between lung

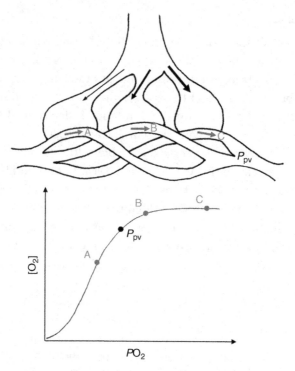

Fig. 4.4 A three-compartment lung model showing how $\dot{V}/\dot{Q}$ heterogeneity between functional lung units can lead to gas-exchange inefficiency. The three lung units are ventilated at different rates (black arrows) but receive equal blood perfusion (gray arrows), leading to different $\dot{V}/\dot{Q}$ ratios. In each unit the gas and blood phase equilibrates and end capillary oxygen content (A, B and C) is dependent on lung PO_2 and the shape of the oxygen dissociation curve (ODC). The 'low' $\dot{V}/\dot{Q}$ ratio contributes more to *mixed* pulmonary venous blood PO_2 (P_{pv}) because the ODC is steep at low PO_2 compared with the units for 'high' $\dot{V}/\dot{Q}$ ratios and high lung PO_2, where ODC is on the flat part. This leads to a reduction in P_{pv} compared with the 'ideal' situation (B). The lung to left atrium difference in PO_2 (P_L–P_{LAt} difference) is further aggravated because *mixed* lung PO_2 is a weighted average between the three compartments, and the hyperventilated unit with the high PO_2 contributes the largest volume, increasing *mixed* lung PO_2 (from Skovgaard and Wang, 2006).

units can increase the P_L–P_{LAt} difference. The functional gas-exchange units receive equal blood perfusion (gray arrows) but are being ventilated at different rates (black arrows), which generate three different $\dot{V}/\dot{Q}$ ratios: A = 'low'; B = 'ideal'; and C = 'high.' As there are no diffusion limitations, PO_2 in the gas and blood phase equilibrates and end capillary O_2 concentration $[O_2]_c$, depends on alveolar PO_2 and the shape of the oxygen dissociation curve (ODC). In the low $\dot{V}/\dot{Q}$ unit there is a large reduction in $[O_2]_c$, compared with the ideal compartment, as the shape of the ODC is steep at low PO_2. However, the high lung PO_2 in the hyperventilated unit does not increase $[O_2]_{c'}$ significantly above $[O_2]_{c'}$ in the ideal compartment, as the ODC is at the flat part at high PO_2. Consequently, the PO_2 in mixed pulmonary venous blood (P_{pv}) is lower than if there were no $\dot{V}/\dot{Q}$ heterogeneity. The P_L–P_{LAt} difference is further aggravated by the fact that mixed lung PO_2 is a weighted average between the three compartments, where the hyperventilated compartment contributes with a greater volume and, therefore, increases PO_2 in the lung.

The lung structure in air-breathing vertebrates is very diverse, ranging from simple lungs to structurally more complex lungs, increasing the oxygen diffusing capacity, but the higher complexity may also increase the possibilities for $\dot{V}/\dot{Q}$ heterogeneity. However, when evaluating $\dot{V}/\dot{Q}$ distributions using MIGET in mammals, birds, and reptiles, Powell and Hopkins (2004) concluded that $\dot{V}/\dot{Q}$ heterogeneity, unexpectedly, was independent of lung complexity. In fact, there was a tendency for some species with structurally simple lungs, such as the tegu lizard, to have high heterogeneity compared with reptiles and mammals with high pulmonary complexity. One way that $\dot{V}/\dot{Q}$ heterogeneity is reduced and arterial oxygenation defended is through local regulation of pulmonary blood flow through hypoxic pulmonary vasoconstriction (von Euler and Liljestrand, 1946).

4.3.9 *Hypoxic pulmonary vasoconstriction*

Hypoxic pulmonary vasoconstriction (HPV) is an adaptive response that diverts pulmonary blood flow from poorly ventilated and hypoxic areas of the lung to more well-ventilated parts. HPV is important for local matching of blood perfusion to ventilation and improves pulmonary gas-exchange efficiency (von Euler and Liljestrand, 1946). The primary site of hypoxic vasoconstriction is the precapillary muscular pulmonary arterioles, where low alveolar oxygen causes constriction of the vasculature (Weir and Archer, 1995). Hypoxic vasoconstriction is an ancient and highly conserved response expressed in the respiratory organs of all vertebrates, including lungs of mammals, birds and reptiles, amphibian skin, and fish gills (Von Euler and Liljestrand, 1946; Faraci et al., 1984; Malvin and Walker, 2001; Smith et al., 2001; Skovgaard et al., 2005).

HPV is a locally mediated response, and the hypoxic constriction persists in isolated and perfused lungs without neurohumoral influences, arterial pulmonary rings, and even in isolated pulmonary arterial smooth muscle cells (PASMCs). Although the mechanism underlying HPV remains elusive, there is general consensus that hypoxia alters the production of reactive oxygen species (ROS) inhibiting voltage-gated K^+ channels and that the resulting depolarization of PASMCs causes contraction as intracellular Ca^{2+} concentration ($[Ca^{2+}]_i$) rises (Moudgil et al., 2005). Numerous controversies, however, surround the role of the endothelium and the involvement of endothelin 1 (ET-1) (Aaronson *et al.*, 2002). Thus, although several studies show that HPV is intrinsic to PASMCs, other studies indicate that the endothelium, possibly through the release of ET-1, is essential for HPV.

The physiological significance of HPV in improving gas exchange in the presence of local hypoxia in the lungs is obvious. However, when the lung is exposed to general hypoxia, following ascent to high altitude, under pathophysiological conditions or when holding breath, HPV is of limited value because all parts of the lungs become equally hypoxic. HPV may even be disadvantageous because general hypoxia increases overall resistance to pulmonary blood flow and leads to a rise in pulmonary arterial blood pressure. The increase in pulmonary arterial pressure and capillary pressure secondary to HPV disturbs pulmonary fluid balance and is partly responsible for the pathophysiology of high-altitude pulmonary edema (HAPE) (Bärtsch, 2007). Moreover, chronic exposure to general hypoxia may cause vascular remodeling and pulmonary hypertension (Bärtsch, 2007). In human populations living at altitude (>3500 m), such as the Tibetans and Andeans, evidence suggests that they exhibit a blunted response to hypoxia (Bärtsch, 2007). The bar-headed goose (*Anser indicus*) crosses the Himalayans twice a year on its migratory route and has been documented flying above the summit of Mount Everest at 8848 m, where inspired PO_2 is as low as 43 mmHg. In these birds, the hypoxic constriction of the vasculature is attenuated, which may be a significant advantage for gas exchange in combined exercise and severe hypoxia (Faraci *et al.*, 1984).

The breathing pattern of many reptiles, particularly aquatic species, is characterized by ventilatory bouts consisting of one or several breaths interspersed with non-ventilatory periods of varying duration, in which lung and blood PO_2 declines as oxygen stores are exhausted (Milsom, 1991). In most non-crocodilian reptiles, the ventricle is anatomically and functionally undivided, so blood pressures are equal in systemic and pulmonary circulations (e.g. Hicks, 1998). Therefore, blood flow distribution between pulmonary and systemic circulations is primarily determined by pulmonary and systemic vascular resistances, respectively (Crossley *et al.*, 1998; Hicks, 1998). When the heart is undivided, HPV during long breath holds induces a bypass of the pulmonary circulation (right-to-left cardiac shunt), which reduces the ability to exploit pulmonary oxygen stores. HPV is present, nevertheless, in some reptiles with a poorly divided ventricle, but the threshold for increased pulmonary vascular resistance is so low (3 kPa in turtles and 6 kPa in tegu lizards) that it remains uncertain whether vascular resistance of the entire pulmonary circulation would be increased during normal breath-hold periods. Thus, hypoxia constricts the pulmonary vasculature in both caimans, which have a fully divided ventricle, and turtles, which have a typical undivided non-crocodilian heart, albeit at very different thresholds (14 vs. 3 kPa). The apparent blunted HPV in, turtles compared with caimans may, therefore, ensure that pulmonary blood flow can be increased during submergence (Crossley et al., 1998; Skovgaard et al., 2005).

4.4   Convection of oxygen by the cardiovascular system

The cardiovascular system is composed of a heart that pumps blood through the arteries to the capillaries, where gas exchange occurs. The blood is subsequently returned to the heart through the veins. In animals with lungs, the pulmonary circulation operates in parallel to the systemic circulation, but it is only within birds and mammals that the systemic and pulmonary circulations are completely separated. Such separation allows for a high systemic pressure, while maintaining a low pressure in the pulmonary circulation. The low pulmonary pressure permits for a thin BGB, which increases the lung diffusive capacity and also protects against fluid loss in the pulmonary capillaries. However, in amphibians and reptiles, the ventricle is not anatomically divided, and both the systemic and the pulmonary arteries are effectively supplied by the same ventricle, implying that the systolic pressures are identical in both circuits. It is characteristic for reptiles to have lower blood pressures than the systemic pressures of birds and mammals, which is often viewed as a necessary means to protect the pulmonary circulation, but may also only be tolerated in the ectothermic vertebrates because the metabolic demands are low compared with those of endotherms. As described in more detail below, the admixture of oxygen-rich and oxygen-poor blood within the heart also has large effects on arterial blood gas composition and can significantly reduce the amount of oxygen that is delivered to the metabolizing tissue.

Cardiac output ($\dot{Q}$) is determined by both the amount of blood that is ejected during each heartbeat (stroke volume, V_S) and heart rate (f_H):

$$\dot{Q} = f_H \times V_S \tag{4.8}$$

and the amount of oxygen delivered by both the cardiovascular system to the metabolizing tissue (systemic oxygen delivery SOD) can accordingly be written as:

$$SOD = f_H \times V_S \times [O_2]_a \tag{4.9}$$

where $[O_2]_a$ is the arterial oxygen concentration. Heart rate both at rest and during exercise is significantly lower in ectothermic vertebrates compared with mammals and birds, whereas V_S is similar when corrected for body mass. Also, $[O_2]_a$ is normally lower in ectothermic vertebrates because the hematocrit tends to be significantly lower. Moreover, blood–oxygen affinity, at a given body mass, tends to be lower in ectothermic vertebrates.

According to the Fick principle, oxygen consumption is a product of cardiac output, the blood capacitance coefficient (β_{blood}), and the difference between arterial and venous PO_2:

$$\dot{V}O_2 = \dot{Q} \times \beta_{blood} \times (P_aO_2 - P_vO_2) \tag{4.10}$$

The blood capacitance coefficient is determined by the carrying capacity of the blood, which depends on the concentration of hemoglobin, and the shape of the oxygen dissociation curve. As the blood capacitance coefficient and PO_2 determine the concentration of oxygen in the blood, the Fick principle can be rewritten as:

$$\dot{V}O_2 = \dot{Q} \times ([O_2]_a - [O_2]_v) \tag{4.11}$$

where $[O_2]_a$ is the arterial concentration of oxygen and $[O_2]_v$ is the venous concentration of oxygen.

As cardiac output is determined by stroke volume and heart rate, the Fick principle can be rewritten again as:

$$\dot{V}O_2 = f_H \times V_S \times ([O_2]_a - [O_2]_v) \tag{4.12}$$

4.4.1   The role of cardiac shunts

In amphibians as well as turtles, lizards, and snakes, the heart is not fully divided and there is the possibility of intraventricular mixing of the oxygenated blood that returns to the left atrium from the pulmonary circulation and the oxygen-poor blood that returns to the right atrium from the systemic veins. This admixture implies that the oxygen content of arterial blood is reduced compared with pulmonary venous return, with proportional reductions in systemic oxygen delivery. Similarly, in many air-breathing fishes, blood from the air-breathing organ is mixed with venous blood, and oxygen levels in the ventral aorta are reduced. Also, in crocodilians, the left aortic arch emerges from the right ventricle, so when systemic blood pressure is low, there is the possibility of recirculation of oxygen-poor blood within the systemic circulation. A practical consequence of these cardiac shunts is that arterial blood gas composition is not necessarily indicative of pulmonary function, and therefore blood needs to be sampled from the pulmonary vein or the left atrium when studies are designed to evaluate pulmonary gas transfer.

The admixture of the bloodstreams within the heart is normally referred to as cardiac shunts, where a Right-to-Left (R–L) shunt denotes systemic venous blood that recirculates within the systemic circulation, while Left-to-Right (L–R) shunts denotes the blood that is recirculated within the pulmonary circulation. The effects of R–L shunts on arterial blood oxygen concentration can be quantified as the weighed average of the oxygen concentrations of the systemic venous blood and the pulmonary vein ($[O_2]_{sv}$ and $[O_2]_{pv}$, respectively):

$$[O_2]_a = (Q_{pul} \times [O_2]_{pv} + \dot{Q}R - L \times [O_2]_{sv})/(\dot{Q}_{pul} + \dot{Q}R - L) \qquad (4.13)$$

where $\dot{Q}_{pul}$ is pulmonary blood flow and $\dot{Q}R-L$ is the right–left shunt flow. Left–right (L–R) shunt does not affect arterial blood oxygen concentration, but increases oxygen levels in the pulmonary artery, which can be described with a similar equation. In reptiles, an in-depth analysis of the effects of cardiac shunts requires that differences in blood composition between the right and left systemic arches are taken into account (see Hicks et al., 1996; Ishimatsu et al., 1996; Hicks, 1998).

In the undivided ventricle, blood with different oxygen concentrations is mixed, and the blood gases behave as if mixed within a closed system. This means that the PO_2 of the resulting mixture is a dependent variable determined by the resulting oxygen saturation of the blood and the blood–oxygen affinity (Wood, 1982; Wood, 1984). Arterial PO_2, therefore, becomes a composite variable that is determined by the amount of shunt flow, the oxygen concentration in the systemic and venous and pulmonary venous blood, as well as the ODC. A consequence of this interaction is that altered cardiac R–L shunt represents a mechanism to alter arterial oxygen levels independent of ventilation (e.g. Wang and Hicks, 1996). The usefulness of R–L cardiac shunt regulating arterial oxygen levels, independently of lung ventilation, is illustrated during digestion in carnivorous reptiles. In these animals the postprandial period is associated with elevated oxygen demands as well as blood alkalinization related to gastric acid secretion, which causes a rise in plasma $[HCO_3^-]$ (a phenomenon known as alkaline tide) (Wang et al., 2001a). In response, lizards and snakes undergo a relative hypoventilation, elevating arterial CO_2 and compensating for the alkaline tide. Simultaneously, these animals reduce R–L shunt, elevating arterial O_2, thus meeting the elevated oxygen demands associated with digestion while regulating acid–base status.

The cardiac shunt patterns are primarily dictated by the outflow resistances in the systemic and pulmonary circulation, such that high resistance in the pulmonary circulation diverts blood flow away from the lungs and, hence, induces R–L shunts, whereas a dilatation of the pulmonary circulation would cause L–R shunts. Although the regulation of the resistances in the systemic and pulmonary circulations is complex and depends on nervous, humoral, and local factors, the pulmonary vascular resistance of most reptiles and amphibians is primarily controlled by smooth muscle surrounding the pulmonary artery, which is innervated by the vagus (Hicks, 1998; Taylor et al., 2009). Thus, apart from slowing the heart, increased vagal tone causes constriction of the pulmonary artery and acts to decrease pulmonary blood and induce R–L cardiac shunts. Often, therefore, heart rate and the cardiac shunt pattern are tightly correlated.

Many reptiles and amphibians are intermittent breathers, in which brief periods of lung ventilation are interspersed with apneas of variable duration. The magnitude of the cardiac shunt flows vary considerably among species and depend on ventilatory state. In general, large R–L shunts prevail during apnea, whereas pulmonary ventilation is associated with small R–L shunts or the development of large L–R shunts. Thus in turtles and some sea snakes, pulmonary blood flow can cease completely during apnea, and very large L–R shunts can occur during ventilation (Hicks, 1998). In other species, such as varanid lizards and pythons, the heart is anatomically compartmentalized such that the potential for cardiac shunts is low and the degree of admixture of bloodstreams is generally small (Burggren and Johansen, 1982; Wang et al., 2003). The functional roles of these cardiac shunts remain elusive (Hicks and Wang, 1996; Hicks, 1998; Wang and Hicks, 2008), but it is evident that the magnitude of the R–L shunt is reduced considerably whenever the demands for oxygen transport are elevated (Wang et al., 2001b). For example, R–L shunts are markedly reduced during exercise in toads and turtles (Hedrick et al., 1999, Krosniunas and Hicks, 2003), and elevated temperature leads to reductions in R–L shunts in rattlesnakes and toads (Wang et al., 1998a; Gamperl et al., 1999; Hedrick et al., 1999). It is possible that the low arterial oxygen levels characteristic of resting animals represent a mechanism to reduce metabolism when demands are low, whereas the ability to reduce shunts is essential when demands are elevated.

4.5   Increased oxygen demand

When metabolism increases because of activity, elevated body temperature, and digestion, vertebrates exhibit rapid responses in ventilation and blood flow to ensure an adequate oxygen delivery to match the increased oxygen demands in the tissue.

All air-breathing vertebrates increase lung ventilation when metabolism is elevated, but the relative roles of f_R and V_T vary among taxonomic classes. In general, during exercise, the increase in minute ventilation tends to be proportional to metabolic increases, such that the ACR does not change relative to resting values. The constancy of ACR insures that lung gases (PO_2 and PCO_2) are maintained at the values measured at rest, and insures adequate diffusional gradients for gas exchange. However, in many species, ACR increases at high exercise intensities, resulting in a significant elevation of lung PO_2 and reduction in lung PCO_2. This relative hyperventilation is illustrated in several species of lizards, snakes, and crocodilians during locomotion. For example, direct measurements of ventilation and arterial blood gases during treadmill exercise in the Savannah monitor lizard (*Varanus exanthematicus*) have consistently documented a relative hyperventilation, where functional ventilation of the lungs increases proportionally more than $\dot{V}O_2$ (Gleeson *et al.*, 1980; Mitchell *et al.*, 1981; Wang *et al.*, 1997; Owerkowicz *et al.*, 1999). Part of this hyperventilation is accomplished by efficient use of buccal ventilation, and prevention of this buccal response leads to significant reductions in $\dot{V}O_{2max}$ (Owerkowicz *et al.*, 1999). A relative hyperventilation may, however, not be a general response in varanid lizards. Frappell *et al.* (2002) recently showed that lung ventilation increases proportionally to $\dot{V}O_2$ during treadmill locomotion in *Varanus mertensi*, which may relate to the 30–35% lower $\dot{V}O_2$max of this species compared with *V. exanthematicus* (Mitchell *et al.*, 1981; Wang *et al.*, 1997; Owerkowicz *et al.*, 1999). In any event, in *V. exanthematicus* the hyperventilation and the associated increase of lung PO_2 during exercise may contribute to overcoming diffusive resistances with the pulmonary circulation (Mitchell *et al.*, 1981; Wang and Hicks, 2004).

The ventilatory response to digestion differs from that during exercise. Thus, in all vertebrates digestion is associated with a large gastric acid secretion, causing a rise in plasma $[HCO_3^-]$ (the alkaline tide). This response is particularly pronounced in amphibians and reptiles that can consume very large prey. In all air-breathing species studied, the alkaline tide is attended by an increase in arterial PCO_2 (P_aCO_2), so that arterial pH remains stable. Thus, the postprandial period of air-breathing vertebrates is characterized by a ventilatory compensation for the metabolic alkalosis caused by gastric acid secretion (Wang *et al.*, 2001a). This rise in P_aCO_2 is caused by a relative hypoventilation where the effective lung ventilation does not rise proportionally to increased metabolic production of CO_2 (Wang *et al.*, 2001a). In Fig. 4.6, the changes in ventilation and ACR during exercise and digestion in the Burmese python (*Python molurus*) are compared and illustrate the relative hyper- and hypoventilation, respectively.

Ventilation also increases when metabolism increases owing to the effects of temperature, but it is characteristic that ACR decreases, which cause arterial PCO_2 to rise with temperature. The relative hypoventilation is considered important for acid–base regulation and explains the fact that arterial pH (pH_a) generally decreases with elevated temperature. Although still somewhat controversial, the decrease in pH_a seems to protect protein ionization and may therefore be an important response to maintain protein function over wide temperature changes. As a consequence, and somewhat paradoxically, the reduction in ACR also means that P_LO_2 decreases with temperature, and given that Hb–O_2 affinity decreases with temperature, it is possible that high

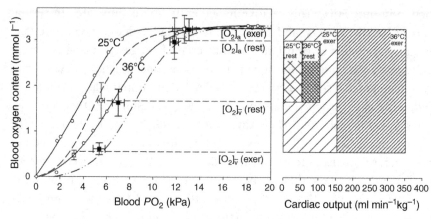

Fig. 4.5 Right: effect of exercise and temperature on arterial and mixed venous PO_2 and O_2 content (25°C, □; 36°C, ■) in relation to hemoglobin oxygen dissociation for *Varanus rosenbergi*. Values are means ± SE. ○, actual measured values of arterial PO_2 and O_2 concentration determined at the appropriate temperature (indicated) from blood taken from animals at rest. The dashed and dot-dashed regression lines have been adjusted according to the venous blood pH during exercise and Bohr effect for each temperature, 25 and 36°C, respectively. Right: graphical representation of the Fick principle showing the relative contributions of cardiac output and arterial and mixed venous oxygen content difference ($[O_2]_a$–$[O_2]_v$) to $\dot{V}O_2$ (enclosed area) during rest and maximum exercise at each temperature (modified from Clark *et al.*, 2005).

temperatures lead to a conflict between the need for acid–base regulation and the need to maintain adequate oxygen delivery (Wang *et al.*, 1998a). The decrease in P_IO_2, nevertheless, is rather minor, and it is possible that acid–base regulation would impair oxygen delivery only at unrealistically high body temperatures.

In terms of the cardiovascular responses to increased metabolism, oxygen supply can be increased through an increase in cardiac output and/or increase in tissue oxygen extraction ($[O_2]_a$–$[O_2]_v$). The degree to which these parameters can increase determines the upper limit for whole body oxygen consumption ($\dot{V}O_{2max}$). Beyond $\dot{V}O_{2max}$, elevated metabolism is supported by anaerobic processes, but given that high rates of lactic acid production cause severe acidosis, this strategy can only be utilized for short periods of time, such as bursts of activities associated with fleeing a predator. Cardiac output is elevated through a rise in heart rate and/or an increase in stroke volume (Clark *et al.*, 2005; Mortensen *et al.*, 2008) (Fig. 4.6). In varanid lizards, as shown in Fig. 4.5, when body temperature is high an increase in cardiac output, caused by a rise in heart rate, is sufficient to meet the elevated oxygen demand. During activity, however, when oxygen consumption increases many-fold and approaches $\dot{V}O_{2max}$, increments in cardiac output alone are insufficient to sustain aerobic

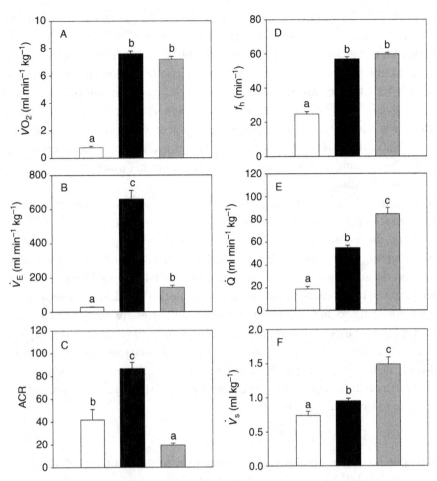

Fig. 4.6 Effects of exercise and digestion on cardiorespiratory parameters in Burmese pythons (*Python molurus*) at 30°C. Exercising pythons were crawling (0.4 km h^{-1}) on a treadmill. Digesting pythons had been fed a meal equivalent to approximately 25% of the snake's body mass and measurements were taken 40 hours after feeding. (A) $\dot{V}O_2$, oxygen uptake; (B) $\dot{V}_E$, minute ventilation; (C) ACR ($\dot{V}_E/\dot{V}O_2$), air convection requirement; (D) f_H, heart rate; (E) $\dot{Q}$, cardiac output; (F) V_S, stroke volume. Resting pythons (white bars); exercising pythons (black bars); and digesting pythons (gray bars). Data are mean ± S.E.M, $N = 6$. Different letters denote significant differences between means ($P < 0.05$). (Modified from Secor *et al.*, 2000.)

metabolism, and oxygen supply to the metabolizing tissue is further increased through a larger tissue oxygen extraction, evident from an increased arterial to venous difference ($[O_2]_a - [O_2]_v$).

The elevated metabolic rate during digestion and the increased need for intestinal absorption and subsequent nutrient transport must be met by an increased blood flow to the metabolizing and digesting tissue. The cardiovascular postprandial response includes an increase in cardiac output, which in pythons increases fourfold, caused by a doubling of heart rate and a rise in stroke volume (Hicks et al., 2000) (Fig. 4.6). The rise in stroke volume is in part caused by a 40% increase in ventricular muscle mass, i.e. the python heart actually grows in response to feeding (Andersen et al., 2005). An increased perfusion of the gastrointestinal organs is also achieved through a redistribution of blood flow from other organs, through a dilation of the mesenteric vascular bed causing a pronounced intestinal hyperemia (Axelsson et al., 1991; Starck and Wimmer, 2005).

During moderate activity, although hyperventilation may increase arterial PO_2, arterial oxygen content remains constant, determined by the shape of the ODC. Thus, the increase in the arterial to venous difference ($[O_2]_a$–$[O_2]_v$) is caused by the lowering of $[O_2]_v$ due to the increased oxygen extraction in the tissue. The increased tissue oxygen extraction is the result of a larger diffusion capacity, mainly caused by capillary recruitment. At rest, not all capillaries in the skeletal muscle are perfused, but during activity more capillaries are opened, increasing vascular conductance and regional blood flow (Krogh, 1919). At rest, the oxygen uptake in the tissue is diffusion limited. However, capillary recruitment reduces diffusion distance of oxygen from the red blood cells to the mitochondria, thereby increasing oxygen diffusing capacity in tissue (D_tO_2) and oxygen uptake according to Fick's law of diffusion:

$$\dot{V}O_2 = D_tO_2 \times (P_aO_2 - P_{mito}O_2) \tag{4.14}$$

where $\dot{V}O_2$ is oxygen uptake, D_tO_2 is tissue oxygen diffusing capacity, P_aO_2 is arterial PO_2, and $P_{mito}O_2$ is mitochondrial PO_2.

The increased conductance in the active muscles is caused by capillary recruitment and a general dilation of the vasculature, which together with a constriction of the vasculature in inactive organs acts to perfuse the muscles with a larger portion of cardiac output, keeping blood pressure unchanged. Moreover, local regulation of blood flow by local factors such as metabolites, oxygen, pH, and regulatory peptides leads to an improved matching of perfusion and oxygen delivery to metabolic demands.

Exercise reduces the efficiency of gas exchange, evident from an increased P_L–P_{LAt} difference (Hopkins et al., 1995; Powell and Hopkins, 2004). During intense exercise in human athletes the P_L–P_{LAt} difference for O_2 can be as large as 40 mmHg, and arterial blood desaturates (Dempsey et al., 1984). This is termed exercise-induced arterial hypoxemia (EIAH), and it has been described in several non-human mammals (Table 4.1). A complete understanding of the underlying mechanisms responsible for EIAH is still debated (Hopkins, 2006).

Table 4.1 *Interspecies comparison of gas-exchange responses to exercise*

	Rest			Maximal exercise				
Species	PO_2 (Torr)	PCO_2 (Torr)	P_A–P_a diff. (Torr)	PO_2 (Torr)	PCO_2 (Torr)	P_A–P_a diff. (Torr)	S_aO_2 (%)	$\dot{V}O_{2max}$ (ml min^{-1} kg^{-1})
Goat	105	37	2	123	26	4	95.0	57
Calf	108	39	0	114	29	9	100.0	37
Rat	95	36	14	108	29	11	93.9	74
Pig	104	43	0	99	37	14	89.7	68
Fox				120	19	12	92.0	216
Dog	97	34	13	101	25	26	92.6	137
Pony	107	37	0	95	25	32	90.3	89
Horse	105	41	4	77	50	28	81.6	144

Modified from Dempsey and Wagner, 1999.

For a definition of the terms used in this table, see the abbreviations list at the beginning of this book.

However, much of the arterial desaturation occurring during exercise can be attributed to diffusion limitations in the lung and $\dot{V}/\dot{Q}$ mismatch, with diffusion limitation becoming increasingly important during heavy, intense exercise (Dempsey and Wagner, 1999). Several mechanisms account for increasing $\dot{V}/\dot{Q}$ heterogeneity during progressive exercise. These have been reviewed in detail (Dempsey and Wagner, 1999) and include: (1) minor structural variations in both conducting airways and in blood vessels; (2) airway irritability resulting in bronchoconstriction and resulting alterations in the distribution of ventilation in the lung; (3) high airflow rates within the airways, which stimulate bronchiole secretions that may alter airflow distributions; (4) secretion of modulators that influence airway and vascular tone, in turn altering ventilation or blood flow distribution; (5) mild interstitial edema that affects the distribution of ventilation or blood flow.

In animals with cardiovascular shunts (amphibians and reptiles), large lung to arterial PO_2 differences (P_L–P_a) can occur at rest and can become progressively larger during exercise. Under resting conditions, the P_L–P_a difference results from cardiac shunts. The R–L shunt adds venous admixture to arterial blood, reducing both O_2 content and PO_2. Thus, small levels of venous shunt can produce large P_L–P_a differences. By contrast, during exercise the P_L–P_a difference dramatically increases. This increase does not result from an arterial desaturation, but rather from an intense hyperventilation during exercise. The ventilatory response is disproportionate to the increase in $\dot{V}O_2$, and thus lung PO_2 increases (Wang and Hicks, 2004).

It is clear that both an increase in $\dot{V}/\dot{Q}$ heterogeneity and diffusion limitation can contribute to the impaired gas exchange; however, the relative contribution differs between species (Hopkins *et al.*, 1995; Seaman *et al.*, 1995; Powell and Hopkins, 2004). Similarly, in varanid lizards, which can attain some of the highest rates of oxygen uptake among reptiles, the lungs appear diffusion limited, and it has been argued that these lizards hyperventilate their lungs during exercise to raise lung PO_2 above resting levels to maintain high arterial PO_2 values (Wang and Hicks, 2004). In the emu, the only bird in which $\dot{V}/\dot{Q}$ distributions have been measured to date, $\dot{V}/\dot{Q}$ heterogeneity does not increase, and any possible diffusion limitation during exercise in birds remains to be quantified (Schmitt *et al.*, 2002).

4.6   Phanerozoic Eon and the evolution of the vertebrate oxygen-transport system

Oxygen levels in the atmosphere provide the diffusive driving force for oxygen, and therefore greatly affect organismal function. Recent models of the Earth's atmospheric composition during the Phanerozoic Eon (the past 550 million years) suggest that oxygen levels might have risen as high as 30% in the Permian, and dropped as low as 12% in the Late Triassic and Early Jurassic (Bergman *et al.*, 2004; Berner, 2006) (Fig. 4.7). Hence, the changing levels of atmospheric oxygen during the Phanerozoic Eon may have influenced the evolutionary history of all animal life (Graham *et al.*, 1995; Huey and Ward; 2005; Berner *et al.*, 2007; Flück *et al.*, 2007). For example, Graham *et al.* (1995) suggested that breathing hyperoxic air may have aided the vertebrate invasion of the land, through reductions in ventilation and ultimate reduction in evaporative water loss. One provocative hypothesis is that rising oxygen levels would

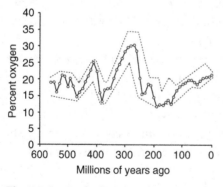

Fig. 4.7 Atmospheric oxygen over the Phanerozoic Eon currently based on geochemical models (e.g. COPSE of Bergman *et al.*, 2004; GEOCARBSULF of Berner, 2006).

have enhanced metabolic capacity, leading to the diversification of the synapsids (Graham *et al.*, 1995). Some mass extinction events appear to be coincident with decreases (sudden or gradual) in the atmospheric O_2 level, e.g. in the Late Devonian (Ward *et al.*, 2006) or in the Late Permian (Erwin, 1993). By contrast, rising O_2 levels in the Cenozoic seem to have had a permissive effect on increasing body size of placental mammals (Falkowski *et al.*, 2005). Interestingly the Triassic, when oxygen levels were close to their lowest levels, is associated with the origins of the major taxa of extant amniotes, with diverse cardiopulmonary morphologies (Perry, 1989; Burggren *et al.*, 1997) and accessory breathing mechanisms (Ruben *et al.*, 1997; Carrier and Farmer, 2000; Claessens, 2004; O'Connor and Claessens, 2004; Brainerd and Owerkowicz, 2006; Klein and Owerkowicz, 2006).

Temporal correlation of these phenomena, however, does not explain mechanistically why and how atmospheric O_2 is responsible for the observed evolutionary trends. Although direct measurements of physiological functions in extinct species is not possible, insights into the effects of the paleoatmosphere can result from studies that investigate the physiological effects of chronic changes in environmental O_2 in extant species. This type of approach has been termed 'experimental paleophysiology' (Berner *et al.*, 2007), but despite repeated calls for 'more paleophysiological studies, from both a fossil interpretation standpoint and a modern experimental standpoint' (Berner 1999; Berner *et al.*, 2003), few experimental studies on chronic exposure to non-normoxic conditions mimicking the hypothesized paleoatmosphere have been conducted (Berner *et al.*, 2007). Only recently have reports started to appear, indicating that exposure to chronic hypoxia and hyperoxia affects growth and metabolism of insects (Harrison, *et al.*, 2006; Kaiser *et al.*, 2007) and vertebrates (Vanden Brooks, 2004; Chan and Burggren, 2005; Owerkowicz *et al.*, 2009). Although such studies can only reveal the effects of inspired oxygen on the developmental trajectories and physiological functions of extant species, these results may provide the basis for making inferences about the broad patterns in vertebrate evolution during the Phanerozoic Eon.

Acknowledgements

The authors have been supported by the National Science Foundation and the Danish Research Council.

References

Aaronson, P. I., Robertson, T. P. and Ward, J. P. T. (2002). Endothelium-derived mediators and hypoxic pulmonary vasoconstriction. *Respir. Physiol. Neurobiol.*, **132**, 107–20.

Andersen, J. B., Rourke, B. C., Caiozzo, V. J., Bennett, A. F. and Hicks, J. W. (2005). Postprandial cardiac hypertrophy in pythons. *Nature*, **434**, 37–8.

Axelsson, M., Fritsche, R., Holmgren, S., Grove, D. J. and Nilsson, S. (1991). Gut blood flow in the estuarine crocodile, *Crocodylus porosus. Acta. Physiol. Scand.*, **142**, 509–16.

Bärtsch, P. (2007). Effect of altitude on the heart and lungs. *Circulation*, **116**, 2191–202.

Bennett, A. F. and Ruben, J. A. (1979). Endothermy and activity in vertebrates. *Science*, **206**, 649–54.

Bergman, N. M., Lenton, T. M. and Watson, A. J. (2004). COPSE: a new model of biogeochemical cycling over Phanerozoic time. *Am. J. Sci.*, **304**, 397–437.

Berner, R. A. (1999). Atmospheric oxygen of the Phanerozoic time. *Proc. Natl. Acad. Sci.*, **96**, 10955–7.

Berner, R. A. (2006) GEOCARBSULF: A combined model for Phanerozoic atmospheric O_2 and CO_2. *Geochim. Cosmochim. Acta*, **70**, 5653–64.

Berner, R. A., Vanden Brooks, J. M. and Ward, P. D. (2007). Oxygen and evolution. *Science*, **316**, 557–8.

Berner, R. A., Beerling, D. J., Dudley, R., Robinson, J. W. and Wildman, R. A., Jr. (2003). Phanerozoic atmospheric oxygen. *Ann. Rev. Earth Planet Sci.*, **31**, 105–34.

Brainerd, E. L. and Owerkowicz, T. (2006). Functional morphology and evolution of aspiration breathing in tetrapods. *Respir. Physiol. Neurobiol.*, **154**, 73–88.

Burggren, W. W. and Johansen, K. (1982). Ventricular hemodynamics in the monitor lizard *Varanus exanthematicus*: pulmonary and systemic pressure separation. *J. Exp. Biol.* **96**, 343–54.

Burggren, W. W. and Moalli, R. (1984). 'Active' regulation of cutaneous gas exchange by capillary recruitment in amphibians: experimental evidence and a revised model for skin respiration. *Respir. Physiol.*, **55**, 379–92.

Burggren, W. and Shelton, G. (1979). Gas exchange and transport during intermittent breathing in chelonian reptiles. *J. Exp. Biol.*, **82**, 75–92.

Burggren, W. W. and West, N. H. (1982). Changing respiratory importance of gills, lungs and skin during metamorphosisin the bullfrog *Rana catesbeiana. Respir. Physiol.*, **47**, 151–64.

Burggren, W. W., Farrell, A. P. and Lillywhite, H. B. (1997). Vertebrate cardiovascular systems. In *Handbook of Comparative Physiology*, ed. W. Dantzler. New York: Oxford University Press.

Carrier, D. R. and Farmer, C. G. (2000). The evolution of pelvic aspiration in archosaurs. *Paleobiology*, **26**, 271–93.

Chan, T. and Burggren, W. W. (2005). Hypoxic incubation creates differential morphological effects during specific developmental critical windows in the embryo of the chicken (*Gallus gallus*). *Respir. Physiol. Neurobiol.*, **145**, 251–63.

Claessens, L. P. A. M. (2004). Dinosaur gastralia: origin, morphology and function. *J. Vertebr. Paleontol.*, **24**, 89–106.

Clark, T. D., Wang, T., Butler, P. J. and Frappell, P. B. (2005). Factorial scopes of cardiac-metabolic variables remain constant with changes in body temperature in the varanid lizard, *Varanus rosenbergi. Am. J. Physiol.*, **288**, R992–7.

Crossley, D. A., Wang, T. and Altimiras, J. (1998). Hypoxia elicits pulmonary vasoconstriction in anaesthetized turtles. *J. Exp. Biol.*, **201**, 3367–75.

Dempsey, J. A. and Wagner, P. D. (1999). Exercise-induced arterial hypoxemia. *J. Appl. Physiol.*, **87**, 1997–2006.

Dempsey, J. A., Hanson, P. G. and Henderson, K. S. (1984). Exercise-induced arterial hypoxemia in healthy human subjects at sea level. *J. Physiol.*, **355**, 161–75.

Dubach, M. (1981). Quantitative analysis of the respiratory system of the house sparrow, budgerigar, and violet-eared hummingbird. *Respir. Physiol.*, **46**, 43–60.

Duncker, H-R. (1972). Structure of avian lungs. *Respir. Physiol.*, **14**, 44–63.

Erwin, D. H. (1993). *The Great Paleozoic Crisis: Life and Death in the Permian.* New York: Columbia University Press.

Falkowski, P., Katz, K., Milligan, A., Fennel, K., Cramer, B., Aubry, M. P., *et al.* (2005). The rise of atmospheric oxygen levels over the past 205 million years and the evolution of large placental mammals. *Science*, **309**, 2202–4.

Faraci, F. M., Kilgore, D. L. and Fedde, M. R. (1984). Attenuated pulmonary pressor response to hypoxia in bar-headed geese. *Am. J. Physiol.*, **247**, R402–3.

Feder, M. E. and Burggren, W. W. (1985). Cutaneous gas exchange in vertebrates: design, patterns, control, and implications. *Biol. Rev.*, **60**, 1–45.

Feder, M. E. and Pinder, A. W. (1988). Ventilation and its effect on 'infinite pool' exchangers. *Am. Zool.*, **28**, 973–83.

Flück, M., Webster, K. A., Graham, J., Giomi, F., Gerlach, F. and Schmitz, A. (2007). Coping with cyclic oxygen availability: evolutionary aspects. *Integr. Comp. Biol.*, **47**, 524–31.

Frappell, P., Schultz, T. and Christian, K. (2002). Oxygen transfer during aerobic exercise in a varanid lizard *Varanus mertensi* is limited by the circulation. *J. Exp. Biol.*, **205**, 2725–36.

Gamperl, A. K., Milsom, W. K., Farrell, A. P. and Wang, T. (1999). Cardiorespiratory responses of the toad (*Bufo marinus*) to hypoxia at two different temperatures. *J. Exp. Biol.*, **202**, 3647–58.

Glass, M. L. (1991). Pulmonary diffusion capacity of ectothermic vertebrates. In *The Vertebrate Gas Transport Cascade*, ed. J. E. P. W. Bicudo. Boca Raton: CRC Press, pp. 154–61.

Gleeson, T. T., Mitchell, G. S. and Bennett, A. F. (1980). Cardiovascular responses to graded activity in the lizards *Varanus* and *Iguana*. *Am. J. Physiol.* **8**, R174–9.

Graham, J. B., Dudley, R., Aguilar, N. and Gans, C. (1995). Implications of the late Paleozoic oxygen pulse for physiology and evolution. *Nature*, **375**, 117–20.

Harrison, J., Frazier, M. R., Henry, J. R., Kaiser, A., Klok, C. J. and Rascon, B. (2006). Responses of terrestrial insects to hypoxia or hyperoxia. *Respir. Physiol. Neurobiol.*, **154**, 4–17.

Hedrick, M. S., Palioca, W. B. and Hillman, S. S. (1999). Effects of temperature and physical activity on blood flow shunts and intracardiac mixing in the toad *Bufo marinus*. *Physiol. Biochem. Zool.*, **72**, 509–19.

Hicks J. W. (1998). Cardiac shunting in reptiles: Mechanism, regulation, and physiological functions. In *Biology of the Reptilia, Vol. 19 (Morphology)*, ed. C. Gans and S. Gaunt. Ithaca: Society for the Study of Amphibians and Reptiles, pp. 425–83.

Hicks, J. W. and Wang, T. (1996). Functional role of cardiac shunts in reptiles. *J. Exp. Zool.* **275**, 204–16.

Hicks, J. W. and Wang. T. (2004). Hypometabolism in reptiles: behavioural and physiological mechanisms that reduce aerobic demands. *Resp. Physiol. Neurobiol.*, **141**, 261–71.

Hicks, J. W. and White, F. N. (1992). Ventilation and gas exchange during intermittent ventilation in the American alligator, *Alligator mississippiensis*. *Respir. Physiol.*, **88**, 23–36.

Hicks, J. W., Wang, T. and Bennett, A. F. (2000). Patterns of cardiovascular and ventilatory response to elevated metabolic states in the lizard, *Varanus exanthematicus*. *J. Exp. Biol.*, **203**, 2437–45.

Hicks, J. W., Ishimatsu, A., Molloi, S., Erskin, A. and Hesiler, N. (1996). The mechanism of cardiac shunting in reptiles: a new synthesis. *J. Exp. Biol.*, **199**, 1435–46.

Hopkins, S. R. (2006). Exercise induced arterial hypoxemia: the role of ventilation-perfusion inequality and pulmonary diffusion limitation. In *Hypoxia and Exercise; Advances in Experimental Medicine and Biology*. Berlin: Springer-Verlag, pp. 17–30.

Hopkins, S. R., Wang, T. and Hicks, J. W. (1996). The effect of altering pulmonary blood flow on pulmonary gas exchange in the turtle *Trachemys (Pseudemys) scripta*. *J. Exp. Biol.*, **199**, 2207–14.

Hopkins, S. R., Hicks, J. W., Cooper, T. K. and Powell, F. L. (1995). Ventilation and pulmonary gas exchange during exercise in the Savanna monitor lizard (*Varanus exanthematicus*). *J. Exp. Biol.*, **198**, 1783–9.

Hopkins, S. R., Stary, C. M., Falor, E., Wagner, H., Wagner, P. D. and McKirnan, M. D. (1999). Pulmonary gas exchange during exercise in pigs. *J. Appl. Physiol,*. **86**, 93–100.

Hochachka, P. W., Lutz, P. L. (2001). Mechanism, origin, and evolution of anoxia tolerance in animals. *Comp. Biochem. Physiol.* **130B**, 435–59.

Hochachka, P. W., Buck, L. T., Doll, C. J. and Land, S. C. (1996). Unifying theory of hypoxia tolerance: molecular/metabolic defense and rescue mechanisms for surviving oxygen lack. *Proc. Natl. Acad. Sci. USA*, **93**, 9493–8.

Huey, R. B. and Ward, P. D. (2005). Hypoxia, global warming, and terrestrial Late Permian extinctions. *Science*, **308**, 398–401.

Ishimatsu, A., Hicks, J. W. and Heisler, N. (1996). Analysis of cardiac shunting in the turtle *Trachemys (Pseudemys) scripta*: application of the three outflow vessel model. *J. Exp. Biol.*, **199**, 2667–77.

Kaiser, A., Klok, J. C., Socha, J. J., Lee, W.-K., Quinlan, M. C. and Harrison J. F. (2007). Increase in tracheal investment with beetle size supports hypothesis of oxygen limit on insect gigantism. *Proc. Natl. Acad. Sci.*, **104**, 13198–203.

Klein, W. and Owerkowicz, T. (2006). Function of intracoelomic septa in lung ventilation of amniotes: lessons from lizards. *Physiol. Biochem. Zool.*, **79**, 1019–32.

Krogh, A. (1919). The supply of oxygen to the tissues and the regulation of the capillary circulation. *J. Physiol.*, **52**, 457–74.

Krogh, A. and Krogh, M. (1910). On the rate of diffusion of CO into the lungs of man. *Skand. Arch. Physiol.*, **23**, 236–47.

Krosniunas, E. H. and Hicks, J. W. (2003). Cardiac output and shunt during voluntary activity at different temperatures in the turtle, *Trachemys scripta*. *Physiol. Biochem. Zool.*, **76**, 679–94.

Maina, J. N. (1998). *The Gas Exchangers: Structure, Function and Evolution of the Respiratory Processes*. Heidelberg: Springer-Verlag.

Maina, J. N. (2006). Development, structure, and function of a novel respiratory organ, the lung-air sac system of birds: to go where no other vertebrate has gone. *Biol. Rev.*, **81**, 545–79.

Maina, J. N. and West, J. B. (2005). Thin and strong! The bioengineering dilemma in the structural and functional design of the blood-gas barrier. *Physiol. Rev.*, **85**, 811–44.

Maina, J. N., King, A. S. and King, D. Z. (1982). A morphometric analysis of the lungs of a species of bat. *Respir. Physiol.*, **50**, 1–11.

Malvin, G. M. (1988). Microvascular regulation of cutaneous gas exchange in amphibians. *Am. Zool.*, **28**, 999–1007.

Malvin, G. M. and Walker, B. R. (2001). Sites and ionic mechanisms of hypoxic vasoconstriction in frog skin. *Am. J. Physiol.*, **280**, R1308–14.

Milsom, W. K. (1991). Intermittent breathing in reptiles. *Annu. Rev. Physiol.*, **53**, 87–105.

Mitchell, G. S., Gleeson, T. T. and Bennett, A. F. (1981). Pulmonary oxygen transport during activity in lizards. *Respir. Physiol.* **43**, 365–75.

Mortensen, S. P., Damsgaard, R., Dawson, E. A., Secher, N. H. and González-Alonso, J (2008). Restrictions in systemic and locomotor skeletal muscle perfusion, oxygen supply and VO_2 during high-intensity whole-body exercise in humans. *J. Physiol.*, **586**, 2621–35.

Moudgil, R., Michelakis, E. D. and Archer, S. L. (2005). Hypoxic pulmonary vasoconstriction. *J. Appl. Physiol.*, **98**, 390–403.

O'Connor, P. M. and Claessens, L. P. A. M. (2004). Basic avian pulmonary design and flow-through ventilation in non-avian theropod dinosaurs. *Nature*, **436**, 253–6.

Owerkowicz, T., Elsey, R. E. and Hicks, J. W. (2009). Atmospheric oxygen affects growth trajectory, cardiopulmonary allometry and metabolic rate in the American alligator (*Alligator mississipiensis*). *J. Exp. Biol.* **212**, 1237–47.

Owerkowicz, T., Farmer, C. G., Hicks, J. W. and Brainerd, E. L. (1999). Contribution of gular pumping to lung ventilation in monitor lizards. *Science*, **284**, 1661–3.

Perry, S. F. (1989). Mainstreams in the evolution of vertebrate respiratory structures. In *Form and Function in Birds, Vol. 4*, ed. A. S. King and J. McMelland. London: Academic Press, pp. 1–67.

Piiper, J. (1961). Unequal distribution of pulmonary diffusing capacity and the alveolar-arterial PO_2 differences: theory. *J. Appl. Physiol.*, **16**, 493–8.

Piiper, J. (1990). Modelling of gas exchange in lung gills and skin. In *Vertebrate Gas Exchange: from Environment to Cell*, ed. R. G. Boutillier. Berlin: Springer-Verlag, pp. 5–44.

Piiper, J. (1993). Medium-blood gas exchange: diffusion, distribution and shunt. In *The Vertebrate Gas Transport Cascade*, ed. J. E. P. W. Bicudo. Boca Raton: CRC Press, pp. 106–120.

Piiper, J. and Scheid, P. (1972). Maximum gas transfer efficacy of models for fish gills, avian lungs and mammalian lungs. *Respir. Physiol.*, **14**, 115–24.

Piiper, J. and Scheid, P. (1975). Gas transport efficacy of gills, lungs and skin: theory and experimental data. *Respir. Physiol.*, **23**, 209–21.

Piiper, J. and Scheid, P. (1977). Comparative physiology of respiration: Functional analysis of gas exchange organs in vertebrates. In *International Review of Physiology: Respiratory Physiology II, Vol. 14*, ed. J. G. Widdecombe. Boston: University Park Press, pp. 220–53.

Powell, F. L. and Gray, A. T. (1989). Ventilation-perfusion relationships in alligators. *Respir. Physiol.*, **78**, 83–94.

Powell, F. L. and Hopkins, S. R. (2004). Comparative physiology of lung complexity: implications for gas exchange. *News Physiol. Sci.*, **19**, 55–60.

Ruben, J. A., Jones, T. D., Geist, N. R. and Hillenius, W. J. (1997). Lung structure and ventilation in therapod dinosaurs and early birds. *Science* **278**, 1267–70.

Scheid, P. and Piiper, J. (1997). Vertebrate respiratory physiology. In *Comparative Physiology, Section 13, Vol. 1*, ed. W. H. Dantzler. New York: American Physiological Society, pp. 309–56.

Schmitt, P. M., Powell, F. L. and Hopkins, S. R. (2002). Ventilation-perfusion inequality during normoxic and hypoxic exercise in the emu. *J. Appl. Physiol.*, **93**, 1980–6.

Seaman, J., Erickson, B. K., Kubo, K., *et al.* (1995). Exercise induced ventilation/perfusion inequality in the horse. *Equine Vet. J.*, **27**, 104–9.

Secor, S. M., Hicks, J. W. and Bennett, A. F. (2000). Ventilatory and cardiovascular responses of pythons (*Python molurus*) to exercise and digestion. *J. Exp. Biol.*, **203**, 2447–54.

Skovgaard, N. and Wang, T. (2006). Local control of pulmonary blood flow and lung structure in reptiles: implications for ventilation perfusion matching. *Respir. Physiol. Neurobiol*, **154**, 107–17.

Skovgaard, N., Abe, A. S. Andrade, D. V. and Wang, T. (2005). Hypoxic pulmonary vasoconstriction in reptiles: a comparative study on four species with different lung structures and pulmonary blood pressures. *Am. J. Physiol.*, **289**, R1280–8.

Smith, M. P., Russell, M. J., Wincko, J. T. and Olson, K. R. (2001). Effects of hypoxia on isolated vessels and perfused gills of rainbow trout. *Comp. Biochem. Physiol.*, **130A**, 171–81.

Starck, J. M. and Wimmer, C. (2005). Patterns of blood flow during the postprandial response in ball pythons, *Python regius*. *J. Exp. Biol.*, **208**, 881–9.

Taylor, C. R. and Weibel, E. R. (1981). Design of the mammalian respiratory system. I. Problem and strategy. *Respir. Physiol.* **44**, 1–10.

Taylor, E. W, D. Andrade, A. S. Abe, Cleo A. C. Leite and T. Wang (2009). The unequal influences of the left and right vagi on the control of the heart and pulmonary artery in the rattlesnake, *Crotalus durissus*. *J. Exp. Biol.* **212**, 145–51.

Vanden Brooks, J. M., (2004). The effects of varying pO$_2$ levels on vertebrate evolution. *J. Vertebr. Paleontol.*, **24**, 124A.

von Euler, U. S. and Liljestrand, G. (1946). Observations on the pulmonary arterial blood pressure in the cat. *Acta Physiol. Scand.*, **12**, 301–20.

Wagner, P. D. (1996). Determinants of maximal oxygen transport and utilization. *Ann. Rev. Physiol.*, **58**, 21–50.

Wagner, P. D., Naumann, P. F. and Laravuso, R. B. (1974a). Simultaneous measurements of eight foreign gases in blood by gas chromatography. *J. Appl. Physiol.*, **36**, 600–5.

Wagner, P. D., Saltzman, H. A. and West, J. B. (1974b). Measurement of continuous distribution of ventilation-perfusion ratios: theory. *J. Appl. Physiol.*, **36**, 588–99.

Wang, T. and Hicks, J. W. (1996). Cardiorespiratory synchrony in turtles. *J. Exp. Biol.*, **199**, 1791–800.

Wang, T. and Hicks, J. W. (2002). An integrative model to predict maximum oxygen uptake of animals with central vascular shunts. *Zoology*, **105**, 45–53.

Wang, T. and Hicks, J. W. (2004). Why Savannah monitor lizards hyperventilate during activity: a comparison of model predictions and experimental data. *Respir. Physiol. Neurobiol.*, **141**, 261–71.

Wang, T. and Hicks, J. W. (2008). Changes in pulmonary blood flow do not affect gas exchange during intermittent ventilation in resting turtles. *J. Exp. Biol.*, **211**, 3759–63.

Wang, T., Abe, A. S. and Glass, M. L. (1998a). Effects of temperature on lung and blood gases in the South American rattlesnake, *Crotalus durissus terrificus*. *Comp. Biochem. Physiol.*, **121A**, 7–11.

Wang, T., Busk, M. and Overgaard, J. (2001a). The respiratory consequences of feeding in amphibians and reptiles. *Comp. Biochem. Physiol.* **128A**, 533–47.

Wang, T., Carrier, D. R. and Hicks, J. W. (1997). Ventilation and gas exchange in lizards during treadmill exercise. *J. Exp. Biol.* **200**, 2629–39.

Wang, T., Smits, A. W. and Burggren, W. W. (1998b). Pulmonary functions in reptiles. In *Biology of the Reptilia, Vol. 19 (Morphology)*, ed. C. Gans and S. Gaunt. Ithaca: Society for the Study of Amphibians and Reptiles, pp. 297–374.

Wang, T., Altimiras, J., Klein, W. and Axelsson, M. (2003). Ventricular haemodynamics in *Python molurus*: separation of pulmonary and systemic pressures. *J. Exp. Biol.* **206**, 4241–5.

Wang, T., Warburton, S. J., Abe, A. S. and Taylor, E. W. (2001b). Vagal control of heart rate and cardiac shunts in reptiles: relation to metabolic state. *Exp. Physiol.* **86**, 777–86.

Ward, P. D., Labandeira, C., Laurin, M., Berner, R. A. (2006). Confirmation of Romer's Gap as a low oxygen interval constraining the timing of initial arthropod and vertebrate terrestrialization. *Proc Natl Acad Sci*, **103**, 16818–22.

Weibel, E. R. and Gomez, D. M. (1962). Architecture of the human lung. *Science*, **137**, 577–85.

Weir, E. K. and Archer, S. L. (1995). The mechanism of acute hypoxic pulmonary vasoconstriction: the tale of two channels. *FASEB J*, **9**, 183–9.

West, J. B. (2003). Thoughts on the pulmonary blood-gas barrier. *Am. J. Physiol.*, **285**, L501–13.

Wood, S. C. (1982). Effect of O_2 affinity on arterial PO_2 in animals with central vascular shunts. *J. Appl. Physiol.* **53**, 1360–4.

Wood, S. C. (1984). Cardiovascular shunts and oxygen transport in lower vertebrates. *Am. J. Physiol.* **247**, R3–14.

Wood, S. C. (1991). Interactions between hypoxia and hypothermia. *Annu. Rev. Physiol.* **53**, 71–85.

PART II SPECIAL CASES

5

Adaptations to hypoxia in fishes

GÖRAN E. NILSSON AND DAVID J. RANDALL

5.1   Hypoxia in the aquatic environment

Both ocean and freshwater environments can challenge the inhabitants with large spatial and temporal variations in oxygen levels. As we pointed out in Chapter 1, oxygen has a low solubility and diffuses slowly in water. Further, the solubility of O_2 in water falls with increases in temperature. At close to 0°C, air-saturated freshwater contains 10.2 ml O_2 per liter, whereas at tropical temperatures (30°C) fresh water can only hold 5.9 ml O_2 per liter when air saturated. These figures are even 20% or so lower in sea water, as salt reduces oxygen solubility (Table 1.1 in Chapter 1).

These physical factors make water breathing more challenging than air breathing, and particularly so when water oxygen levels are below air saturation. The oxygen that enters the water from the atmosphere, or is produced by photosynthesizing algae and phytoplankton, can be rapidly consumed by organisms and chemical oxidation reactions. There is no photosynthetic O_2 production in the dark, and O_2 diffusion is extremely slow in water (see Chapter 1), so oxygen movement to depth depends on convection, i.e. oxygen is carried to depth by water flow rather than diffusion. Surface waters generally have high oxygen content because of both photosynthesis and diffusion of oxygen from air. Aeration is increased by convection and mixing at the surface, a process that is strongly influenced by wind.

Hypoxia is common in several different aquatic habitats. On still days, Hoi Ha Wan, a shallow marine inlet in Hong Kong, becomes hypoxic at depth, as oxygen consumed by zooplankton, coral, fish, and other organisms is not replaced by oxygen in water carried from the surface. On windy days there is no hypoxia at depth in this environment, as the wind mixes the water, carrying sufficient

Respiratory Physiology of Vertebrates: Life with and without Oxygen, ed. Göran E. Nilsson. Published by Cambridge University Press. © Cambridge University Press 2010.

oxygen to depth to prevent hypoxia. Large increases in sediment load in waters around Hong Kong have reduced light penetration and undoubtedly, therefore, depressed photosynthetic rates and oxygen production in the marine environment. No clear estimates of the magnitude of this effect have been determined (Lam, 1999). Oxygen levels in small, productive lakes in Amazonia can be reduced almost to zero during the night and become supersaturated during the day (Val and Almeida-Val, 1995). Oxygen levels in coral lagoons show similar diurnal oscillations in oxygen levels, the magnitude of the oscillation depending on the productivity of the lagoon and the extent of flushing, which is often strongly influenced by the tide (Nilsson et al., 2007a).

Even if cold fresh water can hold a bit more O_2 than warm tropical waters, some of the most extreme examples of long-term anoxia (i.e. no O_2 at all) are found in the far North, in small lakes and ponds that become ice covered for several months during the long winter. Here, the darkness and thick ice effectively block O_2 production and O_2 diffusion from the atmosphere. The anoxia tolerance of the few vertebrates that survive under those extreme conditions will be discussed in Chapter 9.

Vast areas of the open ocean are more or less permanently hypoxic. Photosynthesis and exchange with the atmosphere normally maintain high oxygen levels in surface waters. However, large numbers of organisms live beneath the light zone, feeding on material dropping from above. Oxygen consumption rates are high, and as a result, large regions of the oceans are hypoxic at intermediate depths. At greater depths the oxygen levels rise because the biomass falls, leading to lower O_2 consumption rates, and because ocean currents bring in O_2 in sufficient amounts. The rotation of the Earth and increased water density, due to the low temperature, generate ocean currents that carry oxygen to depths. As the oceans warm close to the equator, the oxygen concentration in the surface waters will decrease due to the low solubility of O_2 in warm seawater, reducing the amount of oxygen carried to depth. A decrease in sinking rate associated with global warming may affect ocean currents and the distribution of oxygen in the oceans.

Some fish, such as herring, form very large groups, and hypoxia can develop within these shoals, such that the fish in the rear are breathing water containing much less oxygen than those at the front. Although they are breathing hypoxic water, they may have a lower metabolic rate than the leaders if they can take advantage of eddies created by the movement of the fish in front of them. In addition, they may change position in the shoal in much the same way as birds flying in V formation.

Sewage from humans and the animals they keep is often released into coastal waters, causing elevated nutrient levels. The associated rise in oxygen use has resulted in a marked increase in the frequency and extent of hypoxia in many coastal waters. Chesapeake Bay, the 'dead zone' in the northern Gulf of Mexico, Tokyo Bay, and the bottom waters of the Baltic Sea are some well-known examples of marine environments that now suffer from hypoxia largely brought about by human activities.

What is clear is that oxygen levels in water are very variable, that aquatic hypoxia is a common event, that the frequency and level of hypoxia is increasing, and, as a result, that fish are more and more often exposed to hypoxia. Little is known of the behavior of fish living in this variable-oxygen environment. Clearly fish may move rapidly through a hypoxic environment when being chased or trying to catch prey. Thus, a fish that is more able to tolerate short periods of hypoxia than its predator/prey would have a distinct advantage. Some fish may be exposed to predictable diurnal oscillations in oxygen levels in the environment, or may be trapped for months in a hypoxic pond under ice, and have evolved adaptations that allow them to survive these somewhat predictable periods of hypoxia. For example, diurnal reductions in metabolism occur in parrotfish that coincide with the putative periods of hypoxia on a coral reef. Other fish may spend much of their lives within the extensive oceanic hypoxic zones.

Most of what is known about the responses of fish to hypoxia is based on laboratory studies, and then, only a few of the more than 30 000 species of fish have been studied in any detail. These studies are largely descriptive, detailing what happens to fish during hypoxia. In most cases the relative contribution of each of the responses to hypoxic survival has not been determined experimentally. Even though only a small number of species have been studied it is clear that some are much more tolerant to hypoxia than others, and responses that are particularly well developed in these fish are deemed to be important for hypoxic survival.

All vertebrates can survive some form of hypoxia. It is the length of time and degree of hypoxia that can be tolerated that varies between tissues and species. Therefore, from a physiological perspective, there is no particular oxygen level that can be defined as hypoxic to all animals. What is experienced as severe, life-threatening hypoxia by some species will present no challenge to others. Most mammals, birds and highly active fishes such as tuna and salmon can survive anoxia only for minutes, whereas some gobies can survive anoxia for hours to days, and some carp tolerate anoxia for months. Striking differences are also seen in the hypoxia tolerance of various tissues. The mammalian brain can survive a few minutes of anoxia, the skin hours or days. The differences are mainly a matter of being able to match energy supply and demand. Terrestrial vertebrates live in a fairly constant oxygen environment, whereas many fish live in a variable and often hypoxic environment. As a result, the responses of fish to hypoxia are likely to be more extensive and developed to a greater degree than those observed in most terrestrial vertebrates.

5.1.1   Ectothermy and hypoxia

It can be argued that being ectothermic ('cold blooded') provides fish with an advantage over mammals and birds during hypoxia. In ectothermic vertebrates, the small amount of heat produced by metabolic processes is not enough to keep their body temperature significantly higher than that of the water. Being warmer than the environment is almost impossible for fish, as their gills are excellent heat exchangers between body and water. (Still, a few highly active fish such as tuna and marlin are partial exceptions as they may increase the temperature of particular organs using counter-current blood flow to thermally insulate parts of their bodies.) When measured at the same body temperature, the resting oxygen consumption of ectothermic vertebrates (fish, amphibians, and reptiles) is only about 10% of that of similarly sized endotherms (birds and mammals). Moreover, the majority of fishes live at temperatures between −2 and 30°C, whereas most endothermic vertebrates keep their body temperature just under 40°C. This low body temperature gives fish an additional reduction in metabolic rate compared with endotherms. Having a low metabolic rate is often considered an advantage in hypoxia, because the primary problem for a hypoxic animal is to maintain cellular energy charge (see Chapter 1). However, to do this, adenosine triphosphate (ATP) use has to be matched to ATP production, and low temperature will also reduce the capacity of ATP production by slowing down respiratory functions at both the organ and the mitochondrial level, as well as suppressing glycolytic enzyme activities. Thus, the advantage of being cold in hypoxia may lie mainly in slowing down the depletion of energy stores and various detrimental processes, rather than facilitating the matching of ATP use with ATP production. Indeed, surviving severe hypoxia and anoxia is not trivial for fish, as illustrated by numerous species that rapidly die under such conditions.

5.1.2   Critical thresholds

The considerable variance in the ability of different fish species to tolerate hypoxia has undoubtedly to do with adaptation to different lifestyles and habitats. A widely used method for characterizing the hypoxia tolerance of fishes is to determine their critical oxygen tension (PO_2crit) or critical oxygen concentration ($[O_2]$crit) using a respirometer to measure the resting rate of

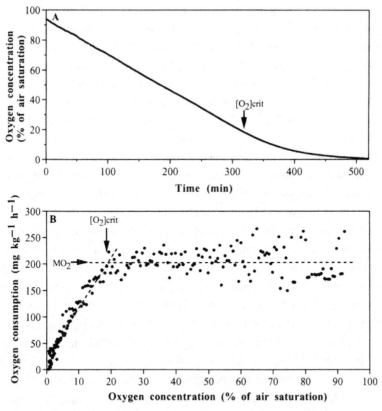

Fig. 5.1 Determination of the critical oxygen concentration ([O$_2$]crit) in fish using closed respirometry. In this example, a goby (*Gobiodon histrio*) is first allowed to acclimatize to the chamber (a 200 ml Perspex cylinder), whereupon water flow through the chamber is closed off and the fall in water oxygen concentration is recorded (using an oxygen electrode). The trace of the recording is shown in A. By taking the weight of the fish and the volume of the chamber into account, the rate of oxygen consumption at the different oxygen levels can be calculated (B). The [O$_2$] crit becomes evident as a sharp break in the curve in B (it can also been seen as a slight deviation from the line in A). At oxygen levels below the [O$_2$]crit, the fish is no longer able to maintain its resting rate of oxygen consumption (VO$_2$ or MO$_2$), and the rate of oxygen uptake becomes more or less linearly dependent on the ambient oxygen level (i.e. the partial pressure of ambient oxygen becomes the main determinant of the influx of oxygen over the gills). Redrawn from Nilsson *et al.*, 2004.

oxygen consumption at different water oxygen levels (Fig. 5.1). The PO$_2$crit and [O$_2$]crit are the lowest partial O$_2$ pressure and lowest O$_2$ concentration, respectively, at which an animal is able to maintain its routine (or resting) rate of oxygen consumption. Physiologists often prefer to discuss oxygen levels in terms of partial pressure, as pressure gradients are what drive oxygen diffusion, while ecologists often are more familiar with using oxygen concentrations. However, $[O_2]$crit in mg or ml O_2 per liter can readily be calculated from a measured PO_2crit if water temperature and salinity are known (e.g. using Table 1.1 in Chapter 1). If $[O_2]$crit is given as a percentage of air saturation, then PO_2crit in mmHg is obtained by multiplying the percentage value by 1.55 (as 100% air saturation normally refers to water equilibrated with a PO_2 close to 155 mmHg, unless the measurements were made well above sea level).

Typically, hypoxia-tolerant species have a lower PO_2crit than hypoxia-sensitive species. Those species that are best adapted to survive hypoxia show PO_2crit values of 6–40 mmHg, whereas species sensitive to hypoxia, such as some salmonids and tuna, tend to have PO_2crit values above 70 mmHg (see Table 5.1). However, even among species that have extremely low PO_2crit values, tolerance to even more severe insults, such as anoxia, varies widely because of apparent differences in anaerobic capacities for ATP production. Thus, whereas an African mormyrid fish, the elephant-nose fish (*Gnathonemus petersii*), has a PO_2crit of 15 mmHg, which is in the same range as that of the North Palearctic crucian carp (*Carassius carassius*), the elephant-nose fish dies virtually immediately if the water oxygen tension falls below PO_2crit (Nilsson, 1996). By contrast, the crucian carp can survive anoxia for days to months, depending on temperature, by being exceptionally well adapted to producing ATP anaerobically. Such cases of extreme anoxia tolerance will be further described in Chapter 9.

The PO_2crit is the point at which oxygen delivery can no longer meet demand. If an animal's oxygen demand increases, its PO_2crit will become higher. Thus, one can expect higher PO_2crit for fish that have been fed than for fasting fish, and for actively reproducing or stressed fish. In addition it is clear that fish can acclimate to hypoxia, and hypoxic exposure may lead to a reduced PO_2crit (as seen in some species listed in Table 5.1).

Although basal metabolic rate rises with temperature, oxygen delivery is largely dependent on diffusion, which increases much less rapidly with temperature than metabolic rate. As a result, PO_2crit will rise with temperature (Fry and Hart, 1948; Schurmann and Steffensen, 1997; Sollid et al., 2005), and it has been suggested that the lethal (critical) temperature (Tc) for many ectothermic animals is reached when their body temperature becomes so high that PO_2crit is reached under normoxic conditions (Lannig et al., 2004; Pörtner et al., 2004). In other words, above Tc the oxygen delivery system can no longer support the basal metabolic rate.

Although the immediate ability of a species to tolerate hypoxia and high water temperature may be fairly well described by its PO_2crit and Tc, other

Table 5.1 *Critical oxygen tensions and concentrations of some fishes*

Species	Habitat	PO$_2$crit (mmHg)	[O$_2$]crit (mg l^{-1})	T (°C)	Reference(s)
Hypoxia-tolerant teleosts					
Toadfish (*Opsanus tau*)	Atlantic coast of North America	29	1.4	22	Ultsch et al. (1981)
Common carp (*Cyprinus carpio*)	European fresh water	30	2.2	10	Beamish (1964)
		30	1.8	20	Beamish (1964)
		27	1.4	25	De Boeck et al. (1995)
Crucian carp (*Carassius carassius*)	European fresh water	12 (6)	1.0 (0.5)	8	Sollid et al. (2003)
		23	1.4	18	Nilsson (1992)
Goldfish (*Carassius auratus*)	Domesticated (orig. Asian fresh water)	25	1.8	10	Beamish (1964)
		40	2.3	20	Beamish (1964)
		74 (36)	4.1 (2.0)	22	Prosser et al. (1957)
European eel (*Anguilla anguilla*)	European fresh water	25	1.4	25	Cruz-Neto and Steffensen (1997)
Elephant-nose fish (*Gnathonemus petersii*)	Tropical African fresh water	15	0.8	26	Nilsson (1996)
Oscar cichlid (*Astronotus ocellatus*)	Amazon	31	1.6	28	Muusze et al. (1998)
Nile tilapia (*Oreochromis niloticus*)	African fresh water	19	1.1	20	Fernandes and Rantin (1989)
		30	1.6	25	Verheyen et al. (1994)
		30	1.4	35	Fernandes and Rantin (1989)
Fragile cardinalfish (*Apogon fragilis*)	Great Barrier Reef	26	1.0	30	Nilsson et al. (2007a)
Humbug damselfish (*Dascyllus aruanus*)	Great Barrier Reef	29	1.2	30	Nilsson et al. (2007a)
Coral goby (*Gobiodon ceramensis*)	Great Barrier Reef	22	0.9	30	Nilsson et al. (2007a)
Hypoxia-sensitive teleosts					
Rainbow trout (*Oncorhynchus mykiss*)	North American fresh water	90	6.0	15	Kutty (1968)
Brook trout (*Salvelinus fontinalis*)	North American fresh water	75	4.9	15	Beamish (1964)
Blunthead cichlid (*Tropheus moorii*)	Lake Tanganyika	47	2.5	25	Verheyen et al. (1994)

Table 5.1 (cont.)

Species	Habitat	PO$_2$crit (mmHg)	[O$_2$]crit (mg l^{-1})	T (C)	Reference(s)
Brichard's cichlid (*Neolamprolgus brichardi*)	Lake Tanganyika	154[1]	8.3	25	Verheyen *et al.* (1994)
Dragonet (*Callionymus lyra*)	Northeastern Atlantic	125	6.8	12	Hughes and Umezawa (1968)
Elasmobranchs					
Epaulette shark (*Hemiscyllium ocellatum*)	Great Barrier Reef	50 (40)	2.2 (1.7)	25	Routley *et al.* (2002)
Bamboo shark (*Hemiscyllium plagiostum*)	Great Barrier Reef	60	2.7	23	Chan and Wong (1977)
Small-spotted catshark (*Scyliorhinus canicula*)	Northeastern Atlantic	60	3.6	7	Butler and Taylor (1975)
		80	3.9	17	Butler and Taylor (1975)

Values within parentheses refer to hypoxia-acclimated individuals.

This is a selection of records from the literature. The studies of Verheyen *et al.* (1994) and Nilsson *et al.* (2007), in particular, contain values from several additional species.

[1] A species that shows a fall in oxygen uptake immediately when exposed to oxygen levels below 100% air saturation.

thresholds are likely to better describe the prospects for long-term survival. Oxygen consumption is closely linked to temperature, and the concept of aerobic scope has been identified as a key factor in population and species survival (Brett and Groves, 1979), particularly in the face of global warming (Pörtner and Knust, 2007). Aerobic scope is the range by which oxygen consumption can be increased above the demand of basal metabolic rate, and it has been found that the aerobic scope starts to decrease above a certain temperature, at which the maximal rate of oxygen delivery cannot be further increased (Fry, 1971; Brett and Groves, 1979). Frederich and Pörtner (2000) termed this the 'pejus temperature' (Tp; pejus = turning worse). The Tp is considerably lower than the Tc and, as Fry (1971), Brett (1979), and Brett and Groves (1979) pointed out, dependent on the physiological state (e.g. reproductive or non-reproductive) of the animal. When the water temperature rises above Tp, the immediate survival of the animal is not threatened, but it will become more and more limited in its ability to perform higher functions necessary for its fitness, such as feeding, growth, and reproduction. Thus, any rise in ambient temperature above Tp will threaten the long-term survival of the population/species, particularly if it is competing for resources with other populations/species with a higher Tp. Pörtner and Knust (2007) presented evidence suggesting that this is already happening to the eelpout (*Zoarces viviparous*) population on the German North Sea coast, where they showed that high summer water temperatures coincide with reduced reproduction in the eelpout. For this species (and possibly many others), the Tp (16.8°C) is only slightly higher than the temperature for optimal growth (15.5°C), and well below the Tc (21.6°C) (Pörtner and Knust, 2007). Similarly, a recent study on the Great Barrier Reef has indicated that some coral-reef fishes will lose virtually all their aerobic scope if ocean temperatures rise by 2–4°C (Nilsson *et al.*, 2009), and during river migration of sockeye salmon (*Oncorhynchus nerka*) in the Fraser river in British Columbia, anomalously high water temperatures in 2004 reduced aerobic scope so much that some populations could not reach their breeding grounds (Farrell *et al.*, 2008).

5.2   Maintenance of oxygen delivery

Fishes utilize several strategies to cope with hypoxia, including mechanisms aimed at maintaining oxygen delivery in the face of reductions in water oxygen levels, upregulation of anaerobic metabolism when oxygen delivery can no longer be maintained, downregulation of energy expenditure, and cellular mechanisms striving to protect tissues against hypoxic damage. The time course of these responses is varied, but increased ventilation is usually immediate, within seconds, whereas the changes in metabolism are slower. Although the nature of protective mechanisms on the cellular level has only recently been examined, the maintenance of oxygen uptake has been fairly well studied for a long time.

Fish use a variety of mechanisms to maintain oxygen delivery to tissues during hypoxia. Some fishes escape hypoxia in the water altogether by directly taking up oxygen from air, adaptations to which the whole of Chapter 6 is devoted. Other fishes skim the surface water, which, because of diffusion from air, is often much more oxygen rich than the water just a millimeter below the water surface (Kramer and McClure, 1982). Many species resort to this option during hypoxia without showing specialized morphological adaptations, one example being the goldfish (*Carassius auratus*) (Burggren, 1982). Others show striking morphological adaptations to this behavior. Possibly the most extravagant specializations to surface water breathing are found in South America, particularly in the Amazon region. Here several species, including members of the genera *Colossoma*, *Brycon*, and *Triportheus*, have a lower lip that develops an edema (i.e. an excessive infiltration of blood and fluid) after an hour or two of hypoxia, making the lip conspicuously extended and perfectly shaped for both oxygenating and transporting the surface water into the mouth (Fig. 5.2) (Branson and Hake, 1972; Braum and Junk, 1982; Winemiller, 1989). Still, the vast majority of species rely on adjusting gill ventilation and perfusion to maintain oxygen delivery in the face of reduced oxygen levels, and this strategy also involves some striking adaptive mechanisms.

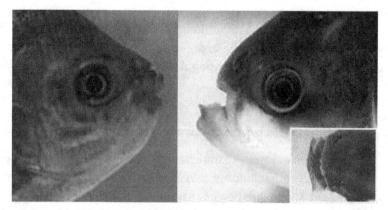

Fig. 5.2 The lower lip of the Tambaqui (*Colossoma macropomum*), an Amazonian fish, expands through edema when exposed to hypoxia (right). Insert shows hypoxic head from above. Courtesy of William Milsom.

2.1 Constitutional adaptation of gills

Fishes that have an active lifestyle that demands high rates of oxygen uptake have larger lamellar surface areas than more sedentary species (Gray, 1954; Bernal *et al.*, 2001). Similarly, fishes that are well adapted to hypoxic conditions often have larger respiratory surface areas than less hypoxia-tolerant relatives (e.g. Fernandes *et al.*, 1994; Chapman *et al.*, 2000; Chapman and Hulen, 2001; Chapman *et al.*, 2002). A large respiratory surface aids oxygen uptake from the water, and one may expect that most fishes should benefit from this. However, the fact that not all fishes have large gills reveals that there are also disadvantages with exposing an extensive surface area to the environment. These are likely to include: (1) increased ion and water fluxes that have to be counteracted by energetically expensive ion pumping (Nilsson, 1986; Gonzalez and McDonald, 1992; Bæuf and Payan, 2001); (2) increased uptake of toxic substances such as ammonia, algal toxins, metal ions, and various anthropogenic toxicants (Wood, 2001); (3) increased exposure to pathogens and parasites; (4) increased risks for bleeding, as the whole cardiac output has to go through the gills (Sundin and Nilsson, 1998a); and (5) impeded feeding capacity, as the gills take up a significant portion of the oral cavity (Schaack and Chapman, 2003). Thus, for fish to afford a large respiratory surface area, there has to be a pay-off, such as a high rate of oxygen uptake allowing endurance swimming, as in mackerel and tuna, or an ability to extract oxygen out of the water even during severe hypoxia.

5.2.2 Gill plasticity in response to hypoxia

Hypoxia may cause adaptive changes in gill morphology through both natural selection and developmental alterations. The best-studied examples of this include some populations of African cichlid and cyprinid fishes, for which an hypoxic environment leads to populations with larger respiratory surface areas compared with conspecific populations living in well-oxygenated habitats (Chapman *et al.*, 2000; Schaack and Chapman, 2003). The increases in gill filament length and lamellar surface area displayed by the 'hypoxic populations' are apparently caused by genetic differences and adaptive changes during development.

It has recently become clear that some fishes have the ability to change their gill morphology in response to a few days of hypoxia exposure (see Nilsson, 2007 for a review). Studies have shown that the lamellae of crucian carp are embedded in an interlamellar cell mass (ILCM) during normoxic conditions or at low temperature, whereas during a few days of hypoxia much of the ILCM dies off, thereby exposing a much larger respiratory surface

A Normoxic crucian carp gill filaments

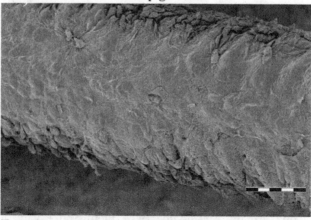

B Hypoxic crucian carp gill filament

Fig. 5.3 The crucian carp remodels its gills during hypoxia. These scanning electron micrographs show gill filaments from crucian carp kept in normoxic water (A) and in hypoxic water (B), both at 8°C. Scale bars are 50 μm. From Sollid *et al.*, 2003.

area (Fig. 5.3). The underlying mechanisms include increased apoptosis and decreased mitosis in the ILCM during hypoxia (Sollid *et al.*, 2003). However, the signals initiating these mechanisms are presently unknown. An apparently identical transformation occurs when crucian carp or goldfish (*Carassius auratus*) are moved from cold to warm water (resulting in an increased metabolic rate), suggesting that the gill remodeling is primarily a response to an increased need for oxygen uptake.

A similar, but more modest, gill remodeling is also displayed by eels (*Anguilla anguilla*) in response to changes in temperature (Tuurala *et al.*, 1998). Moreover, gill remodeling has recently been seen in the Qinghai carp (*Gymnocypris przewalskii*) (Matey *et al.*, 2008) and in the mangrove killifish (*Kryptolebias marmoratus*) (Ong *et al.*, 2007). Such a profound change in gill morphology as that displayed by the crucian carp and goldfish can probably only occur in species with hemoglobins that have extremely high oxygen affinities, as this will allow sufficient rates of oxygen uptake even in the absence of protruding lamellae (Sollid *et al.*, 2005). The crucian carp and goldfish have record-high Hb–O_2 affinities (see Table 3.2 in Chapter 3).

The advantage of having a small respiratory surface area during periods of high oxygen levels or low temperatures is likely to be related to the inherent problems of having large gills (such as costly ion and water fluxes, and uptake of toxic substances, pathogens and parasites), but the importance of each of these factors remains to be clarified (Nilsson, 2007). Gill remodeling is further discussed in Chapter 3 (section 3.1).

5.2.3   Ventilatory and circulatory adjustments

When fishes are exposed to hypoxia, they rapidly show both ventilatory and circulatory changes. Gill ventilation (i.e. the water flow through the gills) is increased by upregulating both the volume and frequency of buccal pumping (e.g. Saunders, 1962; Holeton and Randall, 1967a; Randall *et al.*, 1967). When fishes are exposed to a continuous fall in water oxygen levels, a progressive increase in ventilation is seen until [O_2]crit has been reached, whereupon ventilation falls (Fig. 5.4). The fall in buccal pumping could be due either to an inability of the fish to maintain ventilation when tissue ATP levels start to fall or an adaptive response aimed at suppressing ATP use when only anaerobic glycolysis is available for ATP production.

The circulatory response to hypoxia includes changes that will increase the functional respiratory surface area, i.e. the lamellar area that is perfused with blood. This is done by increasing both the extent by which each lamella is perfused with blood and the total number of perfused lamellae (often called lamellar recruitment) (Booth, 1979a; Soivio and Tuurala, 1981). Compared with other vertebrates, many fishes have a remarkable ability to increase the stroke volume of the heart, sometimes up to threefold (Farrell and Jones, 1992), and during hypoxia fishes generally reduce heart rate while increasing stroke volume (Holeton and Randall, 1967b), thereby maintaining cardiac output during hypoxia (see Farrell, 2007 for a review). This hypoxic bradycardia is largely a cholinergic response mediated by the vagus nerve, with the exception of the hypoxia-tolerant epaulette shark (*Hemiscyllium ocellatum*), which displays hypoxic bradycardia that is unaffected by acetylcholine-receptor blockers (Stensløkken *et al.*, 2004). The larger stroke volume of the heart during hypoxia increases the magnitude of the blood pressure pulse, which in turn may

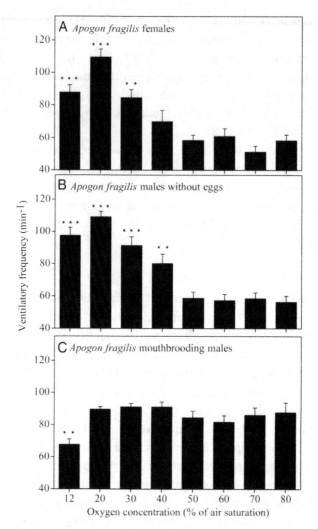

Fig. 5.4 Fishes typically increase their gill ventilation during hypoxia, here illustrated by a species of cardinal fish (*Apogon fragilis*) from the Great Barrier Reef. As often seen in fishes, ventilatory frequency is significantly increased at low oxygen levels (A, B), but ventilation falls below the critical oxygen concentration (which for this species is between 10 and 20% of air saturation). Male cardinal fishes care for the eggs through mouthbrooding, and mouthbrooding males of *A. fragilis* (C) are already ventilating their brood and gills at the maximum rate in normoxia and are therefore unable to increase it further during hypoxia. From Östlund-Nilsson and Nilsson, 2004. Asterisks mark significant differences from the ventilatory frequency at 80% air saturation.

function to increase the extent of filling of the secondary lamellae, augmenting oxygen uptake across the gills (Randall, 1982). The change in blood flow may not only increase the functional surface area for gas exchange but could also thin the epithelium, reducing diffusion distances between blood and water, thereby enhancing gill diffusing capacity. Stroke volume approximates the blood volume of the gills, and a synchrony between heart beat and breathing has been observed repeatedly. Such a synchronization may function to allow blood to be exchanged in the gill during the lowest water flow rates, while remaining in the gill during the highest water flow rates (Randall and Smith 1967). This synchrony also reduces oscillations in lamellar transmural pressure and, therefore, oscillations in the thickness of the lamellar blood sheet. Moreover, the longer diastolic residence time of blood in the heart lumen during hypoxic bradycardia has been suggested to increase the oxygenation of the myocardium (Farrell, 2007) (see section 4 below).

Hormones and neurotransmitters, including catecholamines that fishes release from chromaffin tissue during hypoxia (Butler et al., 1978; Wahlqvist and Nilsson, 1980), mediate numerous adjustments that promote hypoxic survival. In order to further open up the lamellar vasculature, the mean intralamellar blood pressure can be increased through vasoconstriction on the efferent (outgoing) side of the gill vasculature and through vasodilation of afferent (incoming) lamellar arterioles (Davis, 1972; Booth, 1978; Farrell et al., 1980; Taylor and Barrett, 1985). By contrast, in well-oxygenated water, much of the gill blood flow may pass through channels embedded in the body of the gill filaments relatively far from the water (Pärt et al., 1984). Cholinergic innervation of the gill appears to have a role for the microvascular changes occurring in the fish gill during hypoxia, including vasoconstriction of efferent filamental arterioles (Sundin and Nilsson, 1997). Catecholamines acting on β-receptors may mediate a vasodilation of the afferent vasculature (Pettersson, 1983).

Recently, hydrogen sulfide (H_2S) has emerged as a likely candidate for regulating blood flow during hypoxia (Olson, 2008). It is synthesized in tissues by two cytosolic pyridoxyl-5′-phosphate-dependent enzymes, of which cystathionine-β-synthase appears to be responsible for H_2S production in the vasculature. H_2S is constantly oxidized in tissues, particularly by the mitochondria. The rate of H_2S oxidation falls with falling oxygen levels, causing tissue levels of H_2S to rise during hypoxia, thereby making it an 'oxygen sensor.' H_2S has been found to be active as either a vasoconstrictor or vasodilator of both the systemic and branchial vasculature. Because experimental H_2S treatment mimics the effects of hypoxia on vascular resistance, Olson et al. (2006) suggested that H_2S could explain much of the circulatory changes seen in fish and other vertebrates during hypoxia.

Lamellar blood flow, like alveolar blood flow in the mammalian lung, follows sheet flow dynamics: increases in intra-lamellar pressure could cause an increase the thickness of the blood sheet but not the height or length of the

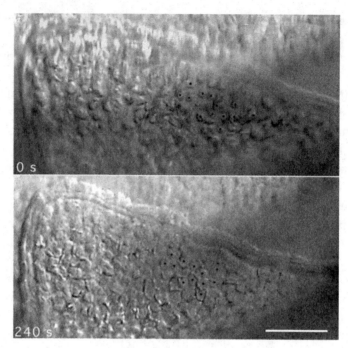

Fig. 5.5 Pillar cell contraction in gill lamellae induced by injection of the peptide endothelin. The images, taken through an epi-illumination microscope, show a gill lamella in a living cod before (upper) and 240 s after (lower) the injection of endothelin. A group of pillar cells is marked with black spots so that the effects of the contraction can be more easily seen. Scale bar = 100 µm. From Stensløkken et al., 1999.

lamellae (Farrell et al., 1980). The collagen in the pillar cells holds the sheet together, with actin and myosin threads organizing the collagen along lines of stress (Booth, 1979b; Randall and Daxboeck, 1984). More recently it has been suggested that functional respiratory surface area may be regulated by changing the thickness of the vascular space inside the lamellae through contraction or relaxation of the pillar cells within the lamellae (Fig. 5.5) (Sundin and Nilsson, 1998b; Stensløkken et al., 2006, Kudo et al., 2007; Sultana et al., 2007). Increases in buccal and opercular cavity pressure during hypoxia also offset rises in lamellar blood pressure and reduce potential increases in the lamellar blood sheet associated with hypoxia (Randall and Daxboeck, 1984). Restrictions of the water outflow from the gills, for example as seen in tuna, may serve to raise opercular pressure and so thin the lamellar blood sheet, augmenting gas transfer. It is presently unclear to what extent these potential mechanisms for regulating lamellar blood-sheet thickness are involved in enhancing gas transfer during hypoxia in fish.

5.3   Defense of the hypoxic brain

Vertebrate tissues show a varying susceptibility to hypoxia. The brain is often very sensitive, due to its constantly high energy demands, whereas the gut, skin, muscle, and liver can withstand fairly prolonged periods of hypoxia or even anoxia. As pointed out in Chapter 1, the ATP pool in a fish brain is turned over about once every minute. The fish brain appears to be prioritized during hypoxia, and increased blood flow in the brain or reduced cerebrovascular resistance have been measured in anoxic crucian carp (Nilsson *et al.*, 1994), in hypoxic common carp (*Cyprinus carpio*) (Yoshikawa *et al.*, 1995), and in hypoxic epaulette sharks (*Hemiscyllium ocellatum*) (Söderström *et al.*, 1999). Conversely, blood flow can be nearly turned off to organs that have a small obligatory need of energy and a function that can be halted in hypoxia, such as the digestive tract. In cod (*Gadus morhua*), a drastic decrease in the blood flow to the visceral organs has been measured during hypoxia (Axelsson and Fritsche, 1991). Muscles are often quiescent in hypoxia, and the liver metabolism is reorganized to support anaerobic metabolism.

Hypoxia can lead to brain swelling, one of the most feared effects of cardiac arrest, stroke, or head trauma in humans. Brain swelling is life threatening in animals that have a brain that fits tightly inside the cranium, as in mammals. Thus, brain swelling in mammals increases tissue pressure inside the cranium and reduces blood flow, which in turn exacerbates the problems of hypoxia. If the intracranial pressure exceeds the blood pressure, no blood can reach the brain. This is a point of no return as there is no way for the organism to restore the delivery of oxygen to the brain. Magnetic resonance imaging (MRI) has revealed that the brain of common carp suffers from cellular edema, net water gain, and a volume increase (by 6.5%) during 2 hours of anoxia. The swelling reached 10% during 100 minutes of subsequent re-oxygenation, but the common carp finally recovered from this insult, proving that the changes were reversible and suggesting that the oversized brain cavity that carp possesses allows brain swelling during energy deficiency without a resultant increase in intracranial pressure and global ischemia (Van der Linden *et al.*, 2001). As hypoxia is a common event in water, selective forces may promote a loose-fitting cranium to allow brain swelling. Indeed, the brain of many fish resides in a rather expansive chamber, where there appears to be room for the brain to swell, presumably without raising pressure and reducing blood flow. Not all fish are hypoxia tolerant, and the relative size of the cranium and brain also varies in fish. For example, the carp brain appears to have more room than the trout brain (personal observations that await a systematic quantitative study). The carp is hypoxia tolerant, whereas the trout is hypoxia sensitive. Although hypoxia tolerance is not simply a matter of cranium to brain size, part of the defense mechanisms may include tolerable brain swelling. A large cranial cavity may have its drawbacks. Salmonids swim up waterfalls, and this may lead to significant mechanical shocks to the brain, such that a more snug fit for the brain has evolved, compared with the hypoxia-tolerant, placid carp.

Fish that are hypoxia tolerant are also often ammonia tolerant (Walsh *et al.*, 2007), suggesting that common mechanisms may be in operation. For one thing, ammonia toxicity also causes brain swelling in mammals. Several theories have been proposed to address the mechanisms of acute ammonia toxicity in mammalian brains, including glutamine accumulation leading to astrocyte swelling (Felipo and Butterworth, 2002). Some fishes accumulate high levels of glutamine in their brains and other tissues – levels that would cause hepatic encephalopathy in mammals (Randall and Ip, 2006) – but this does not seem to be a problem in fish. The glutamine synthetase inhibitor methionine sulfoximine (MSO); at a dosage protective for mammals, does not protect fish against acute ammonia toxicity (Tsui *et al.*, 2004; Ip *et al.*, 2005). It appears that detoxification of ammonia to glutamine is crucial to ammonia tolerance in fish, but, unlike in mammals (Brusilow, 2002), glutamine synthesis and accumulation in the brain does not appear to be a major cause of death following ammonia intoxication. One factor may be that brain swelling during hyperammonia and hypoxia in fish is not detrimental. If the brain can swell, then the animal can better tolerate both high ammonia and hypoxia.

5.4   The fish heart in hypoxia

The heart has a particularly troublesome position in the circulation during hypoxia, as it receives much or all of its oxygen from the venous blood returning to the heart. Venous blood always contains less oxygen than arterial blood, and during hypoxia it may be almost completely depleted of oxygen. Moreover, the acidosis that normally accompanies hypoxia is a serious threat to heart function, as H^+ competes with Ca^{2+} for binding to troponin in the myocytes (Gesser and Jorgensen, 1982). Here, catecholamines appear to perform a protective function in some species by increasing intracellular Ca^{2+} levels in heart tissue (Farrell, 1985; Farrell *et al.*, 1986). Another mechanism that would function to protect the heart during hypoxia is based on the hypoxia-induced opening of arterio-venous (AV) flow through anastomoses in the gills. These allow oxygenated arterial blood to flow back to the venous side of the heart, thereby increasing the PO_2 in the heart, although the contribution is probably small. In a study on hypoxic cod (Sundin and Nilsson, 1992), AV flow from the gills was 8% of cardiac output. All elasmobranchs and some teleosts have coronary arteries that bring oxygenated blood from the gills back to the heart (Davie and Farrell, 1991). However, whereas coronary arteries are generally found in highly active species, such as salmonids, tuna and swordfish, there are numerous examples of hypoxia-tolerant species that lack a coronary blood supply (including cyprinids), so their hearts have to be adapted to make do with very little oxygen. The heart of the crucian carp, which appears to lack a coronary blood supply, is able to maintain or even increase cardiac output after several days without any oxygen, which may be related to the ability of this fish to avoid acidosis (Stecyk et al., 2004) (see Chapter 9).

Hypoxia-tolerant vertebrates reduce energy expenditure by the heart during severe hypoxia by reducing cardiac output (Stecyk and Farrell, 2006; Farrell and Stecyk, 2007; Stecyk and Farrell, 2007), such that energy use matches aerobic and anaerobic energy production. It has been suggested that vertebrates that show a considerable hypoxia tolerance are aided by having a heart power output that is low enough to be sustained by their glycolytic capacities (Farrell and Stecyk, 2007). Moreover, it may also be that a major function of the bradycardia combined with increased stroke volume, which is displayed by many fishes in hypoxia, is to save the heart. These changes may be reducing heart workload and aiding myocardial oxygenation by increasing the blood residence time in the ventricle and by stretching the myocardium so that oxygen diffusion distances in the tissue become shorter (Farrell, 2007).

5.5   Hematological adaptations to hypoxia

Among fishes, the most active species with top swimming performance show poor hypoxia tolerance, whereas the most hypoxia-tolerant species all appear to be relatively sluggish and sedentary. Salmonids with highly active lifestyles have a PO_2crit of around 75–90 mmHg (Table 5.1). Similarly, skipjack tuna (*Katsuwonus pelamis*) die when water $[O_2]$ falls below 60% of air saturation (Gooding et al., 1981). The underlying reason is probably that the high maximal rate of oxygen uptake ($\dot{V}O_{2max}$) displayed by very active fish species preclude hypoxia tolerance because of the opposing demands that a high $\dot{V}O_{2max}$ and hypoxia tolerance put on the oxygen-carrying properties of hemoglobin (see Burggren et al., 1991 for a review). Maintained oxygen uptake in hypoxia requires hemoglobins with a high O_2 affinity, but this means that most of the oxygen remains bound to the hemoglobin even at relatively low partial pressures of oxygen. The most extreme examples of high-affinity hemoglobins are found in crucian carp and goldfish, in which half saturation of the hemoglobin occurs at an oxygen tension (P_{50}) of 0.8 mmHg (crucian carp at 10°C; Sollid *et al.*, 2005) to 2.6 mmHg (goldfish at 26°C; Burggren, 1982) (see Table 3.2 in Chapter 3). Consequently, in the tissues of hypoxia-tolerant fishes O_2 has to be downloaded at a low partial pressure, leading to a small pressure gradient from blood into the mitochondria and therefore a slow O_2 delivery. In goldfish, venous PO_2 is as low as 2.2 mmHg during normoxic conditions, and it falls to 0.7 mmHg during moderate hypoxia, when water PO_2 is close to PO_2crit, and the AV [O_2] difference is still maintained (Burggren, 1982). Consequently, to allow high rates of oxygen delivery, hemoglobins of highly active and hypoxia-sensitive fish, such as salmon, have relatively low O_2 affinities (Burggren *et al.*, 1991; Jensen *et al.*, 1998 (see Table 3.2 in Chapter 3)).

However, fishes have the ability to modulate the Hb–O_2 affinity to some extent during hypoxia (see also Chapter 2, section 4.1). Like other vertebrates, fish hemoglobins are sensitive to (for example) temperature, H^+, PCO_2, and organic phosphates, and a decrease in any of these variables will increase the oxygen affinity, and vice versa. If given the opportunity, fishes will move to cooler water during hypoxia (Rausch *et al.*, 2000; Bicego *et al.*, 2007), which will not only reduce whole-body ATP use due to the Q_{10} effect, but also increase blood–oxygen affinity.

Unlike in mammals, hypoxic hyperventilation has only a minor effect on carbon dioxide excretion and blood pH in fish (Holeton and Randall, 1967a). This is because blood carbon dioxide levels are very low and the gill diffusing capacity is high, such that changes in ventilation have little effect on carbon dioxide levels in the blood: the rate-limiting step in carbon dioxide excretion in teleosts is the bicarbonate transfer from plasma into the red blood cell (Tufts and Perry, 1998). Hypoxia in fish is more often accompanied by a reduction in blood pH due to increased anaerobic metabolism. The increase in blood H^+ reduces Hb–O_2 affinity by inducing Bohr and Root shifts, changes that appear maladaptive in the hypoxic situation (Root and Bohr shifts are explained in Chapter 3, section 5.) In teleosts, catecholamines appear to play a protective function in this situation, as they activate a β-adrenergic Na^+/H^+ exchanger in the erythrocyte membrane that strives to increase intracellular pH (see Nikinmaa and Salama, 1998 for a review). However, in elasmobranchs, a β-adrenergic Na^+/H^+ exchanger appears to be lacking (Tufts and Randall 1989), and catecholamine release in response to hypoxia is more variable (Perry and Gilmour, 1996). In general, changes in blood–oxygen affinity during hypoxia are less well studied in elasmobranches than in teleosts.

By contrast, the organic phosphates may have a more clear-cut role in enhancing hypoxia tolerance. The levels of the most important modulatory organic phosphates in teleost erythrocytes (ATP and guanosine triphosphate

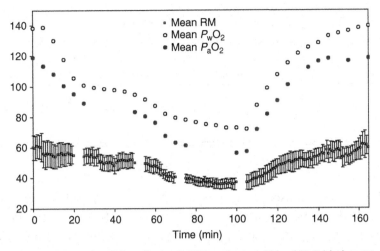

Fig. 5.6 Oxygen levels in red muscle (RM) and arterial blood (P_aO_2) in intact trout subjected to variations in water oxygen tensions (P_wO_2). From McKenzie *et al.* (2004).

[GTP]) fall during hypoxia, which leads to increased Hb–O_2 affinity. When it comes to mechanisms increasing blood–oxygen affinity, it appears that the catecholamine-induced alkalinization of the erythrocytes functions as a fast acute response to hypoxia, whereas the fall in organic phosphates, which takes hours to fully develop, is more important during chronic hypoxia (see Nikinmaa and Salama, 1998 for a review). Changes in organic phosphates, however, are rapid enough to track the diurnal changes in oxygen content of water in Amazonian fish (Val, 2000).

The increase in hemoglobin affinity will aid oxygen uptake from the water during hypoxia but is detrimental for unloading oxygen in the tissues. How then is oxygen delivery to the tissues maintained during hypoxia? First it should be recognized that there are very few measurements of tissue oxygen tensions in fish, compared with the large numbers of blood measurements.

In trout subjected to hypoxia with implanted oxygen probes in the red muscle, McKenzie *et al.* (2004) observed that arterial PO_2 decreased with the fall in water PO_2, but the difference appeared to be smaller during hypoxia, reflecting the increase in gill diffusing capacity. Muscle oxygen tensions also dropped during hypoxia, but not nearly as much as those in arterial blood (Fig. 5.6). In trout, unlike in mammals, tissue PO_2 was found to be higher than venous oxygen levels (Table 5.2), even during hypoxia.

One explanation could be that muscle blood flow is very high compared with that in other tissues, and the muscle is quiescent, such that oxygen extraction from the blood is low. Blood leaving the muscle is then mixed with deoxygenated venous blood from other tissues to present a mixed venous blood tension

Table 5.2 *Muscle oxygen levels are below mixed venous levels in mammals but midway between arterial and mixed venous levels in trout*

PO_2 (mmHg)	Human, Rat, Dog	Trout
Arterial	100	100
Venous	40	
Tissue	25–35	61

Data from McKenzie *et al.* (2004).

that is lower than muscle oxygen tension. Another explanation is that oxygen transfer to the tissues is not directly related to PO_2 differences between arterial blood and muscle, but is due to a hemoglobin Root-off shift, in which acidification of the blood drives oxygen from hemoglobin raising blood PO_2 in the muscle capillaries in a manner basically similar to that postulated for the swimbladder (a mechanism reviewed by Pelster and Randall, 1998). Blood leaving trout gills is in a non-equilibrium state (Gilmore, 1998). Plasma bicarbonate is hydrated as blood flows away from the gills, thereby gradually raising pH. The carbon dioxide formed from bicarbonate hydration will enter and acidify the red blood cell, causing a Root-off shift and raising blood PO_2. Because of the small volume of the arterial system, blood is probably only resident there for a few seconds, and the carbonate system of blood entering the muscle is probably not in equilibrium. Carbonic anhydrase on the muscle endothelium would catalyze the reaction and enhance the Root-off shift and the elevation in PO_2. The arterial blood oxygen levels reported by McKenzie *et al.* (2004), however, will not be those for blood leaving the gills but for blood that has probably reached equilibrium in the measuring system. Thus the non-equilibrium state of blood leaving the gills (Gilmore 1998), while enhancing tissue oxygenation, is likely to have only a minor effect on oxygen delivery to the tissues. Carbon dioxide entering the blood from the tissues could cause a Root shift if the CO_2 transfer is more rapid and precedes the bulk of oxygen transfer (Brauner and Randall, 1996).

If the Root shift dominates in oxygen unloading at the tissues, then responses that enhance arterial oxygen content, such as the decrease in organic phosphate levels, which increases Hb–O_2 affinity, will be selected because uptake at the gills will be enhanced without detrimental effect on oxygen delivery in the tissues. Still, anemia in fish causes a rise in organic phosphates, similar to the response in mammals, thereby reducing blood–oxygen affinity (Nikinmaa and Salama, 1998). Thus, during anemic conditions when oxygen transfer to the tissues is compromised, the general, and presumably adaptive, response in vertebrates appears to be to reduce Hb–O_2 affinity to make oxygen more available to the tissues.

During hypoxia, an increase in hematocrit due to red blood cell swelling and the release of red blood cells from the spleen may occur within minutes/hours, but there is a surprising variability in this response, both within and between species (see Gallaugher and Farrell, 1998 for a review). Hypoxia is associated with an increase in urine flow and a decrease in plasma volume, perhaps due to the release of cardiac peptides (Tervonen *et al.*, 2002), also contributing to the increase in hematocrit seen during hypoxia. There is no apparent tendency for hypoxia-tolerant fishes in general to have higher blood hemoglobin levels than less hypoxia-tolerant species (Nilsson and Östlund-Nilsson, 2008). However, there is evidence for an increase in blood erythrocyte and hemoglobin content in hypoxic rainbow trout (*Oncorhynchus mykiss*) due to erythropoietin-induced erythropoiesis, which occurs within days/weeks (Lai *et al.*, 2006). Several hypoxia-inducible transcription factors (HIFs) have been described in fish (see Chapter 2 for a detailed description of HIF function), and the increase in erythropoietin during hypoxia is presumably HIF related.

5.6   Reducing energy expenditure

If ambient oxygen levels fall below the PO_2crit, oxygen delivery to the tissues is compromised, and for the fish to survive, energy expenditure has to be reduced and/or anaerobic metabolism must be upregulated (Boutilier *et al.*, 1987). Whereas the PO_2crit reflects the ability of the fish to extract oxygen from water, subsequent reductions in energy expenditure and, therefore, oxygen uptake, reflect reorganization of the behavior and physiology of the fish in response to hypoxia. The effect of these changes will probably reduce the PO_2crit of that fish. Indeed, hypoxia acclimation has been found to reduce both resting oxygen consumption and PO_2crit in goldfish (Prosser *et al.*, 1957), speckled trout (*Salvelinus fontinalis*) (Shepard, 1955), and epaulette shark (Routley *et al.*, 2002).

Metabolic depression in response to environmental stress has been reported in both invertebrates and vertebrates, including hypoxic fish (Van Waversveld *et al.*, 1989; Johansson *et al.*, 1995; Van Ginneken *et al.*, 1996; Muusze *et al.*, 1998; Nilsson and Renshaw, 2004). Much work has been devoted to understanding metabolic depression by means of suppressing ATP production and ATP-consuming processes (e.g. ion pumping, protein synthesis, etc.) in a coordinated fashion, and many reviews have been published on this subject (e.g. Hand and Hardewig, 1996; Hochachka *et al.*, 1996; Storey and Storey, 2004). These reviews tend to concentrate on biochemical mechanisms associated with metabolic depression and ignore behavioral and physiological strategies, such as moving to a lower temperature, reduced activity, and inhibition of feeding and reproduction.

Because PO_2crit is the lowest level of oxygen that allows a sustained *resting* metabolic rate, an animal that finds itself in an environment where PO_2 is close to its PO_2crit will have no scope for additional activity. Thus activities that are not immediately needed for survival, such as feeding and reproduction, have to be suppressed. Moreover, like many animals, hypoxic fish save energy by reducing swimming activity (Nilsson *et al.*, 1993) and/or by lowering body temperature (by moving to colder water) (Schurmann and Steffensen, 1997). The epaulette shark, for example, becomes virtually comatose during exposure to anoxia (Renshaw *et al.*, 2002).

Reduction of food intake and retardation of growth in fish during hypoxia has been reported many times (Secor and Gunderson, 1998; Pichavant *et al.*, 2000; Taylor and Miller 2001; Zhou *et al.*, 2001; Foss *et al.*, 2002; Bernier and Craig, 2005). In carp there was an initial hypoxic inhibition of feeding, but after several days the 'hypoxic group' began to feed again, but only at a reduced rate of 1% body weight (Wang *et al.*, 2008). Food transfer in the gut and food conversion are not affected; the overall process appears to be essentially unchanged, only inhibited at very reduced levels of food intake. At low rates of food intake food conversion is reduced because the energy taken up is used to cover the cost of feeding rather than being converted into growth. Inhibition of feeding results in an immediate and considerable saving in energy use during hypoxia.

Reproduction is also inhibited during hypoxia. Gonads do not mature, and sexual activity is curtailed in carp (Wu *et al.*, 2003). Thomas *et al.* (2005) exposed Atlantic croaker (*Micropogonias undulatus*) to long-term hypoxia and found dramatic suppression at all levels of the reproductive axis, including GnRH gene expression. Exposing Gulf killifish (*Fundulus grandis*) to hypoxia for 1 month significantly reduced growth and reproduction: hypoxia-exposed females produced significantly fewer eggs and initiated spawning later than control fish (Landry *et al.*, 2007). Hypoxia reduces growth and increases teratologies in zebrafish embryos (Shang and Wu, 2004), and it has also been shown to increase the proportion of males in laboratory zebrafish colonies (Shang *et al.*, 2006). Hypoxia inhibits mating and the surge in luteinizing hormone, which in turn reduces egg maturation and spawning in carp (Wang *et al.*, 2008).

In cardinalfishes (Apogonidae), hypoxia inhibits reproduction in a very direct way: it forces the mouthbrooding males to prioritize their own survival by spitting out the egg clutch (which fills up much of the oral cavity), thereby immediately improving gill oxygen uptake (Östlund-Nilsson and Nilsson, 2004). Moreover, during hypoxia cardinalfish males with the largest egg clutches spit out the clutch at a higher ambient oxygen level than males with smaller clutches, indicating a trade-off between brood size and hypoxia tolerance.

Anoxia decreases protein synthesis in crucian carp, with the liver showing a much larger decrease (~95%) than muscle and heart (~50%), whereas protein synthesis in the brain remains unchanged (Smith *et al.*, 1996). Changes in gene and protein expression are varied, and the relatively few available studies utilizing mRNA microarrays (Gracey *et al.*, 2001; Van der Meer *et al.*, 2005; Gracey 2007; Ju *et al.*, 2007) and proteomic approaches (Bosworth *et al.*, 2005; Smith *et al.*, 2009) paint a complex picture, with some common themes, such as a suppression of genes involved in aerobic metabolism. Interestingly, a proteomic study on zebrafish indicated that changes in protein expression are less widespread than changes in mRNA expression (Bosworth *et al.*, 2005). Pyruvate dehydrogenase activity is reduced in the muscle of the common killifish during hypoxia exposure (Richards *et al.*, 2008), and AMP-activated protein kinase (AMPK) activity is rapidly increased in goldfish liver within 0.5 hours of hypoxia exposure with no changes in total AMPK protein amount, indicating that the changes in AMPK activity are due to post-translational phosphorylation of the protein (Jibb and Richards, 2008). Similar changes in AMPK activity, also involving brain and heart, are seen in anoxic crucian carp (Stensløkken *et al.*, 2008) (see Chapter 9, section 5.1).

It has been suggested that ion-channel arrest in membranes would contribute to energy savings, but a microarray study on the hypoxia-tolerant goby *Gillichthys mirabilis* did not pick up any genes that indicated channel arrest. It can be noted that not even the master of anoxia tolerance, the crucian carp, appears to show any widespread channel arrest (see Chapter 9). Channel arrest can occur in a wide range of circumstances and during hypoxia could be a response to, rather than a cause of, reduced energy turnover in cells.

Adenosine is released from energy-compromised cells as a result of a net breakdown of ATP, ADP, and AMP. Adenosine has been proposed to contribute to the regulation of depressed metabolism, and its elevation has been reported during times of energy deficiency in fish, including hypoxia (Renshaw *et al.*, 2002). The actions of adenosine in vertebrates include stimulation of glycogenolysis and tissue blood flow to fuel glycolysis, and suppression of neuronal excitability and neurotransmitter release to reduce ATP use. In other words, adenosine aids survival in hypoxia and anoxia by reducing ATP consumption to match production and hence results in metabolic depression (Nilsson and Renshaw, 2004). There is evidence for a role for adenosine in metabolic depression in teleosts (rainbow trout and crucian carp), elasmobranchs (epaulette shark), and hagfish (Nilsson, 1991; Bernier *et al.*, 1996; Renshaw *et al.*, 2002), and for stimulating brain blood flow in anoxic crucian carp (Nilsson *et al.*, 1994).

5.7   Hypoxic tissue damage

Even if many fishes are well adapted to encounter hypoxia, one may expect some degree of tissue damage to occur, especially during severe, near lethal, hypoxia. Indeed, an increased number of apoptotic cells have been detected in the central nervous system of sturgeon (*Acipenser shrenckii*) recovering from 6- to 30-hour-long hypoxia exposures (Lu *et al.*, 2005). It is possible that fishes can survive some degree of tissue damage during hypoxia, as fishes are well known to have large regenerative capacities, even for an organ such as the brain. A reason for this is that the body and brain of fishes continues to grow during much of their life. Zones with proliferating cells are, for example, much more abundant in the fish brain than in the mammalian brain (Sørensen *et al.*, 2007).

Low oxygen levels induce DNA damage and apoptosis in mammalian cell lines (Thompson, 1998; Bras *et al.*, 2005). However, in vivo responses to DNA damage are known for only a few mammals, and very little is known about such responses in fish. In vivo studies of the liver of common carp exposed to hypoxia (Poon *et al.*, 2007) have suggested extensive DNA damage during the first days of hypoxic exposure, as indicated by terminal transferase-mediated dUTP nick-end labeling (TUNEL). TUNEL labeling was very high (found in around 60% of the liver cells) during hypoxia, especially after 4 days of exposure to aquatic hypoxia at 0.5 mg O_2 l^{-1}. The level of TUNEL staining was reduced after about a week of hypoxic exposure but was maintained at a higher level during the 42 days of hypoxia than is seen in normoxic livers. Such extensive DNA damage often leads to programmed cell death or apoptosis. Indeed, TUNEL is often used to indicate apoptosis.

If the TUNEL signal was indicative of rates of apoptosis in the in vivo hypoxic carp liver, then, in the face of low rates of cell proliferation, the carp liver should have been reduced in size after 6 weeks of hypoxia; but both the size of the liver and the number and size of liver cells did not change significantly during this period (Poon *et al.*, 2007). In addition, there was no change in cell proliferation, no increase in caspase-3 activity, and no increase in single-stranded DNA, leading to the conclusion that there was no increase in apoptosis in the liver during hypoxia. There was upregulation of some anti-apoptotic factors (Bcl-2, Hsp70, p27) and downregulation of some pro-apoptotic genes (Tetraspanin 5 and Cell death activator). The liver cells appeared to enter cell cycle arrest, presumably to allow repair of damaged DNA. As there was no change in cell proliferation and cell number, the damaged cells were not entering apoptosis and must have recovered during prolonged hypoxia (Poon *et al.*, 2007). Thus, it may be that fishes have the capacity to counteract apoptosis during hypoxia in some tissues.

DNA damage during hypoxia could be related to increased production of reactive oxygen species (ROS). A large increase in the gene expression and level of uncoupling proteins (UCPs) has been observed in vivo in common carp liver, but not kidney, during hypoxia (Hung, 2005, unpublished Ph.D. thesis). In mammals it is clear that UCPs are inserted into the inner mitochondrial membrane and will shortcircuit the proton gradient generated by NADH oxidation (see Krauss *et al.*, 2005 for a review). This is important in regulating heat production in mammals, but why are there uncoupling proteins in fish, when clearly the fish liver is not being used to generate heat? The rate of production of ROS is related to mitochondrial membrane potential, at least in the rat (Korshunov *et al.*, 1998), and one possibility is that UCP in fish liver functions to lower mitochondrial membrane potential and, therefore, ROS production during hypoxia.

5.8   Hypoxia tolerance and size

Finally, in this chapter we will consider how hypoxia tolerance is influenced by a factor that shows an incredible span in fishes: body size. This will also allow us to summarize some of the major mechanisms guiding hypoxic responses in fishes that we have discussed in this chapter.

Many fish species, the common carp being one example, weigh only one or a few milligrams early in life and still reach weights of over 10 kg, an increase in body mass covering more than six orders of magnitude. If all species of adult bony fishes are considered, their body mass covers a range of eight orders of magnitude, and if we also include elasmobranchs, fishes cover nine orders of magnitude in body mass. The smallest of all vertebrates are two recently described teleosts: *Paedocypris progenetica* from swamps in Sumatra, and *Schindleria brevipinguis*, from the Great Barrier Reef. Both mature at a body length of around 8 mm (Watson and Walker, 2004; Kottelat *et al.*, 2006), when they can be estimated to weigh 10–20 mg. Among teleosts, the largest species include the sunfish (*Mola mola*) and the beluga sturgeon (*Huso huso*), which reach 2000 kg (Frimodt, 1995). Among elasmobranchs, the whale shark (*Rhincodon typus*) can reach 34 000 kg (Chen *et al.*, 1999).

Not surprisingly, there have been many suggestions on how body size influences the hypoxia tolerance of fishes. The study of how body size influences biological functions is termed scaling (see Schmidt-Nielsen, 1984, for an introduction), and the scaling of hypoxia tolerance in fishes was recently the subject of a review in which some general conclusions could be drawn (Nilsson and Östlund-Nilsson, 2008). The two major conclusions are that: (1) body size per se is unimportant for the ability to survive hypoxia aerobically (when a fish relies on taking up the little oxygen there is); but that (2) body size becomes very important in severe hypoxia and anoxia, when ATP has to be produced anaerobically.

A major reason for the first conclusion is that metabolic oxygen demands, on the one hand, and respiratory surface area, on the other, are scaling in the same way in fishes: both relate to body size, with a scaling exponent estimated to lie between 0.76 and 0.90 (see Nilsson and Östlund-Nilsson, 2008 for references). Thus, while mass-specific metabolic rate falls with increasing body size among fishes (as in most organisms), so does the surface area of the gills. In other words, gill surface area appears to be closely matched to the metabolic demands of fish. Other factors that could influence oxygen uptake, such as hematocrit, Hb–O_2 affinity, cardiac output, and oxygen diffusion distances in the gills, show little or no dependence on body size in fish (Nilsson and Östlund-Nilsson, 2008). Therefore, if we consider the ability of fish to take up oxygen from the water during hypoxic conditions, there are no strong arguments for assuming that body size per se should be important. Indeed, a relatively large survey of [O_2]crit of post-settlement damselfishes from the Great Barrier Reef (covering a size range of 10 mg to 40 g) supports the idea that body size per se is essentially irrelevant to how well a fish can take up oxygen from the water (Fig. 5.8A).

Still, there are several examples of fish species in which smaller individuals are either better or worse at taking up oxygen during hypoxic conditions, and the reason for this is most likely that size often correlates with differences in habitat or lifestyle. A striking example of this is provided by early life stages of coral-reef damselfishes, which at the end of their planktonic larval stage become super-performers when it comes to swimming: they are the fastest swimmers in existence, being able to sustain speeds of up to 50 body lengths per second for hours and days (Bellwood and Fisher, 2001). This means that they also have the highest mass-specific rate of oxygen uptake of any fishes, and most likely have hemoglobins with a low oxygen affinity to allow rapid downloading in the tissues (Nilsson et al., 2007b). Not surprisingly, during this short early phase of life, they show a very poor ability for taking up oxygen during hypoxia (having an [O_2]crit of 40–60% of air saturation), but this changes within

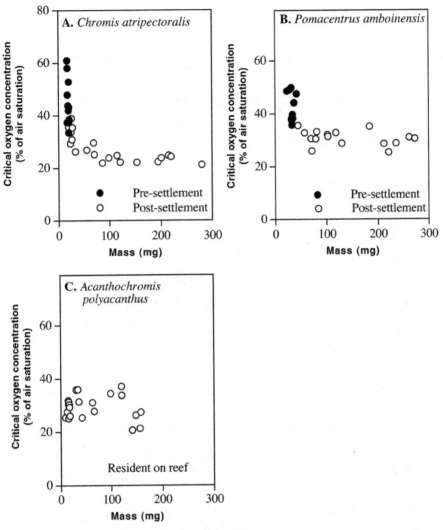

Fig. 5.7 Change in critical oxygen concentration ([O$_2$]crit) with body mass of pre-settlement larvae and post-settlement juveniles of three species of damselfish from Lizard Island, Great Barrier Reef. (A) *Chromis atripectoralis* (weighing 17–280 mg); (B) *Pomacentrus amboinensis* (weighing 23–280 mg); and (C) *Acanthochromis polyacanthus* (weighing 9–157 mg). *Acanthochromis* is unusual as it lacks a planktonic larval stage (the young being subject to parental care). It is clear that at the pre-settlement stage, *C. atripectoralis* and *P. amboinensis* have a significantly higher [O$_2$]crit, which falls rapidly upon settling on the reef. Once on the reef, they need hypoxia tolerance to survive, as they spend the night sheltering in the coral matrix, which can be severely hypoxic at night. As the pre-settlement larvae were caught in the vicinity of the reef, some of them may already have started the transition to a life on the reef, explaining the large variability in [O$_2$]crit seen in this group. Data from Nilsson *et al.*, 2007b.

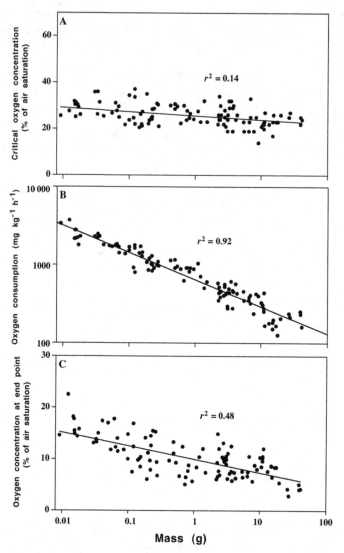

Fig. 5.8 Scaling of hypoxia tolerance with body mass in damselfishes (Pomacentridae) represented by 117 individuals belonging to 15 species, all from the coral reef at Lizard Island, Great Barrier Reef. (A) Critical oxygen concentration ([O₂]crit) is virtually independent of body mass. (B) As in other animals, mass-specific metabolic rate (measured as oxygen consumption) falls with body mass. (C) The near-lethal oxygen concentration when fishes lose their equilibrium is higher for small individuals. Data from Nilsson *et al.* (2007a) and from Nilsson and Östlund-Nilsson (2008).

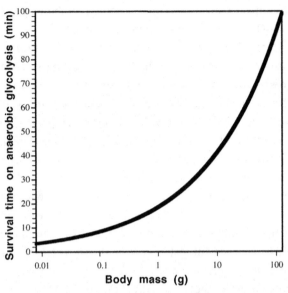

Fig. 5.9 Big is better in anoxia, because a small fish will more rapidly poison itself with lactic acid or run out of glycogen. The semi-log plot illustrates the expected relationship between body mass and survival time on anaerobic glycolysis in anoxia. The general exponential shape of the curve should apply to any fish, where the survival time is primarily dependent on how fast they empty their glycogen stores or fill up with lactate and H$^+$. The values in this case are based on the scaling of metabolic rate of damselfishes in the size range from 10 mg to 100 g, and on the assumptions that metabolic rate is maintained in anoxia and that either a lactate level of 20 mmol kg^{-1} is lethal or that the glycogen store contains 10 mmol glycosyl units kg^{-1} (Nilsson and Östlund-Nilsson, 2008). Metabolic depression, a differently sized glycogen store, or other limits for lactate poisoning, would affect the values on the y-axis but not change the shape of the curve as long as these factors are independent of body mass. This is because the shape is determined by the relationship between metabolic rate and body size.

days when they settle on the reef (Fig. 5.7). Here, they need to be hypoxia tolerant to survive, as they spend the night sheltering in the coral matrix, which can become severely hypoxic as the sun goes down and photosynthesis stops (Nilsson *et al.*, 2007a). The Oscar cichlid (*Astronotus ocellatus*) of the Amazon is another example in which larger individuals have a lower [O$_2$]crit than their smaller relatives (Sloman *et al.*, 2006). By contrast, in the largemouth bass (*Micropterus salmonides*) and yellow perch (*Perca flavescens*), the larger individuals tend to avoid hypoxic water more than small ones (Burleson *et al.*, 2001; Robb and Abrahams, 2003). In all these cases, the differences are most likely the result of life-stage differences in habitat preference (and the selective forces at work in these habitats).

However, when it comes to surviving on anaerobic metabolism (glycolysis) at oxygen levels below [O$_2$]crit, including anoxia, the scaling of metabolic rate gives larger fishes a clear advantage over smaller ones. For example, in damselfish kept in a closed respirometer in which [O$_2$] is falling steadily, small individuals lose their equilibrium at a higher [O$_2$] (around 15% of air saturation for a 10 mg fish) than large individuals (5% of air saturation for a 40 g fish) (Fig. 5.8C). The reason for this is most likely the higher mass-specific metabolic rate of small fish (exemplified for damselfishes in Fig. 5.8B). If they are at all able to compensate for the loss of aerobic respiration with anaerobic glycolysis during severe hypoxia or anoxia, the high metabolic rate of a small fish means that it will rapidly use up the glycogen stores or poison itself with anaerobic end products (lactate and H$^+$), or both (Fig. 5.9). Also in the case when the glycolytic rate is not high enough to maintain ATP levels in anoxia or severe hypoxia, ATP is likely to be used up faster in a small fish than in a large one due to its higher rate of ATP consumption.

There are exceptions: most notably the crucian carp, which is able to turn lactate into ethanol, which is released into the water. This allows long-term anoxic survival and makes it meaningful to store very large amounts of glycogen, as lactic acid poisoning is no longer an issue. Indeed, the crucian carp has record-high glycogen levels in its tissues, and body size does not seem to be a major issue for its anoxic survival. Much of Chapter 9 will be devoted to this champion of anoxia tolerance.

Acknowledgements

We are grateful to Ms Mandy Dung for editing the manuscript.

References

Axelsson, M. and Fritsche, R. (1991). Effects of exercise, hypoxia and feeding on the gastrointestinal blood flow in the Atlantic cod *Gadus morhua*. *J. Exp. Biol.* **158**, 181–98.

Bæuf, G. and Payan, P. (2001). How should salinity influence fish growth? *Comp. Biochem. Physiol.* **130C**, 411–23.

Beamish, F. W. H. (1964). Respiration of fishes with special emphasis on standard oxygen consumption. III. Influence of oxygen. *Can. J. Zool.*. **42**, 355–66.

Bellwood, D. R. and Fisher, R. (2001). Relative swimming speeds in reef fish larvae. *Mar. Ecol. Prog. Ser.* **211**, 299–303.

Bernal, D., Dickson, K. A., Shadwick, R. E. and Graham, J. B. (2001). Analysis of the evolutionary convergence for high performance swimming in lamnid sharks and tunas. *Comp. Biochem. Physiol.* **129A**, 695–726.

Bernier, N. J., Craig, P. M. (2005) CRF-related peptides contribute to the stress response and the regulation of appetite in hypoxic rainbow trout. *Am. J. Physiol.* **289**, R982–90.

Bernier, N., Harris, J., Lessard, J. and Randall, D. (1996). Adenosine receptor blockade and hypoxia-tolerance in rainbow trout and Pacific hagfish. I. Effects on anaerobic metabolism. *J. Exp. Biol.* **199**, 485–95.

Bicego, K. C., Barros, R. C. and Branco, L. G. (2007). Physiology of temperature regulation: comparative aspects. *Comp. Biochem. Physiol.* **147A**, 616–39.

Booth, J. H. (1978). The distribution of blood flow in the gills of fish: application of a new technique to rainbow trout (*Salmo gairdneri*). *J. Exp. Biol.* **73**, 119–30.

Booth, J. H. (1979a). The effect of oxygen supply, epinephrine and acetylcholine on the distribution of blood flow in trout gills. *J. Exp. Biol.* **83**, 31–9.

Booth, J. H. (1979b). Circulation in trout gills: The relationship between branchial perfusion and the width of the lamellar blood space. *Can. J. Zool.* **57**, 2185–93.

Bosworth, C. A. 4th, Chou, C. W., Cole, R. B. and Rees, B. B. (2005). Protein expression patterns in zebrafish skeletal muscle: initial characterization and the effects of hypoxic exposure. *Proteomics* **5**, 1362–71.

Boutilier, R. G., Dobson, G., Hoeger, U. and Randall, D. J. (1987). Acute exposure to graded levels of hypoxia in rainbow trout (*Salmo gairdneri*): metabolic and respiratory adaptations. *Respir. Physiol.* **71**, 69–82.

Branson, B. A. and Hake, P. (1972). Observations of an accessory breathing mechanism in *Piaractus nigripinnis* (Cope). *Zoologischer Anzeiger* **189**, 292–7.

Bras, M., Queenan, B. and Susin, S. A. (2005). Programmed cell death via mitochondria: different modes of dying. *Biochemistry (Mosc.)* **70**, 231–9.

Braum, E. and Junk, W. J. (1982). Morphological adaptation of two Amazonian characoids (Pisces) for surviving in oxygen deficient waters, *Int. Revue Res. Hydrobiol.* **67**, 869–86.

Brauner, C. J. and Randall, D. J. (1996). The interaction between oxygen and carbon dioxide movements in fishes. *Comp. Biochem. Physiol.* **113A** 1, 83–90.

Brett, J. R. (1979) Environmental factors and growth. In *Fish Physiology Vol. 8, Bioenergetics and Growth* (eds. W. S. Hoar, D. J. Randall and J. R. Brett). San Diego: Academic Press, pp. 599–675.

Brett, J. R. and Groves, T. D. D. (1979) Physiological energetics. In *Fish Physiology Vol. 8, Bioenergetics and Growth*, ed. W. S. Hoar, D. J. Randall and J. R. Brett. San Diego: Academic Press, pp. 280–352.

Brusilow, S. W. (2002). Hyperammoniemic encephalopathy. *Medicine* **81**, 240–9.

Butler, P. J. and Taylor, E. W. (1975). The effect of progressive hypoxia on respiration in the dogfish (*Scyliorhinus canicula*) at different seasonal temperatures. *J. Exp. Biol.* **63**, 117–30.

Butler, P. J., Taylor, E. W., Capra, M. F. and Davison, W. (1978). The effect of hypoxia on the levels of circulating catecholamines in the dogfish *Scyliorhinus canicula*. *J. Comp. Physiol. B* **127**, 325–30.

Burggren, W. W. (1982). 'Air gulping' improves blood oxygen transport during aquatic hypoxia in the goldfish *Carassius auratus*. *Physiol. Zool.* **55**, 327–34.

Burggren, W., McMahon, B. and Powers, D. (1991). Respiratory functions of blood. In *Environmental and Metabolic Animal Physiology*, ed. C. L. Prosser. New York: Wiley-Liss, pp. 437–508.

Burleson, M. L., Wilhelm, D. R. and Smatresk, N. J. (2001). The influence of fish size on the avoidance of hypoxia and oxygen selection by largemouth bass. *J. Fish Biol.* **59**, 1336–1349.

Chan, D. K. O. and Wong, T. M. (1977). Physiological adjustments to dilution of the external medium in the lip shark, *Hemiscyllium plagiosum* (Bennet). III. Oxygen consumption and metabolic rates. *J. Exp. Zool.* **200**, 97–102.

Chapman, L. J. and Hulen, K. G. (2001). Implications of hypoxia for the brain size and gill morphometry of mormyrid fishes. *J. Zool.* **254**, 461–72.

Chapman, L. J., Galis, F. and Shinn, J. (2000). Phenotypic plasticity and the possible role of genetic assimilation: hypoxia-induced trade-offs in the morphological traits of an African cichlid. *Ecol. Lett.* **3**, 387–93.

Chapman, L. J., Chapman, C. A., Nordlie, F. G. and Rosenberger, A. E. (2002). Physiological refugia: swamps, hypoxia tolerance and maintenance of fish diversity in the Lake Victoria region. *Comp. Biochem. Physiol.* **133A**, 421–37.

Chen, C. T., Liu, K. W. and Young, S. J. (1999). Preliminary report on Taiwan's whale shark fishery. In *Elasmobranch Biodiversity, Conservation and Management, Proc. Int. Seminar and Workshop in Sabah, Malaysia*, ed. S. L. Fowler, T. Reid and F.A. Dipper. IUCN, Gland, Switzerland, pp. 162–7.

Cruz-Neto, A. P. and Steffensen, J. F. (1997). The effects of acute hypoxia and hypercapnia on oxygen consumption of the freshwater European eel. *J. Fish Biol.* **50**, 759–69.

Davie, P. S. and Farrell, A. P. (1991). The coronary and luminal circulations of the myocardium of fishes. *Can. J. Zool.* **2**, 158–164.

Davis, J. C. (1972) An infrared photographic technique useful for studying vascularization of fish gills. *J. Fish. Res. Board Can.* **29**, 109–11.

De Boeck, G., De Smet, H. and Blust, R. (1995). The effect of sublethal levels of copper on oxygen consumption and ammonia excretion in the common carp, *Cyprinus carpio. Aquatic Toxicol.* **32**, 127–41.

Farrell, A. P. (1985). A protective effect of adrenaline on the acidotic teleost heart. *J. Exp. Biol.* **116**, 503–8.

Farrell, A. P. (2007). Tribute to P. L. Lutz: a message from the heart–why hypoxic bradycardia in fishes? *J. Exp. Biol.* **210**, 1715–25.

Farrell, A. P. and Jones, D. R. (1992). The heart. In *Fish Physiology. Vol. 12A*, ed. W. S. Hoar, D. J. Randall and A. P. Farrell. San Diego: Academic Press, pp. 1–88.

Farrell A. P. and Stecyk, J. A. W. (2007). The heart as a working model to explore themes and strategies for anoxic survival in ectothermic vertebrates. *Comp. Biochem. Physiol.* **147A**, 300–12.

Farrell, A. P., MacLeod, K. and Chancey, B. (1986). Intrinsic mechanical properties of the perfused rainbow trout heart and the effects of catecholamines and extracellular calcium under control and acidotic conditions. *J. Exp. Biol.* **125**, 319–45.

Farrell, A. P., Sobin, S. S., Randall, D. J. and Crosby, S. (1980). Intralamellar blood flow patterns in fish gills. *Am. J. Physiol.* **239**, R428–36.

Farrell, A. P., Hinch, S. G., Cooke, S. J., *et al.* (2008). Pacific Salmon in hot water: applying aerobic scope models and biotelemetry to predict the success of spawning migrations. *Physiol. Biochem. Zool.* **81**, 697–708.

Felipo, V. and Butterworth, R. F. (2002). Neurobiology of ammonia. *Prog. Neurobiol.* **67**, 259–79.

Fernandes, M. N. and Rantin, F. T. (1989). Respiratory responses of *Oreochromis niloticus* (Pisces, Cichlidae) to environmental hypoxia under different thermal conditions. *J. Fish Biol.* **35**, 509–19.

Fernandes, M. N., Rantin, F. T., Kalinin, A. L. and Moron, S. E. (1994). Comparative study of gill dimensions of 3 Erythrinid species in relation to their respiratory function. *Can. J. Zool.* **72**, 160–5.

Foss A., Evensen T. H. and Oiestad V. (2002). Effects of hypoxia and hyperoxia on growth and food conversion efficiency in the spotted wolfish *Anarhichas minor* (Olafsen). *Aquaculture Res.* **33**, 437–44.

Frederich, M. and Pörtner, H. O. (2000). Oxygen limitation of thermal tolerance defined by cardiac and ventilatory performance in spider crab, *Maja squinado*. *Am. J. Physiol.* **279**, R1531–8.

Frimodt, C. (1995). *Multilingual Illustrated Guide to the World's Commercial Coldwater Fish*. Oxford: Fishing News Books.

Fry, F. E. J (1971) The effect of environmental factors on the physiology of fish. In *Fish Physiology Vol. 6*, ed. W. S. Hoar and D. J. Randall. New York: Academic Press, pp. 1–98.

Fry, F. E. J. and Hart, J. S. (1948). The relation of temperature to oxygen consumption in the goldfish. *Biol. Bull.* **94**, 66–77.

Gallaugher, P. and Farrell, A. P. (1998). Hematocrit and blood oxygen-carrying capacity. In *Fish Physiology Vol. 17, Fish Respiration*, ed. S. F. Perry and B. L. Tufts. New York: Academic Press, pp. 185–227.

Gesser, H. and Jorgensen, E. (1982). pHi, contractility and Ca-balance under hypercapnic acidosis in the myocardium of different vertebrate species. *J. Exp. Biol.* **96**, 405–12.

Gilmore, K. M. (1998) Causes and consequences of acid-base disequilibria. In *Fish Physiology Vol. 17, Fish Respiration*, ed. S. F. Perry and B. L. Tufts. New York: Academic Press, pp. 321–348.

Gonzalez, R. J. and McDonald, D. G. (1992). The relationship between oxygen consumption and ion loss in a freshwater fish. *J. Exp. Biol.* **163**, 317–32.

Gooding, R. M., Neill, W. H. and Dizon, A. E. (1981). Respiration rates and low-oxygen tolerance limits in skipjack tuna, Katsuwonus pelamis. *Fisheries Bull.* **79**, 31–48.

Gracey A. Y. (2007). Interpreting physiological responses to environmental change through gene expression profiling. *J. Exp. Biol.* **210**, 1584–92.

Gracey A. Y., Troll, J. V. and Somero, G. N. (2001). Hypoxia-induced gene expression profiling in the euryoxic fish *Gillichthys mirabilis*. *Proc. Natl. Acad. Sci. USA* **98**, 1993–8.

Gray, I. E. (1954). Comparative study of the gill area of marine fishes. *Biol. Bull.* **107**, 219–55.

Hand, S. C. and Hardewig, I. (1996). Downregulation of cellular metabolism during environmental stress: mechanisms and implications. *Ann. Rev. Physiol.* **58**, 539–63.

Hochachka, P. W., Buck, L. T., Doll, C. J. and Land, S. C. (1996). Unifying theory of hypoxia tolerance: molecular metabolic defense and rescue mechanisms for surviving oxygen lack. *Proc. Natl. Acad. Sci. USA* **93**, 9493–8.

Holeton, G. F. and D. J. Randall (1967a). The effect of hypoxia upon the partial pressure of gases in blood and water afferent and efferent to the gills of rainbow trout. *J. Exp. Biol.* **46**, 317–27.

Holeton, G. F., and D. J. Randall (1967b). Changes in blood pressure in the rainbow trout during hypoxia. *J. Exp. Biol.* **46**, 297–305.

Hughes, G. M. and Umezawa, S.-I. (1968). On respiration in the dragonet *Callionymus lyra* L. *J. Exp. Biol.* **49**, 565–82.

Ip, Y. K., Peh, B. K., Tam, W. L. Wong, W. P. and Chew, S. F. (2005). Effects of intra-peritoneal injection with NH4Cl, urea or NH4Cl + urea on nitrogen excretion and metabolism in the African lungfish *Protopterus dolloi*, *J. Exp. Zool.* **303A**, 272–82.

Jensen, F. B., Fago, A. and Weber, R. E. (1998). Hemoglobin structure and functions. In *Fish Physiology, Vol. 17, Fish Respiration*, ed. S.F. Perry and B.L. Tufts. San Diego: Academic Press, pp. 1–40.

Jibb, L. A. and Richards, J. G. (2008) AMP-activated protein kinase activity during metabolic rate depression in the hypoxic goldfish, *Carassius auratus. J. Exp. Biol.* **211**, 3111–22.

Johansson, D., Nilsson, G. E. and Törnblom, E. (1995). Effects of anoxia on energy metabolism in crucian carp brain slices studied with microcalorimetry. *J. Exp. Biol.* **198**, 853–9.

Ju, Z., Wells, M. C., Heater, S. J. and Walter, R. B. (2007). Multiple tissue gene expression analyses in Japanese medaka (*Oryzias latipes*) exposed to hypoxia. *Comp. Biochem. Physiol.* **145C**, 134–44.

Korshunov, S. S., Korkina, O. V., Ruuge, E. Kskulachev, V. P. and Starkov, A. A. (1998). Fatty acids as natural uncouplers preventing generation of O_2^- and H_2O_2 by mitochondria in the resting state. *FEBS Lett.* **435**, 215–18.

Kottelat, M., Britz, R., Hui, T. H. and Witte, K.-E. (2006). *Paedocypris*, a new genus of Southeast Asian cyprinid fish with a remarkable sexual dimorphism, comprises the world's smallest vertebrate. *Proc. R. Soc. Lond. B, Biol. Sci.* **273**, 895–9.

Kramer, D. L. and McClure, M. (1982). Aquatic surface respiration, a widespread adaptation to hypoxia in tropical fishes. *Env. Biol. Fish.* **7**, 47–55.

Krauss, S., Zhang, C. Y. and Lowell, B. B. (2005). The mitochondrial uncoupling-protein homologues. *Nature Rev. Mol. Cell Biol.* **6**, 248–61.

Kudo, H., Kato, A. and Hirose, S. (2007). SourceFluorescence visualization of branchial collagen columns embraced by pillar cells. *J. Histochem. Cytochem.* **55**, 57–62.

Kutty, M. N. (1968). Respiratory quotients in goldfish and rainbow trout. *J. Fish. Res. Bd. Can.* **25**, 1689–728.

Lai, J. C, Kakuta, I., Mok, H. O., Rummer, J. L. and Randall, D. (2006). Effects of moderate and substantial hypoxia on erythropoietin levels in rainbow trout kidney and spleen. *J. Exp. Biol.* **209**, 2734–8.

Lam, K. K. Y. (1999). Hydrography, nutrients and phytoplankton, with special reference to a hypoxic event, at an experimental artificial reef at Hoi Ha Wan, Hong Kong. *Asian Mar. Biol.* **16**, 35–64.

Landry C. A., Steele S. L., Manning S. and Cheek, A. O. (2007). Long term hypoxia suppresses reproductive capacity in the estuarine fish, *Fundulus grandis*. *Comp. Biochem. Physiol.* **148A**, 317–23.

Lannig, G., Bock, C., Sartoris, F. J. and Pörtner, H. O. (2004). Oxygen limitation of thermal tolerance in cod, *Gadus morhua* L., studied by magnetic resonance imaging and on-line venous oxygen monitoring. *Am. J. Physiol.* **287**, R902–10.

Lu, G., Mak, Y. T., Wai, S. M., *et al.* (2005). Hypoxia-induced differential apoptosis in the central nervous system of the sturgeon (*Acipenser shrenckii*). *Microscopy Res. Tech.* **68**, 258–63.

Matey, V., Richards, J. G., Wang, Y. X., *et al.* (2008). The effect of hypoxia on gill morphology and ionoregulatory status in the Lake Qinghai scaleless carp, *Gymnocypris przewalskii*. *J. Exp. Biol.* **211**, 1063–74.

McKenzie, D. J., Wong, S, Randall, D. J., Egginton, S, Taylor E. W. andd Farrell, A.P. (2004). The effects of sustained exercise and hypoxia upon oxygen tensions in the red muscle of rainbow trout. *J. Exp. Biol.* **207**, 3629–37.

Muusze, B., Marcon, J., Van den Thillart, G. and Almeida-Val, V. (1998). Hypoxia tolerance of Amazon fish, respirometry and energy metabolism of the cichlid *Astronotus ocellatus*. *Comp. Biochem. Physiol.* **120A**, 151–6.

Nikinmaa, M. and Salama, A. (1998). Oxygen transport in fish. In *Fish Physiology, Vol. 17 Fish Respiration*, ed. S. F. Perry and B.L. Tufts. San Diego: Academic Press, pp. 141–84.

Nilsson, S. (1986). Control of gill blood flow. In *Fish Physiology: Recent Advances*, ed. S. Nilsson and S. Holmgren. London: Croom Helm, pp. 87–101.

Nilsson, G. E. (1991). The adenosine receptor blocker aminophylline increases anoxic ethanol production in crucian carp. *Am. J. Physiol.* **261**, R1057–60.

Nilsson, G. E. (1992). Evidence for a role of GABA in metabolic depression during anoxia in crucian carp (*Carassius carassius* L.). *J. Exp. Biol.* **164**, 243–59.

Nilsson, G. E. (1996). Brain and body oxygen requirements of Gnathonemus petersii, a fish with an exceptionally large brain. *J. Exp. Biol.* **199**, 603–7.

Nilsson, G. E. (2007). Gill remodeling in fish – a new fashion or an ancient secret? *J. Exp. Biol.* **210**, 2403–9.

Nilsson, G. E. and Östlund-Nilsson, S. (2008). Does size matter for hypoxia tolerance in fish? *Biol. Rev.* **83**, 173–89.

Nilsson, G. E. and Renshaw, G. M. C. (2004). Hypoxic survival strategies in two fishes: extreme anoxia tolerance in the North European crucian carp and natural hypoxic preconditioning in a coral-reef shark. *J. Exp. Biol.* **20**, 3131–9.

Nilsson, G. E., Hobbs, J.-P. A. and Östlund-Nilsson, S. (2007a). A tribute to P. L. Lutz: Respiratory ecophysiology of coral-reef teleosts. *J. Exp. Biol.* **210**, 1673–86.

Nilsson, G. E., Hylland, P. and Löfman, C. O. (1994). Anoxia and adenosine induce increased cerebral blood flow in crucian carp. *Am. J. Physiol.* **267**, R590–5.

Nilsson, G. E., Rosén, P. and Johansson, D. (1993). Anoxic depression of spontaneous locomotor activity in crucian carp quantified by a computerized imaging technique. *J. Exp. Biol.* **180**, 153–63.

Nilsson, G. E., Crawley, N., Lunde, I. G. and Munday, P. L. (2009). Elevated temperature reduces the respiratory scope of coral reef fishes. *Global Change Biol.* **15**, 1405–12.

Nilsson, G. E., Hobbs, J.-P. A., Munday, P. L. and Östlund-Nilsson, S. (2004). Coward or braveheart: extreme habitat fidelity through hypoxia tolerance in a coral-dwelling goby. *J. Exp. Biol.* **27**, 33–9.

Nilsson, G. E., Östlund-Nilsson, S., Penfold, R. and Grutter, A. S. (2007b). From record performance to hypoxia tolerance – respiratory transition in damselfish larvae settling on a coral reef. *Proc. R. Soc. Lond. B, Biol. Sci.* **274**, 79–85.

Olson, K. R. (2008). Hydrogen sulfide and oxygen sensing: implications in cardiorespiratory control. *J. Exp. Biol.* **211**, 2727–34.

Olson, K. R., Dombkowski, R. A., Russell, M. J., *et al.* (2006). Hydrogen sulfide as an oxygen sensor/transducer in vertebrate hypoxic vasoconstriction and hypoxic vasodilation. *J. Exp. Biol.* **209**, 4011–23.

Ong, K. J., Stevens, E. D. and Wright, P. A. (2007). Gill morphology of the mangrove killifish (*Kryptolebias marmoratus*) is plastic and changes in response to terrestrial air exposure. *J. Exp. Biol.* **210**, 1109–15.

Östlund-Nilsson, S and Nilsson, G. E. (2004). Breathing with a mouth full of eggs: respiratory consequences of mouthbrooding in cardinalfishes. *Proc. R. Soc. Lond. B, Biol. Sci.* **271**, 1015–22.

Pärt, P. Tuurala, H., Nikinmaa, M. and Kiessling, A. (1984). Evidence for a non-respiratory intralamellar shunt in perfused rainbow trout gills. *Comp. Biochem. Physiol.* **79A**, 29–34.

Pelster, B. and Randall, D. J. (1998) The physiology of the Root effect. In *Fish Physiology, Vol. 17 Fish Respiration*, ed. S. F. Perry and Tufts, B. L. San Diego: Academic Press., pp. 321–48.

Perry, S. F and Gilmour K. M. (1996). Consequences of catecholamine release on ventilation and blood oxygen transport during hypoxia and hypercapnia in an elasmobranch *Squalus acanthias* and a teleost *Oncorhynchus mykiss. J. Exp. Biol.* **199**, 2105–18.

Pettersson, K. (1983). Adrenergic control of oxygen transfer in perfused gills of the cod, Gadus morhua. *J. Exp. Biol.* **102**, 327–335.

Pichavant, K., Person-Le-Ruyet, J., Le Bayon, N., *et al.* (2000) Effects of hypoxia on growth and metabolism of juvenile turbot. *Aquaculture 188*, 103–44.

Poon, W. L., Hung, C. Y., Nakano, K. and Randall, D. J. (2007). An in vivo study of common carp (*Cyprinus carpio* L.) liver during prolonged hypoxia. *Comp. Biochem. Physiol. D* **2**, 295–302.

Pörtner, H. O. and Knust, R. (2007). Climate change affects marine fishes through the oxygen limitation of thermal tolerance. *Science* **315**, 95–7.

Pörtner, H. O., Mark, F. C. and Bock, C. (2004). Oxygen limited thermal tolerance in fish? – Answers obtained by nuclear magnetic resonance techniques. *Respir. Physiol. Neurobiol.* **141**, 243–60.

Prosser, C. L., Barr, L. M., Pinc, R. D. and Lauer, C. Y. (1957). Acclimation of goldfish to low concentrations of oxygen. *Physiol. Zool.* **30**, 137–41.

Randall, D. J. (1982). The control of respiration and circulation in fish during exercise and hypoxia. *J. Exp. Biol.* **100**, 275–88.

Randall, D. J. and Daxboeck, C. (1984). Oxygen and carbon dioxide transfer across fish gills. In *Fish Physiology, Vol. XA*, ed. W. S. Hoar and D. J. Randall. New York: Academic Press, pp. pp. 263–314.

Randall, D. J. and Ip Y. K. (2006). Ammonia as a respiratory gas in water and air-breathing fishes. *Respir. Physiol. Neurobiol.* **154**, 216–25.

Randall, D. J. and Smith, J. C. (1967). The regulation of cardiac activity in fish in a hypoxic environment. *Physiol. Zool.* **40**, 104–13.

Randall, D. J., Holeton, G. F. and Stevens, E. Don. (1967). The exchange of oxygen and carbon dioxide across the gills of rainbow trout. *J. Exp. Biol.* **46**, 339–48.

Rausch, R. N., Crawshaw, L. I. and Wallace, H. L. (2000). Effects of hypoxia, anoxia, and endogenous ethanol on thermoregulation in goldfish, *Carassius auratus*. *Am. J. Physiol.* **278**, R545–55.

Renshaw, G. M. C, Kerrisk, C. B. and Nilsson, G. E. (2002). The role of adenosine in the anoxic survival of the epaulette shark, *Hemiscyllium ocellatum*. *Comp. Biochem. Physiol.* **131B**, 133–41.

Richards, J. G., Sardella, B. A. and Schulte, P.M. (2008). Regulation of pyruvate dehydrogenase in the common killifish, *Fundulus heteroclitus, during hypoxia exposure*. *Am. J. Physiol.* **295**, R979–90.

Robb, T. and Abrahams, M. V. (2003). Variation in tolerance to hypoxia in a predator and prey species: an ecological advantage of being small? *J. Fish Biol.* **62**, 1067–81.

Routley, M. H., Nilsson, G. E. and Renshaw, G. M. C. (2002). Exposure to hypoxia primes the respiratory and metabolic responses of the epaulette shark to progressive hypoxia. *Comp. Biochem. Physiol. A* **131**, 313–21.

Saint-Paul, U. (1988). Diurnal routine O_2 consumption at different O_2 concentration by *Colossoma macropomum* and *Colossoma brachypomum* (Teleostei, Serrasalmidae). *Comp. Biochem. Physiol.* **89A**, 675–82.

Saunders, R. L. (1962). The irrigation of the gills in fishes. II. Efficiency of oxygen uptake in relation to respiratory flow activity and concentrations of oxygen and carbon dioxide. *Can. J. Zool.* **40**, 817–62.

Schaack, S. and Chapman, L. J. (2003). Interdemic variation in the African cyprinid *Barbus neumayeri*: correlations among hypoxia, morphology, and feeding performance. *Can. J. Zool.* **81**, 430–40.

Schmidt-Nielsen, K. (1984). *Scaling: Why is Animal Size so Important?* Cambridge: Cambridge University Press.

Schurmann, H. and Steffensen, J. F. (1997). Effects of temperature, hypoxia and activity on the metabolism of juvenile Atlantic cod. *J. Fish Biol.* **50**, 1166–80.

Secor D. H. and Gunderson T. E. (1998). Effects of hypoxia and temperature on survival, growth, and respiration of juvenile Atlantic sturgeon, *Acipenser oxyrinchus. Fish. Bull.* **96**, 603–13.

Shang, E. H. H and Wu, R. S. S. (2004) Aquatic hypoxia is a teratogen and affects fish embryonic development *Environ. Sci. Tech.* **38**, 4763.

Shang, E. H. H., Yu, R. M. K. and Wu, R. S. S. (2006). Hypoxia affects sex differentiation and development, leading to a male-dominated population in zebrafish (*Danio rerio*). *Environ. Sci. Tech.* **40**, 3118–22.

Shepard, M. P. (1955). Resistance and tolerance of young speckled trout (*Salvelinus fontinalis*) to oxygen lack, with special reference to low oxygen acclimation. *J. Fish Res. Bd. Can.* **12**, 387–433.

Smith, R. W., Cash, P., Ellefsen, S. and Nilsson, G. E. (2009). Proteomic changes in the crucian carp brain during exposure to anoxia. *Proteomics*, **9**, 2217–29.

Smith, R. W., Houlihan, D. F., Nilsson, G. E. and Brechin, J. G. (1996). Tissue specific changes in protein synthesis rates in vivo during anoxia in crucian carp. *Am. J. Physiol.* **271**, R897–904.

Sloman, K. A., Wood, C. M., Scott, G. R., *et al.* (2006). Tribute to R. G. Boutilier: the effect of size on the physiological and behavioural responses of oscar, *Astronotus ocellatus*, to hypoxia. *J. Exp. Biol.* **209**, 1197–205.

Söderström, V., Renshaw, G. M. C. and Nilsson, G. E. (1999). Brain blood flow and blood pressure during hypoxia in the epaulette shark (*Hemiscyllium ocellatum*), a hypoxia tolerant elasmobranch. *J. Exp. Biol.* **202**, 829–35.

Soivio, A. and Tuurala, H. (1981). Structural and circulatory responses to hypoxia in the secondary lamellae of *Salmo gairdneri* gills at two temperatures. *J. Comp. Physiol. B* **145**, 37–43.

Sollid, J., Weber, R. E. and Nilsson, G. E. (2005). Temperature alters the respiratory surface area of crucian carp *Carassius carassius* and goldfish *Carassius auratus. J. Exp. Biol.* **208**, 1109–16.

Sollid, J., De Angelis, P. Gundersen, K. and Nilsson, G. E. (2003). Hypoxia induces adaptive and reversible gross-morphological changes in crucian carp gills. *J. Exp. Biol.* **206**, 3667–73.

Sørensen, C., Øverli, Ø., Summers, C. H. and Nilsson, G. E. (2007). Social regulation of neurogenesis in teleosts. *Brain Behav. Evol.* **70**, 239–46.

Stecyk, J. A. W. and Farrell, A. P. (2006). Regulation of the cardiorespiratory system of common carp (*Cyprinus carpio*) during severe hypoxia at three seasonal acclimation temperatures. *Physiol. Biochem. Zool.* **79**, 614–27.

Stecyk, J. A. W. and Farrell, A. P. (2007). Effects of extracellular changes on spontaneous heart rate of normoxia- and anoxia-acclimated turtles (*Trachemys scripta*). *J. Exp. Biol.* **210**, 421–31.

Stecyk, J. A. W., Stensløkken, K.-O., Farrell, A. P. and Nilsson, G. E. (2004). Maintained cardiac pumping in anoxic crucian carp. *Science* **306**, 77.

Stensløkken, K.-O., Sundin, L. and Nilsson, G. E. (1999). Cardiovascular and gill microcirculatory effects of endothelin-1 in Atlantic cod: evidence for pillar cell contraction. *J. Exp. Biol.* **202**, 1151–7.

Stensløkken, K.-O., Sundin, L. and Nilsson, G. E. (2006). Endothelin receptors in teleost fish: cardiovascular effects and branchial distribution. *Am. J. Physiol.* **290**, R852–60.

Stensløkken, K.-O., Sundin, L., Renshaw, G. M. C. and Nilsson, G. E. (2004). Adenosinergic and cholinergic control mechanisms during hypoxia in the epaulette shark (*Hemiscyllium ocellatum*), with emphasis on branchial circulation. *J. Exp. Biol.* **207**, 4451–61.

Stensløkken, K.-O., Ellefsen, S., Stecyk, J. A. W., Dahl, M. B., Nilsson, G. E. and Vaage, J. (2008). Differential regulation of AMP-activated kinase and AKT kinase in response to oxygen availability in crucian carp (*Carassius carassius*). *Am. J. Physiol.* **295**, R1403–14.

Storey, K. B. and Storey, J. M. (2004). Metabolic rate depression in animals: transcriptional and translational controls. *Biol. Rev.* **79**, 207–33.

Sultana, N., Nag, K., Kato, A. and Hirose, S. (2007). Pillar cell and erythrocyte localization of fugu ETA receptor and its implication. *Biochem. Biophys. Res. Comm.* **355**, 149–55.

Sundin, L. and Nilsson, S. (1992). Arterio-venous branchial blood flow in the Atlantic cod *Gadus morhua*. *J. Exp. Biol.* **165**, 73–84.

Sundin, L. and Nilsson, G. E. (1997). Neurochemical mechanisms behind gill microcirculatory responses to hypoxia in trout: in-vivo microscopy study. *Am. J. Physiol.* **272**, R576–85.

Sundin, L. and Nilsson, G. E. (1998a). Acute defence mechanisms against haemorrhage from mechanical gill injury in rainbow trout. *Am. J. Physiol.* **275**, R460–5.

Sundin, L. and Nilsson, G. E. (1998b). Endothelin redistributes blood flow through the lamellae of rainbow trout gills: evidence for pillar cell contraction. *J. Comp. Physiol. B* **168**, 619–23.

Taylor, E. W. and Barrett, D. J. (1985). Evidence of a respiratory role for the hypoxic bradycardia in the dogfish *Scyliorhinus canicula* L. *Comp. Biochem. Physiol.* **80A**, 99–102.

Taylor, J. C. and Miller, J. M. (2001). Physiological performance of juvenile southern flounder, *Paralichthys lethostigma*, in chronic and episodic hypoxia. *J. Exp. Mar. Biol. Ecol.* **258**, 195–214.

Tervonen, V., Ruskoaho, H., Lecklin, T., Ilves, M. and Vuolteenaho, O. (2002). Salmon cardiac natriuretic peptide is a volume-regulating hormone. *Am. J. Physiol.* **283**, E353–61.

Thomas, P., Rahman, S., Kummer, J. and Khan, I., (2005). *Neuroendocrine Changes Associated with Reproductive Dysfunction in Atlantic Croaker after Exposure to Hypoxia.* Baltimore: Society of Environmental Toxicology and Chemistry, p. 59.

Thompson, E. B. (1998) Special topic: apoptosis, *Ann. Rev. Physiol.* **60**, 525–32.

Tsui, T. K. N., Randall, D. J., Hanson, L., Farrell, A. P., Chew, S. F. and Ip, Y. K. (2004) Dogmas and controversies in the handling of nitrogenous wastes: ammonia tolerance in the oriental weatherloach, *Misgurnus anguillicaudatus*. *J. Exp. Biol.* **207**, 1977–83.

Tufts, B. L. and Perry, S. F. (1998) Carbon dioxide transport and excretion. In *Fish Physiology, Vol. 17 Fish Respiration*, ed. S. F. Perry and B. L. Tufts. New York: Academic Press, pp. 229–81.

Tufts, B. L. and Randall, D. J. (1989) The functional significance of adrenergic pH regulation in fish erythrocytes. *Can. J. Zool.* **67**, 235–8.

Tuurala, H., Egginton, S. and Soivio, A. (1998). Cold exposure increases branchial water-blood barrier thickness in the eel. *J. Fish Biol.* **53**, 451–5.

Ultsch, G. R., Jackson, D. C. and Moalli, R. (1981). Metabolic oxygen conformity among lower vertebrates–the toadfish revisited. *J. Comp. Physiol.* **142**, 439–43.

Val, A. L. (2000). Organic phosphates in the red blood cells of fish. *Comp. Biochem. Physiol.* **125A**, 417–35.

Val, A. L. and Almeida-Val, V. M. F. (1995). *Fishes of the Amazon and their Environment: Physiological and Biochemical Features*. Heidelberg: Springer Verlag.

Van der Linden, A. Verhoye M. and Nilsson, G. E. (2001). Does anoxia induce cell swelling in carp brains? Dymanic in vivo MRI measurements in crucian carp and common carp. *J. Neurophysiol.* **85**, 125–33.

Van der Meer, D. L., Van den Thillart, G. E., Witte, F., *et al.* (2005). Gene expression profiling of the long-term adaptive response to hypoxia in the gills of adult zebrafish. *Am. J. Physiol.* **289**, R1512–19.

Van Ginneken, V., Nieveen, M., VanEersel, R., Van den Thillart, G. and Addink, A. (1996). Neurotransmitter levels and energy status in brain of fish species with and without the survival strategy of metabolic depression. *Comp. Biochem. Physiol.* **114A**, 189–96.

Van Waversveld, J., Addink, A. D. F. and Van den Thillart, G. (1989). Simultaneous direct and indirect calorimetry on normoxic and anoxic goldfish. *J. Exp. Biol.* **142**, 325–35.

Verheyen, E., Blust, R. and Decleir, W. (1994). Metabolic rate, hypoxia tolerance and aquatic surface respiration of some lacustrine and riverine African cichlid fishes. *Comp. Biochem. Physiol. A* **107**, 403–11.

Wahlqvist, I. and Nilsson, S. (1980). Adrenergic control of the cardio-vascular system of the Atlantic cod, *Gadus morhua*, during 'stress'. *J. Comp. Physiol.* **137**, 145–50.

Walsh, P. J., Veauvy, C. M., McDonald, M. D., Pamenter, M. E., Buck, L. T. and Wilkie, M. P. (2007). Piscine insights into comparisons of anoxia tolerance, ammonia toxicity, stroke and hepatic encephalopathy. *Comp. Biochem. Physiol.* **147A**, 332–43.

Wang, S-H., Yuen, S. F., Randall, D. J., *et al.* (2008) Hypoxia inhibits fish spawning via LH-dependent final oocyte maturation. *Comp. Biochem. Physiol.* **148C**, 363–9.

Watson, W. and Walker Jr, J. J. (2004). The world's smallest vertebrate, *Schindleria brevipinguis*, a new paedomorphic species in the family Schindleriidae (Perciformes: Gobioidei). *Records of the Australian Museum* **56**, 139–42.

Winemiller, K. O. (1989). Development of dermal lip protuberances for aquatic surface respiration in South American characid fishes. *Copeia*, **1989**, 382–90.

Wood, C. M. (2001) Toxic responses of the gill. In *Target Organ Toxicity in Marine and Freshwater Teleosts, Vol. 1 – Organs*, ed. D. Schlenk and W. H. Benson. London: Taylor & Francis, pp. 1–89.

Wu, R. S. S., Zhou, B. S., Randall, D. J., Woo, N. Y. S. and Lam, P. K. S. (2003). Aquatic hypoxia is an endocrine disruptor and impairs fish reproduction. *Envir. Sci. Tech.* **37**, 1137–41.

Yoshikawa, H., Ishida, Y., Kawata, K., Kawai, F. and Kanamori, M. (1995). Electroencephalograms and cerebral blood-flow in carp, *Cyprinus carpio*, subjected to acute hypoxia. *J. Fish Biol.* **46**, 114–22.

Zhou, B. S., Wu, R. S. S., Randall, D. J. and Lam, P. K.S. (2001) Bioenergetics and RNA/DNA ratios in the common carp (*Cyprinus carpio*) under hypoxia *J. Comp. Physiol. B* **171**, 49–57.

6

Breathing air in water and in air: the air-breathing fishes

JEFFREY B. GRAHAM AND NICHOLAS C. WEGNER

6.1 Introduction

Air breathing is an auxiliary respiratory mode utilized by some fishes when environmental factors such as exposure to hypoxic water or emergence impede aquatic respiration. All of the 28,000 living fish species use gills to exchange O_2 and CO_2 with their aqueous environment. However, nearly 400 species, distributed among 50 families and spanning 17 orders of bony fishes (Osteichthyes), are known to be capable of breathing air. Air breathing enables these fishes to survive in and occupy habitats in which aquatic respiration cannot be used to sustain aerobic metabolism. Among all air-breathing fishes, the principal causal factor associated with this specialization is exposure, at some point during their life history, to either chronic or periodic environmental hypoxia.

A chapter on air breathing in fishes is essential for a book about vertebrate adaptation to hypoxia, because fishes are the basal vertebrates and were also the first vertebrates to breathe air (Graham, 1997). The recent literature contains substantive accounts of the adaptations for air breathing (Graham, 1997; Graham, 1999; Graham, 2006) and emersion from water (Sayer, 2005) in fishes. Using three cases studies, this chapter shows how both hypoxia and aerial O_2 access have shaped the behavior, physiology, and natural history of different fish groups.

6.2 Oxygen and water

With the increasing overlap in disciplines such as comparative physiology, field ecology, and environmental biology, there is a need for precise

Respiratory Physiology of Vertebrates: Life with and without Oxygen, ed. Göran E. Nilsson. Published by Cambridge University Press. © Cambridge University Press 2010.

quantitative terminology describing the properties of water affecting respiration. Chapter 1 showed that water has a much lower O_2 capacitance than air. Air contains 20.95% O_2, or about 210 ml l^{-1}. By contrast, air-saturated fresh water at 25°C holds only 6.3 ml O_2 l^{-1} and air-saturated sea water at 25°C has only 5.0 ml O_2 l^{-1}. The phrase 'air-saturated water' used here is important because it signifies the need to differentiate between the total volume of O_2 that is dissolved in water (i.e. ml O_2 l^{-1}) and the O_2 partial pressure (PO_2) of water, which is the fraction of total atmospheric pressure attributable to O_2. At sea level, the atmospheric pressure is 1 atm (= 760 Torr $\approx$ 760 mmHg = 101.3 kilopascals [Pa = 1 N m^{-2}]). If this atmosphere is 100% saturated with water vapor and its temperature is 25°C, its PO_2 is calculated by:

$$PO_2 = (760 - 24) \times 0.2095 = 154.2 \text{ mmHg} (= 20.6 \text{ kPa} = 0.206 \text{ atm})$$
(6.1)

where 760 is total atmospheric pressure in mm Hg, 24 is the vapor pressure of water in mm Hg at 25°C, and 0.2095 is the O_2 percentage in air. Water at 25°C that is air saturated (i.e., in full diffusive equilibrium with air) has the same O_2 partial pressure (154.2 mm Hg = 20.6 kPa) as air. It is water's low O_2 solubility that reduces its total O_2 content; the relationship between PO_2 and O_2 solubility is described by Henry's law:

$$[O_2] = \alpha PO_2$$
(6.2)

where $[O_2]$ is the O_2 concentration (ml l^{-1}) and α is the O_2 solubility or capacitance coefficient (i.e. the volume of O_2 dissolved in water; ml l^{-1} atm^{-1}). This coefficient is reduced by increased temperature and salinity, as evident from Table 1.1 in Chapter 1. Thus, interpretation of information on water O_2 concentration must be accompanied by data on water temperature and salinity. However, water PO_2 can be readily contrasted with atmospheric PO_2, and this defines both the gradient and direction for O_2 diffusion. It is the relative PO_2 of water and air that gives meaning to the frequently used term 'percent saturation' (= PO_2 water/PO_2 air × 100) and gives rise to easily understood terms describing the amount of O_2 in water relative to air, such as 'normoxic' (i.e. from 100 down to about 70% saturation), 'hyperoxic' (water PO_2 > air, also termed 'supersaturated'), and 'hypoxic' (water PO_2 less than about 70% saturation). In addition, hypoxia levels can be further defined by wording such as 'moderate,' 'severe,' 'extreme', and 'anoxic' (zero O_2). Finally, it is the PO_2-diffusion gradient between water or air and the blood on the other side of a respiratory surface that determines the starting point for the step decreases in PO_2 that occur at each station along the flow cascade of O_2 from the respiratory organ to its ultimate destination, the mitochondria.

6.3   Aquatic hypoxia and air breathing in fishes

6.3.1   Habitat

A recent review of fish respiration (Graham, 2006) chronicles the occurrence of hypoxia in different aquatic habitats, including the ocean depths; the present discussion is therefore limited to hypoxia in shallow-water habitats, where – because air is accessible – air breathing occurs. Oxygen is depleted from the water mainly by biological oxygen demand (BOD; the respiration carried out by the resident biota, including bacterial decomposition). Oxygen is added to the water by the photosynthesis of aquatic plants and by diffusion from the atmosphere, which is aided by convective and mixing processes, such as currents and tidal flows, and by wind- and thermally driven circulation. Depending upon the balance between O_2 sources and sinks, shallow waters can become hyperoxic during the day and hypoxic or anoxic at night. Isolated intertidal pools, for example, can become hyperoxic (i.e. high rates of photosynthetic O_2 production in an enclosed volume) during a daytime low tide, but severely hypoxic during a nocturnal low tide (Congleton, 1980; Nilsson et al., 2007). Similarly, the mixing and permeation of ocean water through a submerged mud burrow during high tide can maintain the O_2 of the burrow water at a level adequate for fish respiration. However, during low tide, when the burrow water stagnates, the high BOD and reducing potential of the mud quickly exhausts the O_2 supply, requiring the fish to come to the mouth of its burrow and breathe air (Gonzales et al., 2006) or to emerge onto land (Lee et al., 2005).

Hypoxia commonly occurs in tropical and temperate swamps and flooded forests that are heavily vegetated and have both low light penetration and low mixing. Tropical waters do not undergo large temperature excursions, but oxygenation in many freshwater habitats is altered seasonally by the rainy (flowing water) and dry (stagnant water) conditions, which can reduce a flowing stream to a series of isolated pools, heated by the sun to 41°C or warmer and choked with organisms (Moritz and Linsenmair, 2007). The flooded papyrus forests of central Africa and adjacent areas are perpetually hypoxic because of the low light penetration of the canopy and a high density of organisms living in the water. Comparable conditions of permanent hypoxia occur in regions of the Amazon, where fish populations exist in a dissolved O_2 content ranging from 0.4 to 2.0 ml O_2 l^{-1} (~5–36% saturation, depending on temperature), with the added stress of 28–31°C or warmer water that has a low pH of 3.5 (Chapman and Liem, 1995; Val and Almeida-Val, 1995; Graham, 2006).

6.3.2   Circumstances of air breathing in fishes

Fishes breathe air while either in water (aquatic air breathers) or on land (amphibious air breathers). Aquatic air breathers gulp or aspirate air at the water surface and deposit it into their air-breathing organ (ABO), from which the O_2 is absorbed across a respiratory epithelium and into the blood for distribution to the body. When the bulk of the O_2 contained in the air breath has been absorbed, the fish exhales and takes another breath.

There are two categories of aquatic air breathers: facultative and continuous. Facultative air breathers normally respire aquatically and use air breathing only when required by environmental conditions, principally hypoxia. Continuous air breathers regularly inspire air, even in normoxic water, and this breathing behavior is usually, but not always, a trait of species living in habitats where hypoxia is either chronic or a frequent occurrence (Seymour et al., 2008).

The 'in-series' or loop pattern of fish circulation (Graham, 1997; Graham, 2006; Farrell, 2007) causes a problem for most air breathers in that O_2 gained in the ABO enters the venous circulation and is thus at risk of dilution by mixing with other venous (hypoxic) bloodstreams returning to the heart. Also, before reaching the body, this blood must pass through the gills, where the possibility exists for the outward diffusion of O_2 to hypoxic water. Accordingly, specializations to lessen contact between the ABO and other venous streams and to decrease the potential for transbranchial O_2 loss are signature features of the heart, gill structure and circulation of the three extant lungfish genera (Neoceratodus, Lepidosiren, Protopterus) and a few other genera such as Polypterus and Erpetoichthys (bichirs and ropefish), Gymnarchus (African knifefish), Amia (bowfin), and Channa (snakehead) (Graham, 1997). Reducing gill area also lessens transbranchial loss potential, and some continuous air breathers have such a small gill area that they are obligatory air breathers (i.e. they drown without air access) (Graham, 1997). Another O_2-conserving mechanism used by some species involves modulation of the relative rates of water flow (V) and blood flow (Q) through the gills over the period that each air breath is held; in this case the V:Q mismatch functions to push the O_2-rich blood through under-ventilated gills with minimal O_2 loss to water (Graham, 2006; McKenzie et al., 2007).

Amphibious air breathers must respire without water contact and thus risk desiccation as well as the interruption of normal gill function in CO_2 release and both ion and acid–base balance (Graham, 1997; 2006). Included among the amphibious air breathers are lungfish, swamp eels (Synbranchidae), loricariid catfish (Hypostomus), and a number of other species occurring in lowland tropical swamps that experience extreme dry season conditions that can completely dehydrate small pools (Graham, 1997; Ip et al., 2004; Sayer, 2005). Many intertidal fishes also become exposed to air during low tide (Graham, 1976; Martin, 1995; Yoshiyama et al., 1995; Hill et al., 1996; Halpin and Martin, 1999; Sayer, 2005), and fishes residing in a congested littoral pool that becomes progressively hypoxic during low tide may initially respond by facultative air breathing and then emerge from the water when conditions become more extreme (e.g. the sculpins *Clinocottus* and *Oligocottus*, the mudsucker goby *Gillichthys*) (Congleton, 1980; Yoshiyama *et al.*, 1995; Gracey *et al.*, 2001; Gracey, 2008). The mangrove killifish (*Kryptolebias* = *Rivulus*) occupies unique, above-the-waterline habitats, such as in leaf litter and in the termite burrows of floating logs (Ong *et al.*, 2007; Taylor *et al.*, 2008). Although amphibious air breathing is, in many cases, associated with hypoxic stress imposed by the disappearance of water from a habitat, the natural behavior of a number of intertidal fishes such as the rockskipper (*Mnierpes* = *Dialommus*) and the mudskippers (*Periophthalmus*, *Periophthalmodon*) extends to volitional emergence in order to exploit the resources occurring at the air–water interface (Graham, 1976, 1997; Sayer, 2005; Graham *et al.*, 2007).

6.4   The air-breathing fish panorama

The modern synthesis of fish air breathing interprets laboratory studies and field data in the larger context of the fossil record, paleoclimatology, and the evolutionary relationships of extant fishes.

6.4.1   *Phylogeny*

Many aspects of vertebrate respiratory adaptation are rooted in the evolutionary history of fishes. Figure 6.1 depicts the history of fish evolution summarized from accounts by Gilbert (1993), Long (1995), Janvier (2007), and others. Fishes are the first vertebrates; the acquisition of a body-supporting vertebral column distinguishes them from the lower chordates from which they evolved in the early Cambrian Period (more than 500 million years ago, mya) (Carroll, 1988). The first fishes were jawless and probably used branchial filtration for feeding and respiration. The Devonian Period (Fig. 6.1), which is referred to as the 'Age of Fishes,' saw the simultaneous occurrence of jawless fishes, the placoderms – which were the first fishes with jaws and the sister group of all other jawed vertebrates (gnathostomes) – the acanthodians (spiny sharks), and both the early chrondrichthyans (sharks) and bony fishes (osteichthyans = Euteleostomi = Osteichthyes) (Nelson, 2006; Janvier, 2007), including the latter's two major groups, the sarcopterygians (lobefins) and the actinopterygians (ray-fins). Also present in Devonian waters was an assemblage of sarcopterygians, the Tetrapodomorpha, that were closely related to lungfish and gave rise to the early tetrapods, which first invaded land in the Late Devonian (360 mya) (Clack, 2002; Graham and Lee, 2004; Janvier, 2007). Whereas chrondrichthyans, osteichthyans, and lungfish all diversified in the Devonian and are still in existence, both the placoderms and acanthodians went extinct before the end of the Paleozoic (Fig. 6.1).

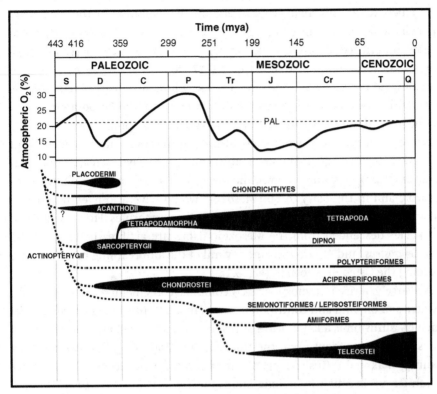

Fig. 6.1 Fish evolutionary record and atmospheric O_2 levels from the beginning of the Silurian (443 mya) to present. Dotted lines show hypothesized phylogenetic relationships among groups. Fossil occurrences and probable diversities of different groups are indicated by relative band thickness. Compiled from Carroll (1988), Gilbert (1993), Helfman *et al.* (1997), Nelson, (2006), and Janvier (2007). O_2 levels (Ward *et al.*, 2006; Berner *et al.*, 2007) are shown relative to present atmospheric level (PAL). Geologic period abbreviations: S, Silurian; D, Devonian; C, Carboniferous; P, Permian; Tr, Triassic; J, Jurassic; Cr, Cretaceous; T, Tertiary; Q, Quaternary.

The Mesozoic Era (251–65 mya) saw the sequential appearance, radiation, diversification, and then contraction of different rayfinned groups, including the Acipenseriformes (sturgeons, *Acipenser*), Semionotiformes (the ancestral group of gars, *Lepisosteus* and *Atractosteus*), and Amiiformes (bowfin, *Amia*). Teleosts (Teleostei), which also originated in the Mesozoic, are the largest subdivision of the extant bony fishes and comprise about 96% of all living fish species (Nelson, 2006); their major radiation occurred in the Cenozoic Era ($\leq$65 mya; Fig. 6.1).

6.4.2   *The paleoatmosphere and air breathing in fishes*

Geochemical data allow approximation of Earth's atmospheric O_2 level over geologic time, and this provides added perspective to the possible role of both hyperoxic and hypoxic atmospheres in the evolution of the biota. The atmospheric O_2 record since early in the Paleozoic (Fig. 6.1) shows that there have been both reductions and increases in O_2 relative to the present atmospheric level (PAL) of 20.95%. Lows of 12–17% O_2 occurred in the Mid–Late Devonian and Early Triassic, and from the Early Jurassic to Early Cretaceous. A steady O_2 increase to levels greater than 30% began in the Carboniferous and continued until the Early Permian, and then O_2 dropped precipitously.

A lower atmospheric O_2 would have affected respiration and metabolism and further exacerbated aquatic-hypoxia effects and could have increased selection for air breathing in groups such as the lungfish. Groups appearing in the Late Paleozoic and in the Early to Mid-Mesozoic would have also lived in a lower atmospheric O_2. With the exception of the Acipenseriformes (Fig. 6.1), most of the species descended from the Mesozoic fishes are air breathers, including *Amia*, *Lepisosteus*, and *Atractosteus*, and several of the air-breathing basal teleosts, including the Osteoglossiformes (Fig. 6.1).

Conversely, a rise in atmospheric O_2 could have augmented aerobic activity in water and on land and contributed to metabolic transitions in organisms in which O_2 diffusion is a key feature of respiration. Insect respiration, for example, is highly dependent upon the tracheal diffusion of O_2, and two features requiring maximal diffusion, gigantism and flight, both occurred during the Carboniferous–Permian hyperoxic atmosphere (Graham *et al.*, 1995; Dudley, 1998; Kaiser *et al.*, 2007). The invasion of land by tetrapods occurred in the Devonian, when atmospheric O_2 was low. However, the major radiation of early tetrapods took place in the Carboniferous, when atmospheric O_2 was rising. Hypotheses linking early tetrapod evolution to atmospheric O_2 have been forwarded by Graham *et al.* (1995), Huey and Ward (2005), Ward *et al.* (2006), and Berner *et al.* (2007).

6.4.3   Diversity of air-breathing fishes

Known air-breathing fishes are listed in Graham (1997), and a discussion of criteria used in the listing process and some possible new inclusions are found in Appendix A of this chapter. All known air-breathing fishes are osteichthyans (Grade Teleostomi, Class Actinopterygii) (Nelson, 2006). The record for extant species indicates that air breathing has evolved independently in a number of taxa. It also suggests that air breathing is likely to have first occurred early, before the group's separation into the rayfins and lobefins. The only suggestion of air breathing at an earlier stage in vertebrate evolution is the report of paired, lung-like structures in fossils of the Devonian placoderm *Bothriolepis*. However, the interpretation of these structures as 'lungs' remains controversial, and additional confirmation is needed (Graham, 1997; Perry,

2007). Many well-preserved *Bothriolepis* fossils exist, and it would be feasible to obtain refined morphological data on these structures using high-resolution X-ray computed tomography.

Because placoderms have long been regarded as ancestral to the chrondrich-thyans, the report of 'lungs' in *Bothriolepis*, together with knowledge that no air-breathing structures occur in extant chondrichthyans, prompted speculation that air breathing evolved in placoderms, was lost in chondricthyans, and reappeared in the osteichthyans (Wells and Dorr, 1985; Liem, 1988; Graham, 1997). However, support for this hypothesis has never been strong (Liem, 1988; Liem, 1989; Perry, 2007) and, given the lesson of independent origin provided by extant air breathers (Graham, 1997), the presence of this specialization in a single placoderm does not define the entire group's character state. Also, the absence of air breathing among extant sharks and rays does not rule out the possibility that this adaptation could have been present in a species such as *Orthacanthus*, a large, eel-like Paleozoic shark that lived in freshwater swamps.

6.4.4   *The lung and the respiratory gas bladder*

Fishes use both lungs and gas bladders as ABOs (Graham, 1997). Characteristics of the fish lung include: (1) development as a ventral outpock-eting of the posterior embryonic pharynx; (2) growth into a paired or a bilobed structure that extends into the posterior, mainly lower, part of the peritoneal cavity; (3) a tube (pneumatic duct) connecting the organ to the pharynx that is guarded by a glottis valve; and (4) perfusion by a pulmonary circulation (the arteries of this originate from efferent branchial arches 3 and 4, while a pulmonary vein returns O_2-rich blood to or near the heart). Among the sarcop-terygians, the extant lungfishes use a lung for air breathing but also use their gills and skin for aquatic respiration. Also, while it is not an air breather, the coelacanth (*Latimeria*; a primitive lobefin related to lungfish and tetrapods) has a vestigial, fat-invested, lung-like organ connected to a pulmonary circulation. Among actinopterygians, only *Polypterus* and *Erpetoichthys* utilize auxiliary lung respiration.

Details for the fish respiratory gas bladder include: (1) development as an outpocketing of the dorsal or lateral wall of the posterior embryonic pharynx; (2) posterior growth as a single or bilobed tube occurring within the dorsal mesentery and occupying the upper sector of the peritoneum; (3) a pneumatic-duct connection (i.e. a physostomous gas bladder) that may be long or short and generally without the glottis; and (4) a blood circulation that is, in most cases, via a non-pulmonary flow and in series with the systemic loop (i.e. in most species supply via the dorsal aorta to the celiac or gas bladder arteries, drainage into the hepatic portal system, or the post cardinal vein [Graham, 1997]).

6.4.5   Lung and gas bladder homology

Figure 6.2 shows the phyletic distribution of lungs and respiratory gas bladders among fishes and tetrapods. No fossil evidence exists for the presence of lungs in the early vertebrates; however, criteria such as the site of differentiation (posterior embryonic pharynx), the pattern of early development (ventral midline lung buds with glottis), and morphology (pulmonary septa and ediculae) all support the homology of the lungs of lungfish and the tetrapods (Fig. 6.2), and it is this homology that firmly anchors comparative vertebrate respiratory physiology in its piscine roots (Romer, 1972; Liem, 1988; Liem, 1989; Graham, 1997; Maina, 2002; Perry, 2007; Torday et al., 2007). Figure 6.2 also shows that, among the actinopterygians, polypterids have lungs, whereas respiratory gas bladders occur in at least eight orders, 11 families, and 19 genera, including both primitive non-teleosts such as *Lepisosteus* and *Amia* and six teleost orders.

In addition to respiration, gas bladders provide other functions, such as buoyancy control, sound reception, and sound production, and comparative studies demonstrate that changes in organ structure and function have occurred in concert with actinopterygian evolution, and particularly as a result of the radiation of teleosts into many diverse ecological niches (Liem, 1988; Liem, 1989; Helfman et al., 1997). In general, evolutionary progression of the gas bladder has been from respiratory to non-respiratory (although some groups appear to have secondarily returned to using the gas bladder as an ABO), and from physostomous (pneumatic duct present) to physoclistous (no duct) (Graham, 1997; Helfman et al., 1997).

The presence of lungs and gas bladders within the rayfinned fishes has been a source of controversy insofar as the evolutionary relationship of the two organs is concerned. Darwin (1859) viewed the gradual transformation of a fully functional organ for buoyancy regulation into a fully functional lung as a prime example of 'descent with change.' Such a transformation would entail both a vertical shift in organ position within the peritoneum (i.e. from dorsal to ventral) and a 180° rotation in the site of pneumatic duct attachment to the pharynx. Later, as the evolutionary evidence pointed to the lung and not the gas bladder as the primitive organ, elaborate scenarios, replete with examples of species representing transformational intermediates, were developed to explain 'descent' in the opposite direction; however, these remain unconvincing (Graham, 1997).

Recent analyses (Perry and Sander, 2004; Perry, 2007; Torday et al., 2007) indicate that, although actinopterygian lungs and gas bladders both undergo embryological differentiation in the posterior pharynx, there are far too many structural differences among them to warrant postulation of a single origin for

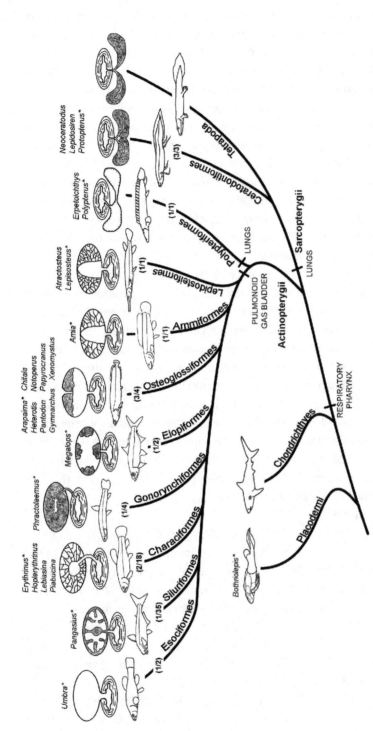

Fig. 6.2 Occurrence and structural diversity of the lungs and pulmonoid gas bladders in ten orders of extant fishes and the tetrapods. Indicated for each fish order is the relative number of families in which a respiratory gas bladder or lung occurs (e.g. species in 1 of 35 siluriform families use a respiratory gas bladder) and a list of the genera having the organ. Transverse-section drawings (anterior view) of one genus (*) in each group show the respiratory organ and the relative complexity of its parenchyma, and its position relative to the pneumatic duct and digestive tract (flattened oval ring with horizontal stippling). (Note: many extant fish lineages in which a physostomous, non-respiratory gas bladder is present are not shown.) Compiled from illustrations contained in Graham (1997), Perry (2007), and Podkowa and Goniakowska-Witalińska (1998).

the lung or to account for a sequential (by means of vertical movement and rotation) transformation of a functional lung to a functional gas bladder. Whereas the classical view has been that there was a single origin of the osteichthyan lung and this became the ABO of the tetrapods and was also the pleisomorphic condition for the actinopterygian gas bladder (Graham, 1997), Perry (2007) suggests that the posterior respiratory pharynx, the region immediately behind the functional gill arches, was a the rudimentary ABO and the common point for the separate origin of the lung and gas bladder ABOs of the osteichthyans. As shown in Fig. 6.2, Perry (2007) postulates at least three separate organ origins and suggests that a determining factor in ABO morphology was the position of embryonic organ-bud formation in the respiratory pharynx; ventral budding formed a lung (and the lungs of sarcopterygians and tetrapods originated independently of the lungs of the polypterids), whereas dorsal or lateral budding formed a gas bladder.

Perry's (2007) view is that the presence of an aerial respiratory capacity is a basal osteichthyan character and that selection pressures operating at different times in evolutionary history determined whether the ABO expressed was a lung or a gas bladder. Viewed in this context, the high level of homology demonstrated for both the ultrastructure of the gas bladder and lungs (e.g. laminated osmophilic bodies and type I and II cells) (Graham, 1997), and the remarkably similar chemical properties of the surfactant proteins in both organs (Power *et al.*, 1999; Daniels *et al.*, 2004), would be expected based on their common origin from pharyngeal tissue. Perry (2007) extends his idea about the undifferentiated respiratory pharynx by suggesting that the site of embryonic pouch formation also determined the organ's pattern of blood flow (i.e. a pulmonary circulation formed with lungs and a serial circulation formed with a respiratory gas bladder). However, this idea is not supported by *Amia*, which has a respiratory gas bladder and a pulmonary circulation (Graham, 1997).

A phyletic survey of the fishes with respiratory gas bladders (Fig. 6.2) suggests the independent origin of this ABO type in different groups. For example, in the catfish family Pangasiidae, the respiratory gas bladder occurs in the genus *Pangasius* but not in the family's other two genera (*Helicophagus, Pangasianodon*). The Pangasiidae is one of 35 families comprising the Order Siluriformes, a group consisting of over 2850 species in 446 genera contained in 35 families (Nelson, 2006). The silurid cladogram (Fig. 6.3) shows pangasiids are not a basal group and that, while there are eight siluriform families with air-breathing species, pangasiids are the only of these having a gas bladder ABO. This indicates that the origin of the respiratory gas bladder in *Pangasius* is an apomorphic character (i.e. derived and not found in an ancestral group) and that the overall diversity of the siluriform air breathers reflects an independent origin of this adaptation.

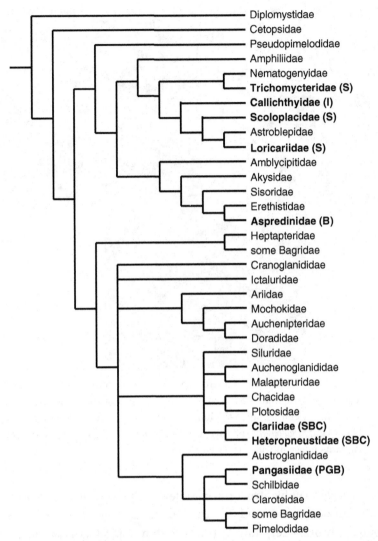

Fig. 6.3 Phylogeny of Siluriformes (modified from de Pinna (1998)) showing the families in which air breathing occurs (bold) and their air-breathing organ (ABO) type in parentheses: B, buccal chamber; I, intestine; PGB; pulmonoid gas bladder; S, stomach; SBC, suprabranchial chamber.

6.4.6   *Epithelial complexity*

Although similar in ultrastructure and cell type, the morphology of the respiratory epithelium of fish lungs and respiratory gas bladders varies considerably (Figs. 6.2, 6.4, 6.5). Lungfishes, for example, have a highly complex, three-dimensional, septated alveolar parenchyma (Grigg, 1965; Maina, 1987; Maina, 2002) (Figs. 6.2 and 6.4). By contrast, the lungs of polypterids (Figs. 6.2

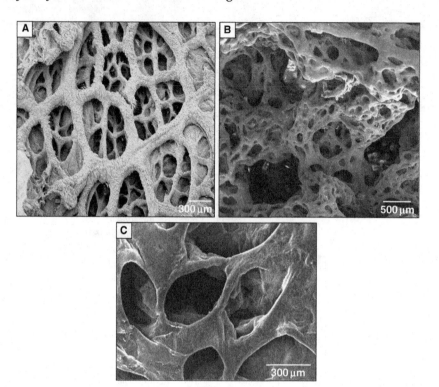

Fig. 6.4 Scanning electron microscope images of respiratory surfaces showing the alveolar-like respiratory parenchyma composed of a cartilaginous matrix covered by a respiratory epithelium in the lung of *Protopterus* (A), and in the respiratory gas bladders of *Megalops* (B) and *Pangasius* (C). (*Protopterus* image provided by J. Maina; *Pangasius* image provided by D. Podkowa.)

and 6.5) lack comparable structural complexity; the respiratory epithelium lines the walls but does not extend into the lumen (Zaccone, *et al.*, 1995). Similarly, the complexity of the respiratory surface within respiratory gas bladders ranges from *Umbra*, which has a flat epithelium that does not expand into the lumen, to the presence of highly septated ediculae in *Megalops*, *Pangasius*, *Phractolaemus*, and others (Graham, 1997; Podkowa and Goniakowska-Witalińska, 1998; Seymour *et al.*, 2008) (Figs. 6.2 and 6.4).

6.4.7   *Other ABOs: air breathing beyond the respiratory gas bladder*

The evolution of the physoclistous gas bladder, which occurred independently in many groups, eliminated the pneumatic duct and largely ended gas-bladder utility in teleost air breathing. However, as teleosts continued to radiate into every habitable body of water, there were new requirements for auxiliary aerial respiration, and these led to development of novel ABO

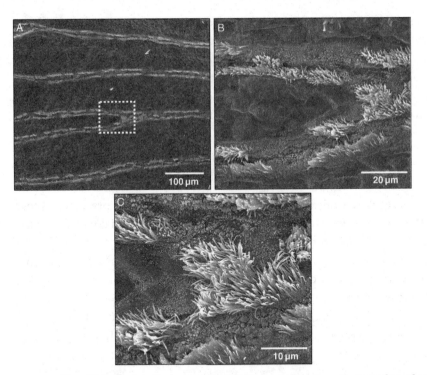

Fig. 6.5 Scanning electron microscope images of the inner lung surface of a 5.7 g *Polypterus senegalis*. (A) Ciliated furrows, one bifurcated, running parallel to the lung's axis contain granular and mucous cells and border the respiratory epithelium. (B, C) Magnifications of the boxed area in A detailing the cilia and the small folds in the respiratory epithelium.

structures. Both the structural details and diversity of these organs are reviewed in Graham (1997) and need only brief survey here. Included in the group of fishes with non-gas bladder ABOs are those that breathe air using gills (*Mnierpes* = Dialommus, rockskippers), modified gills (*Electrophorus*, electric eel, or knifefish), a specialized buccopharyngeal epithelium (*Periophthalmus*, mudskipper), or the skin (many species). In general, amphibious marine air breathers have relatively few respiratory specializations and use their gills and skin for air breathing. The eleotrid *Dormitator* has a unique skin respiration method. In hypoxic water, *Dormitator* hyper-inflates its physoclistous gas bladder and becomes positively buoyant. This emerges its forehead, where a dense cutaneous capillary network then becomes engorged with blood and functions for aerial respiration. Predominantly freshwater groups such as the anabantoids (labyrinth fishes) and channids (snakeheads), which also have a physoclistous gas bladder, have evolved elaborate suprabranchial chambers for air breathing. Among the silurids (Fig. 6.3), most of which have a closed gas bladder, there are several species of clariid catfishes (*Clarias, Heterobranchus, Dinotopterus*) that have elaborate respiratory dendrites growing out from their gill arches into suprabranchial chambers. The closely related *Heteropneustes* has paired, lung-like-projections of its branchial chamber that extend posteriorly into its myotomes. Various groups make use of the esophagus (*Dallia*, blackfish), stomach (Loricariidae, armored catfish), intestine (Callichthyidae, armored catfish; Misgurnidae, loaches) and pneumatic duct (*Anguilla*, eel) as ABOs.

5   Case studies of hypoxia and air breathing in fishes

Three case studies of air breathing in fishes are now presented. These show the link between air breathing and aquatic hypoxia and demonstrate the role that factors such as geologic time, environmental change, and ecological opportunity can have in influencing the development of this specialization. Figure 6.6 frames the connection between hypoxia and air breathing and these factors.

For a non-air-breathing fish living in normoxic water, a key, first-line response to hypoxia is behavioral: the fish readily searches for areas having higher O_2

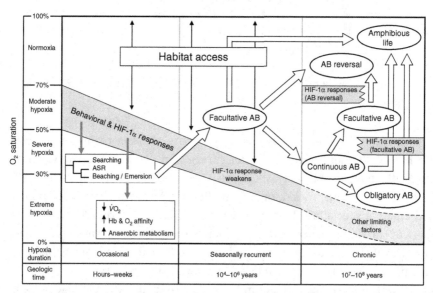

Fig. 6.6 Integrated aspects of environmental hypoxia adaptation and the selective mechanisms operating over different periods of time leading to the evolution of air breathing and terrestriality in different fish groups. Behavioral responses such as ASR that take a fish close to the surface increased the potential for inadvertent air gulping and may have been a major factor in the origin of air breathing. Specialization for hypoxia and air breathing would lead to downregulation of HIF-1α response mechanisms.

concentration, including shallow or surface waters. Many species use aquatic surface respiration (ASR), a behavior in which the mouth is positioned as close to the water surface as possible in order to ventilate the gills with the upper few millimeters of water that remains O_2 rich because of atmospheric diffusion (Graham, 1997; Graham, 2006).

Closely following the behavioral response is activation of the hypoxia-inducible factor (HIF-1α), which switches on genes with protein products that either increase O_2 transfer (i.e. erythropoiesis, Hb affinity increases, angiogenesis, etc.) or trigger metabolic adaptation (through genes controlling anaerobiosis and O_2 consumption rate [$\dot{V}O_2$]). HIF-1α induction is an ancient adaptation that first appeared in eukaryotic cells and thus long preceded the origin of metazoans (Nikinmaa and Rees, 2005; Flück et al., 2007). The efficacy of HIF-1α has been demonstrated for both water-ventilating and air-breathing fishes (Gracey et al., 2001; Gracey and Cossins, 2003; Nikinmaa and Rees, 2005; Gracey, 2008), and, as illustrated in Fig. 6.6, it would play an adaptive role during irregular and brief (i.e. hours to weeks) periods of aquatic hypoxia experienced by a non-air-breathing fishes. In such conditions HIF-1α expression is continually reinforced by natural selection (i.e. individuals with an effective HIF-1α response survive and propagate the next generation) and, among populations of species experiencing more severe hypoxia (i.e. 30–50% saturation) over an extended period, both the suite of adaptive responses encompassed by HIF-1α and the threshold of its onset would probably change (Fig. 6.6).

The origin of air breathing in many groups is probably linked to the combined interactions of behavioral and HIF-1α responses. Numerous behavioral, morphological, and physiological specializations are associated with ASR. For example, some species gulp air, either inadvertently or for the purpose of increasing buoyancy for more efficacious ASR (Burggren, 1982; Gee and Gee, 1995; Graham, 1997; Armbruster, 1998; Graham, 2006). In some cases, the O_2 contained in gulped air is also incorporated in respiration, and it is this, in combination with geologic time spans (probably on the order of 10^4–10^6 years), and exposure to severe (e.g. a combination of duration and low PO_2) and possibly regular (i.e. seasonal) hypoxia events, that has selected for facultative air breathing in groups such as the loricariid catfishes (Fig. 6.6 and Case study 1). The evolution of facultative air breathing expanded habitat accessibility and, although HIF-1α induction functions would complement air breathing, the activation threshold for these and their scope of action could be reduced because of the auxiliary O_2 access (Fig. 6.6).

Strong selection for more proficient air breathing, occurring over a vast expanse of geologic time and driven by environmental change and by the diversification and radiation of groups into different and more variably oxygenated habitats, probably also contributed to the origin of continuous air breathing and ushered in morphological and physiological changes, such as a reduction in gill area to lessen the potential for O_2 loss, leading to obligatory air breathing (Fig. 6.6). The respiratory specializations of different groups, brought on by regular exposure to environmental hypoxia over periods of 10^7–10^8 years, concomitantly reduce the importance of an aquatic hypoxia-induced HIF-1α response (e.g. Case study 2 shows that *Lepidosiren* and *Protopterus* are largely insulated from aquatic O_2 conditions) and expand habitat access to the point at which it is limited by other factors such as the absence of prey and poor water quality (the production of anoxic water and sediment constituents such as H_2S). Conversely, the gradual radiation, over geologic time, of some groups into habitats having greater O_2 access could result in the relaxation of selection pressures for air breathing. This has been documented for some African stream-dwelling clariid catfish (e.g. *Xenoclarias*, *Clariallabes*, *Gymnallabes*, and *Tanganikallabes*) (Graham, 1997), and may have also occurred in the Australian lungfish, *Neoceratodus* (Case study 2).

The transition to land is another dimension of fish air breathing, and Case study 3 explores this in gobies, a diverse group in which hypoxia and aerial respiration, in proximity to open mudflat niches, became catalysts for the origin of amphibious life.

6.5.1   Case study 1. Transition to air breathing: the loricariid model

This case study examines evolutionary and physiological aspects of facultative air breathing in the suckermouthed armored catfishes, family Loricariidae. Figure 6.3 shows loricariids as one of four siluriform families (Loricariidae, Scoloplacidae, Callichthyidae, and Trichomycteridae) that use either the stomach or intestine as an ABO. Air-breathing investigations have been conducted on species in each of these families (Graham, 1997). With regard to the two other families within this clade (Fig. 6.3), Gee (1976) determined that one astroblepid, *Astroblepus longifilis*, did not hold air in its stomach. It is not known whether air breathing occurs in other species of this family (about 54 total) or in *Nematogenys inermis*, the only species in the family Nematogenyidae.

Loricariids number about 92 genera and over 680 species (Nelson, 2006). They range in length from a few centimeters to nearly a meter and are found in a variety of habitats, from fast-flowing streams to floodplain lakes and swamps throughout the tropical regions of South America and in Panama and Costa Rica. Loricariids are mainly substrate dwellers with a dorsoventrally depressed body form. Their common name derives from plate-like scales covering their dorsal surface and a ventral mouth with large, fleshy lips used for feeding by scraping the bacterial and algal slime that coats benthic substrates. Loricariid fossils first occur in the late Paleocene to early Miocene epochs (23 mya).

6.5.1.1 Behavior, morphology and evolution

Details about respiration remain unknown for most loricariids; however, facultative air breathing has been documented for species in at least ten genera, all of which use the stomach as an ABO (Santos *et al.*, 1994; Graham, 1997; Silva *et al.*, 1997; Armbruster, 1998; Takasusuki *et al.*, 1998). Air breathing occurs among five of the six loricariid subfamilies and appears to have originated independently in each group in the course of its diversification and radiation into habitats requiring auxiliary air breathing (Armbruster, 1998). The early stages of this evolution took place after the gas bladder became highly specialized for sound detection (it is small, closed, and encased in the skull). The gas bladder thus could no longer function as an ABO and, when the environmental conditions confronting a diversifying and radiating group required air breathing, a novel ABO, the stomach, was recruited. The stomach's utility as an ABO is enhanced by both its ready contact with an air source via the esophagus and its vascularization. Loricariid utilization of the stomach as an ABO may thus parallel the first air-breathing fishes, which swallowed and deposited air in their posterior pharynx (Perry, 2007; Torday *et al.*, 2007). Also, selection for air swallowing and aerial respiration probably began when, in a hypoxia-driven search for oxygenated water in shallow areas, loricariids made surface contact and took inadvertent air gulps (Fig. 6.6) (Burggren, 1982; Graham, 1997). It may be that the inflexibility of the armor plates on the anterior body precluded selection for expanded supra- or post-branchial chambers similar to the structures serving as ABOs in other silurids (Fig 6.3). Although air breathing would seem to compromise the stomach's role in feeding, laboratory observations showed no effect of air breathing on food-ingestion rate (Graham, 1997). Also, most loricariids utilize the stomach ABO for relatively brief periods (i.e. the tropical dry season), when food is scarce, and this would lessen the potential functional conflict (Armbruster, 1998).

Comparative studies show a range of air-breathing capacities among the loricariids. Some do not breathe air (e.g. *Leptoancistrus*, *Neoplecostomus*); genera such as *Chaetostoma* and *Sturisoma*, which live in fast-flowing waters where stagnation and hypoxia are less likely, will gulp air but are not as proficient air breathers as species living in areas where stream flow is low or seasonal flooding and drying are possible (e.g. *Hypostomus*, *Ancistrus*, *Liposarcus*, *Pterygoplichthys*, *Rhinelepis*) (Gee, 1976; Graham, 1997; Armbruster, 1998).

Loricariid air breathing is associated with different behaviors. Initial exposure to hypoxia elicits searching for areas having more O_2 (Fig 6.6). Small species such as *Rineloricaria* enter shallow areas and partially beach themselves. Even though *Rineloricaria* breathes air, staying in shallow water appears to be crucial, as this species cannot survive if restricted to deeper water and required to swim repeatedly to the surface for air. Beaching thus appears to be the functional equivalent of ASR (Gee, 1976; Graham, 1997; Armbruster, 1998; Graham, 2006; Fernandes-Castilho et al., 2007) and thus facilitates both branchial and aerial respiration. Some intertidal fishes (e.g. cottids, blennies, gobies) will also beach themselves when ambient water becomes hypoxic (Graham, 1976; Graham, 1997; Congleton, 1980; Martin, 1995; Sayer, 2005).

Most loricariids breathe air by rapidly swimming to the surface, gulping air, and swallowing it while returning to the bottom. Once O_2 is depleted, the breath is 'burped' out of the operculae either before or during the ascent for a new breath. When corrected for the absorbed O_2 (there is no volume replacement by CO_2, which is lost to water via the gills), the volume of released gas is nearly the same as that inspired, which means that the stomach ABO is completely emptied after each air breath (Graham, 1983). Loricariids in close proximity to one another will frequently release air at nearly the same time and then rise en masse to air breathe synchronously; this temporal schooling behavior is an anti-predation adaptation (Graham, 1997). In addition to providing O_2, air ingestion may also occur in conjunction with feeding, as near-neutral buoyancy facilitates foraging on vertical surfaces or submerged roots or tree branches. Several loricariid genera also have air-filled diverticulae connected at the esophageal–stomach junction. A respiratory function has been suggested for some of these (Silva et al., 1997); however, their size, shape, position, and wall thickness all suggest a primary function for increasing buoyancy (Armbruster, 1998).

Studies of loricariid stomachs show an ultrastructure comparable to other ABOs (lungs and gas bladders) formed in or by outgrowths of the digestive tube. Investigations also identify respiratory areas that are thin, have a large number of capillaries for O_2 uptake, and have a reduced number of digestive cells (Armbruster, 1998; de Oliveira et al., 2001; Podkowa and Goniakowska-Witalińska, 2002; Podkowa and Goniakowska-Witalińska, 2003). A morphological survey of the digestive tracts of over 40 loricariid genera demonstrated eight graded character states in stomach size and position and venous-blood-drainage pattern related to air breathing. In all cases arterial supply to the stomach is via the celiac artery; however, venous return (O_2-rich blood) to the heart bypasses the hepatic portal circulation and is via the inter-renal vein to the post-cardinal vein (Graham, 1997).

6.5.1.2 Physiology and biochemistry of air breathing

Similar to most other fishes, loricariids respond to progressive hypoxia by increasing gill ventilation in order to sustain their aquatic $\dot{V}O_2$ down to, but not below, a critical PO_2 (PO_2crit) (Graham, 1983; Graham, 1997; Mattias et al.,

1998; Nelson *et al.*, 2007) (see Chapter 5 for a detailed description of PO_2crit). Facultative air breathing commences when the water PO_2 drops to the threshold level triggering this behavior (Graham and Baird, 1982; Mattias *et al.*, 1998; Takasusuki *et al.*, 1998). In most species the air-breathing threshold PO_2 is higher than PO_2crit (i.e. the fish begins air breathing before ambient O_2 declines to the level where routine aquatic $\dot{V}O_2$ cannot be sustained) (Graham, 2006). In most loricariids that have been studied, facultative air breathing continues while PO_2 remains at or below threshold; further reductions in PO_2 increase air-breathing frequency or, if PO_2 rises above the threshold, air breathing ceases. This demonstrates the importance of facultative air breathing in enabling a fish to acquire O_2 when it cannot be obtained by aquatic respiration in hypoxic water.

Initiation of facultative air breathing is frequently associated with the onset of an air intake-mediated cyclic shift in heart rate and ventilation. This onset is gradual (Graham, 1983), and some species appear to be polymorphic with respect to the extent to which it is invoked (Nelson *et al.*, 2007). When a fresh breath is taken the heart accelerates (i.e. air-breath tachycardia) and gill ventilation declines. After a few minutes, when the O_2 content of the breath is reduced, heart rate begins to decline and the gill ventilation rate increases. This shift limits the potential for the transbranchial loss of O_2 that might occur because of in-series circulation. Normal respiration optimizes gas transfer by closely matching the O_2 capacitances (i.e. total contents) of the blood and water flowing on opposite sides of the exchange surface (V:Q matching) (Graham, 2006). A shift in V:Q during air breathing would permit the passage of O_2-rich blood (from the ABO) through the gills during a period of low ventilation, which minimizes the potential for the transbranchial O_2 loss. This appears to be important for loricariids, because they do not have reduced gill areas. The areas of a 100 g *Rhinelepis strigosa* (20 000 mm^2) (Santos *et al.*, 1994) and a 100 g *Hypostomus plecostomus* (9000 mm^2) (Perna and Fernandes, 1996) both lie within the range of most non-air-breathing freshwater fishes and are higher than values for most air-breathing fishes (Palzenberger and Pohla, 1992; Graham, 1997).

If hypoxic conditions require a loricariid to use facultative air breathing for an extended period, it undergoes a series of metabolic changes that increase respiratory efficacy. Major among these is an increase in blood hemoglobin (Hb) concentration and a reduction in the quantity of intra-erythrocytic nucleoside triphosphates (e.g. adenosine and glutamine triphosphate [ATP and GTP]), which causes a left-shift (= increase affinity = lower P_{50} value) in Hb–O_2 affinity (Graham, 1983; Val *et al.*, 1990). The mechanism for this affinity shift is the hypoxia-induced release of catecholamines, which activate Na$^+$–H$^+$ exchangers on the red-cell membrane, causing intracellular alkalization and the entry of water, hence phosphate dilution (Nikinmaa and Rees, 2005; Brauner and

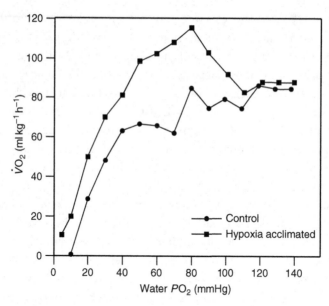

Fig. 6.7 Comparative PO_2 effects on the aquatic $\dot{V}O_2$ (without air access) of control and hypoxia-acclimated armored catfish (*Ancistrus chagresi*) at 25°C. The hypoxia-acclimated group has a higher Hb–O_2 affinity and can sustain a higher $\dot{V}O_2$ in hypoxia. Modified from Graham, 1983.

Berenbrink, 2007). The increase in affinity enables the fish to bind more O_2 in hypoxic water; this lessens the potential for transbranchial loss and reduces PO_2crit and thus makes the fish more proficient in aquatic respiration in hypoxia (Fig. 6.7). Further, after 2 weeks of air breathing the stomach of *Ancistrus* holds 25% larger air breaths (Graham, 1983). Although facultative air breathing for an extended period does not affect the air-breathing threshold, the sum effect of loricariid adaptations for air breathing is to lessen air-breathing frequency (i.e. a larger stomach volume brings more O_2 with each breath and, because of increased Hb affinity, more of this O_2 is used). Even though facultative air breathing would seem to obviate environmental-hypoxia effects, the increases in Hb and Hb–O_2 affinity in loricariids during prolonged hypoxia exposure are consistent with HIF-1α expression. The increases in Hb that occur in natural populations of air-breathing fishes (*Ancistrus*, *Hypostomus*, and *Dormitator*) during the tropical dry season may precondition them for hypoxia and the need to breathe air (if this occurs), and may also be related to an HIF-1α induction mechanism linked to seasonal change (Graham, 1985).

In summary, hypoxia adaptation in most loricariids depends upon facultative air breathing, which is in turn keyed to the overarching influence of aquatic conditions on aquatic respiration. This is reflected in the occurrence of both behavioral and HIF-1α-induced hypoxia adaptations that lessen the need for air breathing by enhancing aquatic respiration.

6.5.2   Case study 2. Ancient air breathers: the lungfish

Elaborate modifications in the lungs, gills, heart, and circulation provide lungfishes (Order Ceratodontiformes) with an air-breathing efficacy that distinguishes them from nearly all other air-breathing fishes (Graham, 1997). This case study examines comparative aspects of hypoxia responses and air breathing among the three extant lungfish genera.

Lungfish fossils first appear in the Devonian (416–359 mya), and several of the earliest forms resemble living species (Long, 1995; Janvier, 2007). Living lungfishes are classified in three separate families and genera, each of which occurs on a different continent: *Protopterus* (Protopteridae, Africa; four species: *P. aethiopicus*, *P. amphibious*, *P. annectens*, *P. dolloi*); *Lepidosiren paradoxa* (Lepidosirenidae, South America); *Neoceratodus forsteri* (Ceratodontidae, Australia). The extant genera are derived from two ancestral lineages that separated in the Late Permian or Early Triassic (290–210 mya) (Long, 1995; Clack, 2002). *Neoceratodus* (Suborder Ceratodontoidei) represents one of these and both *Protopterus* and *Lepidosiren* (Suborder Lepidosirenoidei) the other. Differences between these lineages are exemplified by *Neoceratodus*, which has a more primitive body form with large scales and paddle-shaped fins. *Neoceratodus* is also a facultative air breather, whereas both *Protopterus* and *Lepidosiren* are obligate air breathers. Accounts of the natural history of the three genera were written by Greenwood (1987), Kemp (1987), and Harder *et al.* (1999).

6.5.2.1 Morphology

The lungs of lungfish are comprised of parenchymal septa that subdivide the lumen into small, alveolar-like respiratory chambers or ediculae (Figs. 6.2 and 6.4). Each septum contains smooth muscle and is covered by a dense capillary bed and a thin respiratory epithelium. The lungs, which can be paired or single, fill most of the posterior coelomic cavity and are connected to paired pulmonary arteries that branch from the common epibranchial artery of the third and fourth gill arches; a single pulmonary vein returns oxygenated blood into the heart (Fig. 6.8a). The pneumatic duct connecting the lung and the digestive tract originates on the ventral surface of the pharynx. Heart specializations for maintaining separation between the O_2-rich blood in the pulmonary vein and the O_2-poor blood in other systemic veins include a partially subdivided atrium, a plug and ventral muscular ridge or septum in the ventricle, and a nearly complete septum within the outflow tract (bulbus cordis), which directs the O_2-rich stream into the systemic circulation and the O_2-poor stream into gill

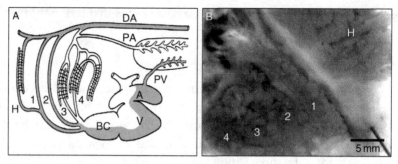

Fig. 6.8 (A) Pulmonary and branchial circulation in *Protopterus*. O_2-rich blood (gray) exits the lung via the pulmonary vein (PV), flows through the heart and then through the non-respiratory first and second gill arches (1, 2) and into the systemic circulation via the dorsal aorta (DA). Partially divided cardiac chambers (A, atrium; V, ventricle; BC, bulbus cordis) maintain separation between the pulmonary blood and the O_2-poor systemic venous blood (white), which flows through gill arches (3, 4) before entering the pulmonary artery (PA). Modified from Satchell (1976), Burggren and Johansen (1987), Graham (1997), and Farrell (2007). (B) Posterior view of the gill arches of *Lepidosiren* showing club-like filaments on all four gill arches and the hyoid (H). Image provided by M. Fernandes and modified after de Moraes *et al.*, 2005.

arches 3 and 4 and then to the lung (Fig. 6.8a; Burggren and Johansen, 1987; Graham, 1997; Icardo *et al.*, 2005a; Icardo *et al.*, 2005b; Farrell, 2007).

6.5.2.2 Comparative morphology and respiration

Differences in the three genera correlate directly with their air-breathing dependence. The ridges and septa within the heart of the faculta-tively air-breathing *Neoceratodus* are less prominent and are unlikely to achieve the same level of pulmonary and systemic separation that occurs in *Lepidosiren* and *Protopterus* (Burggren and Johansen, 1987; Farrell, 2007). Also, the lung of *N. forsteri* is unpaired (an embryonic left-lung bud appears but does not develop), occurs entirely in the dorsal part of the body cavity, has a smaller respiratory surface area, and has a correspondingly longer pneumatic duct that arises on the ventral pharynx and extends dorsally to the lung. *Neoceratodus* also has gills on all of its branchial arches and its estimated total area of about 2500 cm^2 in a 6 kg fish (= 417 cm^2 kg^{-1}) (Hughes, 1976) is comparable to that of other non-air-breathing freshwater fishes (Palzenberger and Pohla, 1992). Except when in hypoxic water, *Neoceratodus* respires aquatically and seldom breathes air. However, if it is exposed to warmer water or forced to be more active, both of which elevate $\dot{V}O_2$, its air-breathing rate increases (Grigg, 1965).

In contrast to *Neoceratodus*, the paired lungs of *Protopterus* and *Lepidosiren* have more septation and are fused anteriorly to form a common chamber that begins under the pharynx. Posterior to this point the lungs separate, proceed dorsally, and extend to nearly the end of the coelom. Owing to the ventral position of the anterior lung, the pneumatic duct is short and nearly vertical (Fig. 6.2). The lungs of *Lepidosiren* and *Protopterus* are morphologically similar, but experiments show that *Lepidosiren* has a greater capacity for aerial O_2 utilization than *Protopterus*, and this is consistent with its greater development of heart specializations that separate the systemic and pulmonary flows (Farrell, 2007).

Both *Lepidosiren* and *Protopterus* normally obtain between 70 and 90% of their O_2 via pulmonary uptake (Amin-Naves *et al.*, 2004; Perry *et al.*, 2005; Amin-Naves *et al.*, 2007). The recent report that *Lepidosiren* could obtain only 60% of its total required $\dot{V}O_2$ via the lung (Abe and Steffensen, 1996a) is not consistent with other respiratory data (Burggren and Johansen, 1987; Farrell, 2007) and does not agree with morphometric analyses (Bassi *et al.*, 2005; de Moraes *et al.*, 2005) showing a high lung diffusing capacity, but skin and gill diffusing capacities that are far too small to support 40% of the routine $\dot{V}O_2$ via aquatic respiration. A reduced efficacy for skin breathing seems to be especially true in light of the finding that in hypoxic water ($PO_2 \leq 3$ kPa) *Lepidosiren* experiences a high rate of transcutaneous diffusive O_2 loss to water (Abe and Steffensen, 1996a).

Anatomical descriptions have consistently reported that gill arches 1 and 2 in *Protopterus* and *Lepidosiren* are totally devoid of gills and that these arches function as conduits for the passage of O_2-rich blood from the bulbus cordis to the dorsal aorta and on to the systemic circulation (Fig. 6.8a) (Burggren and Johansen, 1987; Farrell, 2007). Figure 2.13 in Farrell (2007) illustrates the gill-less condition of arches 1 and 2 in *Protopterus*. However, a recent morphometric study on *Lepidosiren* (de Moraes *et al.*, 2005) documents the presence of small gill filaments on arches 1 and 2 as well as on arches 3 and 4 and the hyoid arch (Fig. 6.8b). This study also showed that the total gill area in *Lepidosiren* (0.65 cm^2 kg^{-1}) is very small and that gill O_2- and CO_2-diffusing capacities (i.e. the transfer rate per mean effective partial pressure gradient between the external medium and blood) are too small to significantly contribute to respiration. These facts raise two problems. First, the presence of gills on all arches in *Lepidosiren* conflicts with all previous reports stating that it and *Protopterus* both have gill-less arches 1 and 2. Secondly, the conclusion of de Moraes *et al.* (2005) that the gills of *Lepidosiren* cannot contribute to aquatic respiration focuses doubt on the essential function readily ascribed to gills of arches 3 and 4: that of preconditioning deoxygenated venous blood about to enter the lung by removing CO_2 and equilibrating its acid–base status (Graham, 1997; Farrell, 2007). In addition, the finding of a low gill diffusing capacity for both O_2 and CO_2 raises further questions about the aquatic respiratory capacity of *Lepidosiren*, in particular its mechanism of CO_2 release (Bassi *et al.*, 2005).

6.5.2.3 Hypoxia effects

Progressive aquatic hypoxia increases the gill ventilation of *Neoceratodus* down to its air-breathing threshold (about 10 kPa), and, once initiated, air-breathing frequency increases with hypoxia (Fritsche *et al.*, 1993; Kind *et al.*, 2002). Similar to loricariids and other facultative air breathers, acclimation of *Neoceratodus* to aquatic hypoxia (~7.8 kPa) results in an increased (left-shifted) Hb–O_2 affinity (Kind *et al.*, 2002); however, the change in P_{50} is very small (0.4 kPa) and may lack physiological significance. Also, because lungfish lack the catecholamine-induced intracellular phosphate-mediated mechanism for shifting Hb–O_2 affinity (Brauner and Berenbrink, 2007), the basis for the slight shift in P_{50} is unknown. Moreover, this shift is not accompanied by significant changes in other blood properties (hematocrit, Hb) that usually accompany a shift in P_{50} (Graham, 1997). This generally low-level response to hypoxia may reflect a low relative stress imposed by the experimental hypoxia level. However, it is more likely to indicate a correspondingly lower HIF-1α-mediated response level made possible by a seamless transition to an efficacious air-breathing mode featuring a lung and pulmonary circulation.

No aspect of the cardiorespiratory function of *Lepidosiren* and *Protopterus*, including O_2 uptake by the lung, the frequencies of air-breathing and aquatic ventilation, or either blood O_2 level or Hb–O_2 affinity or other features, is affected by exposure to aquatic hypoxia (Graham, 1997; Sanchez *et al.*, 2001). This complete absence of homeostatic responses to ambient water hypoxia also occurs in some air-breathing teleosts (*Electrophorus*, *Synbranchus*). Although it has been suggested that the non-response of lungfish to aquatic hypoxia indicates the loss of external O_2 receptors, a more likely explanation is that pulmonary circulation, heart modifications, and specializations in gill microcirculation, together with the capacity to modulate in relative blood flow, enable these fishes to isolate their aerial O_2 supply from contact with hypoxic water (Graham, 1997; Farrell, 2007).

In spite of their independence from ambient hypoxia, both *Lepidosiren* and *Protopterus* do retain an external (water) O_2-sensing capacity. Evidence for this is their reflex responses to the branchial application of nicotine and cyanide and the utilization of hypoxia-activated behaviors such as parental fanning of nests (*Protopterus*) and the use of 'fin gills' (dense, filamentous extensions that appear on the paired fins of nest-tending males and function to oxygenate the nest water) by *Lepidosiren* (Urist, 1973; Graham, 1997). The reduction in air-breathing frequency by both *Lepidosiren* and *Protopterus* in hyperoxic water may also indicate some level of tonic regulatory influence of aquatic O_2 (Sanchez *et al.*, 2001). Moreover, although aquatic PO_2 does not have an important role in controlling the air breathing of *Lepidosiren* or *Protopterus*, internal PO_2 sensors do function for this (Sanchez *et al.*, 2001). Also present are pulmonary chemoreceptors (serotoninergic neuroepithelial cells) (Zaccone *et al.*, 1995) that sense PO_2 (these also occur in *Neoceratodus*) and pulmonary mechanoreceptors, some of which are sensitive to CO_2 or pH (Fritsche *et al.*, 1993; Kind *et al.*, 2002; Perry *et al.*, 2005). The central respiratory control region of lungfish also appears to have both O_2 and CO_2 receptors (Graham, 1997; Sanchez *et al.*, 2001; Graham, 2006).

Consistent with their status as obligatory air breathers, both *Lepidosiren* and *Protopterus* readily respond to aerial hypoxia by increasing air-breathing frequency (Graham, 1997; Sanchez *et al.*, 2001; Perry *et al.*, 2005). Although aerial hypoxia is rarely encountered by any air-breathing fish, such exposure for *Lepidosiren* and *Protopterus* elicits a range of metabolic and stressor effects, such as a reduction in total $\dot{V}O_2$ and increases in circulating catecholamine levels; similar responses also occur in non-air-breathing fishes and tetrapods (Powell, 2003; Bickler and Buck, 2007) and reflect HIF-1α activity. Although catecholamines do not affect lungfish Hb–O_2 affinity (Brauner and Berenbrink, 2007), they can alter blood pressure by affecting both heart rate and ventricular contractility. Catecholamines (stored in the heart, other organs, and blood sinuses) and cholinergic neurons also control the flow-resistances of vascular beds, affecting cardiac output to the lungs of *Lepidosiren* and *Protopterus*. These include the opening of the pulmonary vasomotor segment (located at the base of arches 3 and 4) (Fig. 6.8a), which increases flow into arches 3 and 4, and the closing of the ductus arteriosus (the conduit between the pulmonary artery and dorsal aorta) (Fig. 6.8a), which ensures the stream of blood exiting the gills flows into the lung (Graham, 1997; Perry *et al.*, 2005; Farrell, 2007).

Lungfish hearts lack adrenergic nerve fibers and the heart rate is thus determined by circulating catecholamine levels in combination with vagal (parasympathetic) tone. Because catecholamine levels are stress related, they can potentially increase lungfish heart rate to a level at which the relaxation of vagal tone does not further elevate it (Graham, 2006; McKenzie *et al.*, 2007). This dynamic appears to explain the variable results in air breath-related shifts in lungfish heart activity during air breathing, which is most effectively demonstrated in specimens with low rather than high heart rates (Fritsche *et al.*, 1993; Sanchez *et al.*, 2001; Perry *et al.*, 2005).

6.5.2.4 Estivation

Estivation by *Protopterus* enables it to survive desiccation when extreme dry season conditions evaporate all of the water in its habitat. The fish digs into the drying mud, coils itself with the head up, and secretes a mucus cocoon around its body, but leaves a small breathing hole through the covering mud.

All four *Protopterus* species can estivate; however, this expediency depends on local climate conditions and soil types, as some populations of all four species reside in areas less prone to complete drying and thus may not estivate (Greenwood, 1987). *Lepidosiren* may also become confined to moist burrows during the dry season, and there are early, undocumented reports that it forms a cocoon (Graham, 1997; Harder *et al.*, 1999). *Neoceratodus* does not estivate.

Monumental works documenting the metabolic changes accompanying esti-vation were done by H.W. Smith (1930). Although natural cocoons have been studied, many investigations use laboratory-induced estivation, performed by placing a fish in an aquarium containing several centimeters of mud substrate and, while starving it, slowly allowing the water to evaporate (Fishman *et al.*, 1987; Sturla *et al.*, 2002; Chew *et al.*, 2004).

Estivating *Protopterus* reduces total metabolism by 80–99%, loses body mass, and switches from ammonotelism to ureotelism to conserve water and detoxify ammonia (Graham, 1997). Recent studies with *P. dolloi* confirm this general pattern but show that this species can estivate on the mud surface under a thin mucus covering – and in this circumstance, which affords greater O_2 access, it upregulates its rate of ureotelism (Chew *et al.*, 2004). Studies of *Lepidosiren* also show a 29% $\dot{V}O_2$ reduction in estivating (awake but starved) specimens (Abe and Steffensen, 1996b); however, there have been no metabolic studies on *Lepidosiren* that are comparable to those done with different *Protopterus* species. In a histomorphological comparison of water-dwelling and estivating adult *P. annectens*, Sturla *et al.* (2002) reported that the gills of estivating fish were collapsed and covered in mucus and thus non-functional. These workers also found that the parenchymal septa in the lungs of estivating fish were open, fully vascularized, and filled with red cells and thus functional. Paradoxically, Sturla *et al.* (2002) also reported that the lungs of water-dwelling *P. annectens* were reduced to thin, flat tissue strips that lacked obvious ridges or septa, and were devoid of red cells. Although not specifically stated by these workers, the implication of their findings is that a free-swimming adult *P. annectens* does not have the capacity for air breathing. Although there may be specific or population differences in *P. annectens*, the absence of functional air breathing is not consistent with reports of active air breathing by all the adult *Protopterus* species (Greenwood, 1987). Also, and even though there have been no systema-tic studies of morphometric changes in the lung associated with estivation, another implication of Sturla *et al.* (2002) is that the lung of *P. annectens* is only functional during estivation, a time when its total $\dot{V}O_2$ is reduced, and then its lung atrophies to a non-functional state during the wet season, when the fish, an obligate air breather, is free-swimming in mainly hypoxic water. All of this is highly improbable based on both natural history and experimental data obtained by other workers, and because of the extensive levels of tissue recycling (i.e. apoptosis and regeneration) that this would entail.

In summary, the lungfishes, which are the closest bony fish relatives to the Tetrapodomorpha, reached their greatest diversity in the Devonian. Although air breathing is considered to be common in early osteichthyans, the fossil record indicates that not all lungfishes had this capability; it appears to have become fully developed in the course of the group's radiation into freshwater (Long, 1995; Clack, 2002). The primitive form of the lungfish air-breathing specialization may never be known, but the morphology and physiology of the extant genera permits some inferences. Similarities in the lungs, pulmonary circulation, and heart modifications for parallel-flow separation (Fig. 6.8a) suggest that strong, hypoxia-driven selection for air breathing occurred in the ancestral lungfish group before the Permian-Triassic separation of ceratodontoid and lepidosirenoid suborders. Lungfish cocoons from the Permian further indicate that adaptations for habitat drying, air breathing, and biochemical changes during estivation rapidly ensued after this separation. *Neoceratodus*, which is known from Jurassic Period fossils, has many structural similarities with lungfishes that lived from the late Devonian through the Mesozoic (Carroll, 1988; Long, 1995). The long pathway taken by the pneumatic duct of *Neoceratodus* suggests that subsequent changes in its environment and natural history selected for the vertical migration of its lung to a position favorable for buoyancy. (The long pneumatic duct of this species partly inspired the lung–gas bladder transformation model for actinopterygians (Liem, 1988; Liem, 1989; Graham, 1997).) The early appearance of morphological changes imposed on the efferent circulation of gill arches 3 and 4 by the pulmonary arteries indicates that gill changes related to bimodal respiration were also well integrated into air breathing. However, new findings that *Lepidosiren* gills are not the morphological twin of those in *Protopterus* and have too small an area to function in aquatic gas transfer raise questions about their role in bimodal gas exchange. Although many air-breathing fishes have a reduced gill surface area, gills remain vitally important for ion and nitrogen regulation, and no aquatic air-breathing fish can function without them. However, the concept in place for *Lepidosiren* and *Protopterus*, that the gills on arches 3 and 4 are important for aquatic respiration (Graham, 1997; Farrell, 2007), is challenged by the finding of a diminished respiratory capacity for *Lepidosiren*. New information about the gill-arch modifications in fossil lungfish, as well as a clarification of the function of *Lepidosiren* gills, are crucial areas of discovery needed to determine whether the pleisomorphic state for bimodal lungfish gills is most closely represented by the condition in *Neoceratodus* or in *Protopterus*.

6.5.3   Case study 3. Air breathing and the gobies: hypoxia and life on land

This case study compares the air-breathing specializations of species in three subfamilies of the Gobiidae. It demonstrates the persistent selective pressure of environmental hypoxia in the origin of air breathing and in the development of air breathing-related behaviors affecting natural history and niche expansion. With about 2000 species, the Gobiidae is the largest family within the teleost Order Perciformes. Goby fossils date back to the Paleocene (about 60 mya), and the radiation of gobies into niches having diverse environmental O_2 regimes has resulted in a continuum in respiratory specialization, from non-air breathers living subtidally, to facultative air breathers frequenting shallow waters, to amphibious air breathers. The capacity to breathe air has evolved independently in genera in at least four of the five goby subfamilies (Gobiinae, Gobionellinae, Amblyopinae, Oxudercinae) (Fig. 6.9). The use of the buccal chamber as an ABO is common to these groups, and many also utilize their skin (Graham, 1997; Nilsson *et al.*, 2007), which has numerous specializations for gas transfer and desiccation resistance (Zhang *et al.*, 2000; Park, 2002; Zhang *et al.*, 2003). There are also many diverse behavioral and metabolic adaptations related to air breathing in gobies (Gracey *et al.*, 2001; Ip *et al.*, 2006; Gracey, 2008).

6.5.3.1 Gillichthys

The long jaw mudsucker *Gillichthys mirabilis* (Gobionellinae) normally lives in crab burrows or other recesses in shallow bays and estuaries in the temperate zone of the North American Pacific coast (Todd and Ebeling, 1966). When exposed to aquatic hypoxia it gulps air, which is held in the mouth in close contact with areas where O_2 will be absorbed, such as the gills, a vascular region on the roof of the mouth, and a dense bed of capillaries on the tongue. Severe hypoxia causes *G. mirabilis* to emerge from water. Stiff filaments

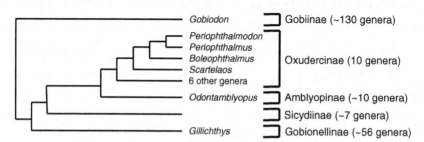

Fig. 6.9 Phylogenetic relationship of the five subfamilies of the Gobiidae (modified from Murdy [1989] and Thacker [2003]). Parenthetic numbers indicate genera diversity. Genera (*italics*) listed next to each clade are known to be air breathers.

probably enable its gills to hold their shape in air and may also enable their function in aerial respiration (Graham, 1997). In air, the buccal capillary beds become engorged with blood. *Gillichthys*, however, has only a rudimentary capacity for aerial respiration and a very limited ability to move on land. Microarray analyses show that the global gene expression of *Gillichthys* responds to hypoxia and air exposure by means of HIF-1α-mediated changes in the levels of numerous transcripts encoding for proteins involved in carbohydrate metabolism, protein synthesis, growth, and other metabolic factors (Gracey *et al.*, 2001; Gracey and Cossins, 2003; Gracey, 2008). This is the first documentation of parallel hypoxia- and emersion-induced changes in the genetic expression of an air-breathing fish.

6.5.3.2 Odontamblyopus

The eel gobies (Amblyopinae) occupy burrows in soft, muddy substrates in tropical and subtropical bays and estuaries. Air breathing in this group was first detailed for *Taenioides rubicundus*, which gulps air and holds a volume in its mouth that is sufficient to cause it to float at the surface (Graham, 1997). Gonzales *et al.* (2006, 2008) described a similar air-breathing mechanism in *Odontamblyopus lacepedii*, and their laboratory and field observations provide context to the otherwise enigmatic surface-floating behavior described for air-breathing *Taenioides*. The burrows of *O. lacepedii* occur intertidally and subtidally. However, this species is not amphibious, and fish living in the intertidal zone remain confined to their burrows during low tide. Both the absence of ocean-water mixing at low tide and the mud's high BOD cause the isolated burrow water to become extremely hypoxic (Ishimatsu *et al.*, 1998; Gonzales *et al.*, 2006; Ishimatsu *et al.*, 2007).

When its burrow water PO_2 drops to between 1.0 and 3.1 kPa (mean = 2.8 kPa), *O. lacepedii* commences facultative air breathing. The inspired air is held in the buccal chamber, where it envelops the gills and contacts vascular beds on the inner wall of each operculum and other buccopharyngeal surfaces probably used for aerial O_2 uptake (Gonzales *et al.*, 2008). As in loricariids and *Neoceratodus*, the air-breathing frequency of *Odontamblyopus* increases with greater hypoxia. Another loricariid similarity is that *O. lacepedii* completely empties its ABO at exhalation. Its ABO volume increases with body mass; however, the mass-scaling exponent is only 0.6. When expressed as a percentage of total mass, the ABO volume reported for *Odontamblyopus* is about 6.1% (= 0.061 ml g^{-1} × 100, range 5.5–10.0%). A volume ratio of 6% is sufficient for positive buoyancy (Gee and Gee, 1991; Gee and Gee, 1995) and would cause an untethered fish to float at the water surface with the tip of its snout out of water (as was observed for *Taenioides*). However, provided it remains in a burrow and breathes air only at low tide, *Odontamblyopus* would not normally float at the surface. In laboratory tests, a fish exposed to hypoxia in burrow-like tubes located a few centimeters below the water surface would extend out the tube far enough to gulp air at the surface, but would then retract itself back into the tube (Gonzales *et al.*, 2008).

6.5.3.3 Mudskippers and their allies

There are ten genera and about 38 species in the goby subfamily Oxudercinae, and these occur in tropical and subtropical estuaries and inter-tidal mudflats of the Old World (Murdy, 1989). In terms of amphibious behavior, the oxudercines are readily separated into two groups: the more basal non-mudskippers (six genera), and the mudskippers (four genera). Two of the non-mudskipper genera, *Pseudapocryptes* and *Apocryptes*, are not amphibious and have air-breathing behaviors similar to those of *Taenioides* and *Odontamblyopus* (i.e. they breathe air during low tide while remaining in their burrows). This air-breathing behavior may be typical of all six of the non-mudskipper oxudercine genera, none of which are amphibious.

Mudskippers number approximately 27–29 species. Listed with their number of species (Murdy, 1989) and in order of increased terrestrial capability, the four genera are: *Scartelaos* (four species), *Boleophthalmus* (five), *Periophthalmus* (15–17), and *Periophthalmodon* (three). Mudskippers couple aerial respiration to water emergence and to a constellation of behavioral, sensory, physiological, and locomotor specializations that elevate amphibious capability to levels unmatched by any other fishes (Graham *et al.*, 2007; Ishimatsu *et al.*, 2007). All mudskippers readily breathe air using their buccopharyngeal epithelium, skin, and, in some species, gills (Graham, 1997; Kok *et al.*, 1998; Zhang *et al.*, 2000; Park, 2002; Zhang *et al.*, 2003).

The gills of *Scartelaos* and *Boleophthalmus* are similar to those of the non-mudskipper oxudercines and many other fishes and thus show little specialization for air exposure (Graham *et al.*, 2007). By contrast, gills modified for life out of water are found in the more amphibious *Periophthalmus* and *Periophthalmodon*, the species of which live in the high intertidal zone, are highly active on land, and have several respiratory and metabolic specializations for an amphibious life (Graham, 1997; Ishimatsu *et al.*, 1999; Takeda *et al.*, 1999; Ip *et al.*, 2006; Graham *et al.*, 2007). The gill filaments of *Periophthalmus* are relatively short and twisted, which opens the space between them and may have advantages both for gas transfer and for preventing the gills from coalescing while in air (Mazlan *et al.*, 2006; Graham *et al.*, 2007). Although other respiratory surfaces are important, the gill structure of *Periophthalmus* species appears adequate for aquatic respiration (i.e. no obligate air-breathing species are known). By contrast, the gill lamellae of *Periophthalmodon* are enclosed in an epithelial matrix that impedes gas diffusion and makes the organ unsuited for either aquatic or aerial respiration. This matrix retains water and is rich in chloride cells and probably supports gill function in acid–base regulation and ammonia excretion (Randall et al., 2004; Ip et al., 2006; Graham et al., 2007). Their gill structure thus makes *Periophthalmodon* species obligate air breathers and also highly dependent upon other respiratory surfaces such as the skin and buccopharyngeal epithelium. For example, a *Periophthalmodon* held in water without air access cannot saturate its blood with O_2, and fish denied air for an extended period reduce $\dot{V}O_2$, initiate an asphyxic bradycardic response, switch to glycolysis and concentrate lactate in their muscle and blood, and are at risk of drowning (Ishimatsu et al., 1999; Takeda et al., 1999; Ip et al., 2006). Finally, the structural and functional modifications of *Periophthalmodon* gills are permanent and cannot be altered by habitat conditions. This is different from the cyprinodont *Kryptolebias*, which, during prolonged air exposure, deactivates its gills by embedding them in a cellular matrix, but then reactivates them when it returns to water (Ong et al., 2007).

The mudskipper buccopharyngeal chamber is also important for bimodal respiration. Chamber volumes (14–17% of body mass) are much larger than in most other gobies (4%) (Gee and Gee, 1991), including *Odontamblyopus* (6.1%) (Gonzales et al., 2008), and are largest in the more amphibious *Periophthalmus* and *Periophthalmodon* (Graham et al., 2007). (Note: in subsequent use of the genus-species nomenclature, *Periophthalmus* will be abbreviated as *Ps.* and *Periophthalmodon* as *Pn.*). Mudskippers also have greater numbers of capillaries on their buccal chamber surfaces: *Ps. magnuspinnatus*, for example, has 59.1 capillaries per mm along the length of its inner opercular wall compared with only 14.5 per mm for *O. lacepedii* (Park, 2002; Park et al., 2003).

The skin of mudskippers also serves for gas exchange: in some species nearly 50% of the total $\dot{V}O_2$ is transcutaneous. Skin specializations for aerial respiration include small (or no) scales and a high density of capillaries occurring within the epidermis where they are close to air. In *Ps. magnuspinnatus*, epidermal capillaries occur within 1.5 μm of air. By contrast, the skin capillaries of *Odontamblyopus* are confined to the dermis, which is 275 μm from the skin surface (Park, 2002; Park et al., 2003). (This difference would be expected because the skin of *Odontamblyopus* is usually immersed in hypoxic water.) Although epidermal capillaries are rare among fishes, they are common on the heads and dorsal body surfaces of a number of mudskippers and other amphibious fishes, including *Kryptolebias* (Graham, 1997; Zhang et al., 2000; Zhang et al., 2003). In mudskippers, the amount of body surface covered by capillaries is largest in *Periophthalmus* and *Periophthalmodon*, and although *Periophthalmus* has more skin capillaries, there are considerable interspecific differences in the position of capillaries, their density, and the air–blood diffusion distance (Park, 2002; Zhang et al., 2000; Zhang et al., 2003).

6.5.3.4 Hypoxia and mudskippers

This section compares the metabolic and behavioral responses of *Periophthalmodon* and *Periophthalmus* to exercise and burrow hypoxia. All mudskippers are effective amphibious air breathers, but this does not prevent them from encountering hypoxia. With their high activity on land, mudskippers potentially experience functional hypoxia (i.e. an O_2 debt resulting from exercise). Also, because of their burrow use, mudskippers must tolerate environmental hypoxia.

Activity on land. Whether it is the result of running on land or swimming vigorously, functional hypoxia is very likely to be experienced by all air-breathing fishes (Graham, 2006). A recurring question about fish air breathing has been whether aerial O_2 access increases either aerobic scope (i.e. more O_2 = more work) or the capacity to recover from functional hypoxia more rapidly (i.e. more O_2 = faster debt recovery, lactate clearance, and glucose restoration). Chasing and prodding have been used to increase the activity levels of *Periophthalmodon*, *Periophthalmus*, and other mudskippers (Ip et al., 2006). Both *Periophthalmodon* and *Periophthalmus* elevate aerial $\dot{V}O_2$ during exercise on land (Graham, 1997; Kok et al., 1998; Ip et al., 2006; Chew et al., 2007). Exercised *Ps. argentilineatus* incurred an O_2 debt and increased post-exercise aerial $\dot{V}O_2$ by 3.1 times. Also, *Ps. chrysospilos* chased to exhaustion had a depleted creatine phosphate, a reduced energy charge (i.e. low ATP, elevated ADP and AMP), and a sixfold increase in muscle lactate (Ip et al., 2006). Air-exercised *Pn. schlosseri* also increased post-exercise aerial $\dot{V}O_2$ by 2.5 times but, reflecting the reduced gill function and the need for obligatory air breathing by this species, exercised fish that were placed in normoxic water without air access could not repay their O_2 debt until given air access (Takeda et al., 1999). Based on its gill structure, it is likely that *Periophthalmus* species can repay an O_2 debt using aquatic respiration; however, it is unknown whether this can also be done in air.

In an effort to answer the air-access and metabolic scope question, Wells et al. (2007) used forced swimming tests with the aquatic-air-breathing Pacific tarpon (*Megalops cyprinoides*). They found that lessening the PO_2 of the tarpon's air supply increased its anaerobic scope (i.e. a greater buildup of lactate); however, after the exercise bout was completed, the tarpon's O_2 debt (i.e. various metabolic costs, including that of either oxidizing lactate or converting it back to glycogen) was handled by increased aquatic respiration. A similar experiment comparing the lactate production and $\dot{V}O_2$ of *Periophthalmus* and *Periophthalmodon* exercised in hypoxia might reveal differences in their anaerobic and aerobic poising related to differences in their ABO structure. A variation on this theme, comparing these same parameters in mudskippers exercised in hyperoxia, would directly answer the question about the capacity of air breathing to increase aerobic scope.

Burrowing. Unlike *Periophthalmodon*, which occupies permanent burrows in the highest reaches of the intertidal zone, most species of *Periophthalmus* occupy burrows mainly during the breeding season. Mudskippers thus use burrows for reproduction, and to compensate for the prevalent condition of burrow hypoxia, all four mudskipper genera use air-depositing behavior, in which they transport gulps of air from the mud surface and release them into the burrow air chambers. This is needed for the respiration of developing eggs and may also benefit the adult fish during periods of burrow confinement (Ishimatsu et al., 1998; Ishimatsu *et al.*, 1999; Lee *et al.*, 2005; Graham *et al.*, 2007; Ishimatsu et al., 2007).

In addition to its utility in air breathing, burrowing, and display, the large buccopharyngeal chamber volume of mudskippers appears to be an important adaptation for air-transport behavior. Field observations of the air-containing chamber in the burrow of *Ps. modestus* confirm that developing eggs are located in the chamber's air phase, where they are tended by the male. Time-series data for burrows in which fish activity and air-chamber PO_2 were monitored (Ishimatsu *et al.*, 2007) show that the chamber O_2 level declines during high tide (i.e. when the burrow is isolated from the air) and that during low tide the male *Ps. modestus* makes numerous air-gulp and air-deposition trips and restores O_2 to near air levels (Fig. 6.10). Experiments further demonstrated that an abrupt reduction of air-chamber PO_2 by the injection of N_2 gas was immediately sensed by the male, which accelerated its air-gulping rate and re-oxygenated the chamber in the short period before the rising tide covered the burrow entrance (Fig. 6.10).

In summary, for the gobies, a central feature of their air breathing is the role of hypoxia in selecting for this specialization in different groups, but with the common attribute of the buccal chamber as an ABO. *Gillichthys*, *Odontamblyopus*, and several oxudercines are facultative air breathers that gulp air. A model for the evolutionary sequence leading to land invasion by the mudskippers is suggested by *Odontamblyopus*, for which range extensions of normally subtidally dwelling species into the intertidal zone initially selected for hypoxia-driven air breathing. This was followed by speciation events leading to the colonization of the intertidal zone and the acquisition of amphibious behavior, which enabled niche expansion onto the mudflat surface. In contrast to other gobies, the air breathing and emergence of mudskippers are normal actions that are highly integrated with locomotor, sensory, and behavioral adaptations for amphibious

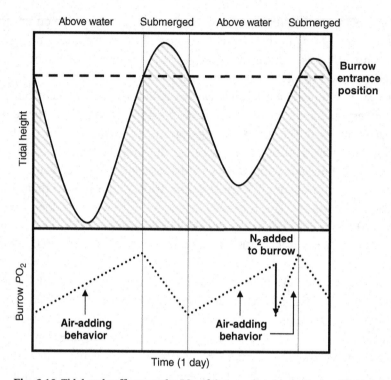

Fig. 6.10 Tidal-cycle effects on the PO_2 of the egg chamber in the mudflat burrow of the mudskipper, *Periophthalmus modestus*. At low tide the burrow entrance is open to the air and the male fish, which guards the nest, transports gulps of air into the egg chamber to elevate its PO_2. During high tide the burrow entrance is submerged and egg respiration, combined with the effect of anoxic mud and possibly the respiration of the guarding fish, reduces the egg chamber O_2 content, which is then restored, by air-gulp transport, during the next low tide. The experimental addition of N_2 into the egg chamber causes the male fish to increase its frequency of air-deposition behavior in order to raise burrow PO_2 before the next incoming tide covers the burrow entrance. Modified from Ishimatsu *et al.* (2007).

life. Nevertheless, the different mudskipper genera display both similarities and differences in their terrestrial adaptations. Whereas *Periophthalmus and Peri-ophthalmodon* are highly amphibious, the gills of *Periophthalmodon* have minimal respiratory function, and these species are obligatory air breathers. By contrast, the gills of *Periophthalmus* appear functional for both aerial and aquatic respiration, and *Periophthalmus* species have more epidermal capillaries. One marked metabolic difference is that *Periophthalmodon* species can utilize amino acids rather than glycogen as an energy source during aerial respiration.

Although *Gillichthys* also emerges from hypoxic water, Gracey's (2008) microarray studies show that hypoxia and emergence both elicit stress responses that trigger changes in the expression of genes coding for proteins governing metabolism. These findings make it likely that microarray analyses of genetic expression can be used to explore mudskipper air breathing and emergence physiology and the link between phenotypic expression and goby genomics (Gracey and Cossins, 2003). Although air-breathing mudskippers can largely avoid aquatic hypoxia by living out of water, the requirements of terrestrial activity subject them to exercise-induced functional hypoxia (Takeda et al., 1999). Also, burrow confinement, whether for refuge during high tide or for rearing eggs, subjects mudskippers to aquatic hypoxia and requires both sensory and behavioral specializations to ensure survival (Lee et al., 2005; Ishimatsu et al., 2007).

6.6   Summary and conclusions

Air breathing has independently evolved in many fishes. The diversity of these fishes and their different types of ABOs and air-breathing behaviors all signify the efficacy of natural selection in achieving a nearly perfect solution to the common occurrence of environmental hypoxia in shallow-water habitats throughout the long history of fish evolution. The three case studies demonstrate the central role of aquatic hypoxia in driving the independent origin of air breathing and for, over the expanse of geologic time, either intensifying or relaxing selection for air breathing and related adaptations, in concert with changes in environment factors and with the pattern of expansion and radiation taken by a particular group. Specifically, the capacity of fishes to breathe air was an important precondition for the evolution of amphibious life, both in the early tetrapods and in many extant air-breathing fishes such as the mudskippers. Although air breathing seems to present the 'perfect answer' to environmental hypoxia, not all fishes utilize this adaptation. Because the independent evolution of air breathing can be viewed as an ongoing process, it is likely that some fishes, because of environmental change or their ecological radiation into less favorable habitats, may now be undergoing the initial selective processes leading to the acquisition of this capacity.

Appendix A: New findings about air-breathing in fishes

Many fishes appear to have a potential or requirement for air breathing (Graham, 1997; Martin and Bridges, 1999). Lampreys, for example, can breathe air and endure prolonged aerial exposure in the laboratory, and field observations show occasional amphibious movements during upstream migration. Similarly, even though air-breathing has not been documented for any shark or ray, the sand tiger shark, *Odontaspis taurus*, regularly gulps air to increase buoyancy, but there are no data linking this behavior to aerial respiration. In these and many other cases, research is required to distinguish the behavioral and functional utility of air breathing for a species as opposed to a capacity to survive exposure to air or hypoxia, or to ingest air for the purpose of increasing buoyancy. Listed here are species for which air breathing or terrestrial behavior (which is probably accompanied by amphibious breathing) have been described. For some of these the level of information available warrants inclusion in the list of air-breathing fishes published by Graham (1997). For others additional research is needed.

Order Petromyzontiformes

Family Petromyzontidae

The lamprey *Geotria australis* (Subfamily Geotriinae) becomes amphibious for brief periods when negotiating barriers to upstream migration. Also, this lamprey can use both its gills and skin for aerial respiration and has aerial O_2 consumption rates comparable to its aquatic respiration (Potter *et al.*, 1997).

Order Osteoglossiformes

Family Mormyridae

Brevimyrus niger is in the freshwater African family Mormyridae, which numbers about 200 species in about 18 genera. This group's common name, elephant fishes, derives from the long proboscis of some species. Mormyrids are weakly electric, using direct-current fields for sensory location and intraspecific communication.

Although not included in Graham (1997), air breathing in *B. niger* was first reported by Benech and Lek (1981), who deduced this based on field observations indicating its long-term survival with other air-breathing fishes in drying ponds and also reported this fish to breathe air in an aquarium. Moritz and Linsenmair (2007) also documented long-term survivorship of this species in drying pools and showed photos of a two-phase air-gulping maneuver. They suggested that ingested air is held in the gas bladder but did not observe any specializations of this organ for air breathing. Another weakly electric African freshwater fish *Gymnarchus niloticus*, a monotypic species in the closely related family Gymnarchidae, is also an air breather and has a well-developed respiratory gas bladder (Fig. 6.2). When confirmed by additional studies, aerial respiration in *Brevimyrus*, together with its use of the gas bladder as an ABO, would demonstrate the occurrence of air breathing in all four osteoglossiform families.

Order Siluriformes

Family Scoloplacidae

Lithoxus lithoides. The Scoloplacidae is closely related to the Loricariidae and is thus within the cluster of siluriform families (Fig. 6.3) that utilize a segment of their digestive tract as an ABO (Case study 1). Armbruster (1998) was the first to document scoloplacid air breathing. He observed periodic air gulping by *Lithoxus* in hypoxic water and also described features of its stomach related to its ABO function.

Order Cyprinodontiformes

Family Fundulidae

Fundulus heteroclitus heteroclitus. Terrestrial activity by both *Fundulus nottii* and *F. majalis* has been documented (Graham, 1997). For the mummichog, *F. heteroclitus heteroclitus*, a salt-marsh inhabitant that can be passively exposed to air during low tide, Halpin and Martin (1999) used respirometry to verify its capacities for aerial respiration and for maintaining a high respiratory exchange ratio ($\dot{V}CO_2$: $\dot{V}O_2$). Also, although the total $\dot{V}O_2$ of fish in air is less than in water, fish exposed to air for one hour and returned to water did not develop an O_2 debt, indicating that metabolic needs in air can be sustained by aerial respiration. Halpin and Martin further noted that the three amphibious *Fundulus* species are on separate clades, suggesting the independent appearance of this capability and its likely occurrence in other species.

Order Perciformes

Family Plesiopidae

Acanthoclinus fuscus occurs in the upper littoral zone in New Zealand, where it is occasionally exposed to air at low tide. Hill *et al.* (1996) measured very similar aerial and aquatic respiration rates for specimens of this species, ranging from 2 to 100 g. They also found that all *A. fuscus* tested in progressive hypoxia ultimately emerged from water; however, not until the PO_2 was extremely low (0.8 kPa). There are no ABO details.

Family Cichlidae

Sarotherodon aureus. Many cichlids are proficient at ASR (Graham, 1997; 2006); however, air breathing has not been documented for any species. Claims that air breathing occurs in some of the tilapia species stem from their capacity to survive in very small puddles. Also, the 'inability to breathe air because of a covering of surface-growing plants' (Ross, 2000) is considered to be the cause of occasional large-scale mortalities for *S. aureus* in aquaculture ponds. However, descriptions by Ross (2000) suggest a more derived form of ASR: '...tilapia will gulp air at the water surface when dissolved oxygen falls... Atmospheric oxygen dissolves in the buccal water and passes quickly into the gills.' Although not air breathing per se, the aspiration of air to augment aquatic respiration by oxygenating the branchial ventilatory stream can be regarded as an advanced form of ASR that was possibly incipient to air breathing in certain lineages (Burggren, 1982; Graham, 1997) (Case study 1).

Oreochromis alcalicus grahami. This species can tolerate a range of osmotic, alkaline and hypoxic conditions, and Maina *et al.* (1996) report it to be an air breather, having both a respiratory gas bladder and a structure similar to a pneumatic duct that connects this organ to the esophagus. Additional studies are needed to verify this observation and confirm aerial respiration. No other perciform fishes are known to have either a respiratory gas bladder or any connection between the gas bladder and the digestive tract. If these anatomical details are confirmed for *Oreochromis*, they would represent the de novo origin of structural specializations for air breathing that are analogous to those found in more basal bony fish groups.

Family Tripterygiidae

Forsterygion sp. lives in the littoral zone of New Zealand where, with *Acanthoclinus fuscus* (above), it is occasionally exposed to air at low tide. Hill *et al.* (1996) reported the ability of this fish to breathe air, but noted that the aerial $\dot{V}O_2$ of larger specimens (10–20 g) was less than their aquatic rates. Similar to *A. fuscus*, *Forsterygion* sp. did not emerge in progressive hypoxia until aerial PO_2 was extremely low (0.7 kPa), and a few of the fish so tested did not leave water. No ABO information is available.

Family Gobiidae

Gobies are a diverse group for which air breathing has been well documented. Two new air-breathing genera have been reported since Graham (1997), and information for one of these, *Odontamblyopus lacepedii* (subfamily Amblyopinae) is presented in Case study 3.

Gobiodon. Nilsson *et al.* (2007) surveyed the hypoxia responses and air-breathing capacities of gobies living between branches of coral colonies and found that seven species in the genus *Gobiodon* (subfamily Gobiinae, Fig. 6.9) could breathe air when their host coral was exposed to air during low tide. Four of these species, *G. axillaris*, *G. erythrospilus*, *G. histrio*, and *G. unicolor*, could carry out aerial respiration for up to 4 h with a $\dot{V}O_2$ similar to that in water. However, the other three species, *G. acicularis*, *G. ceramensis*, and *G. okinawae*, could only breathe air for an hour. In general, these breathing durations correlate with the depth distributions of the species: greater air-breathing capacity occurs in species occupying corals that are more likely to be exposed to air. All of these species are small and lack scales, suggesting that cutaneous respiration may be an important element in air breathing.

Family Scorpaenidae

Caracanthus unipinna. The behavior and circumstances of air breathing in *C. unipinna* are the same as described for the species of *Gobiodon*, and Nilsson *et al.* (2007) determined that this fish could also breathe air for up to 4 h. This species also has no scales and presumably uses its skin for respiration.

Acknowledgements

This work was partially supported by the U.S. National Science Foundation (IBN 9604699 and IBN 0111241) and the UCSD Academic Senate. N.C. Wegner was supported by the Nadine A. and Edward M. Carson Scholarship awarded by the Achievement Rewards for College Scientists (ARCS), Los Angeles, CA, USA. SEM studies were partially subsidized by the Scripps Institution of Oceanography Unified Laboratory Facility, and we thank Evelyn York for technical assistance. We thank Drs Marissa Fernandes, Dagmara Podkowa, and John Maina for providing images used in Figs. 6.4 and 6.8b.

References

Abe, A. S. and Steffensen, J. F. (1996a). Bimodal respiration and cutaneous oxygen loss in the lungfish *Lepidosiren paradoxa*. *Rev. Bras. Biol.*, **56**, 211–16.

Abe, A. S. and Steffensen, J. F. (1996b). Lung and cutaneous respiration in awake and estivating South American lungfish, *Lepidosiren paradoxa*. *Rev. Bras. Biol.*, **56**, 485–9.

Amin-Naves, J., Giusti, H. and Glass, M. L. (2004). Effects of acute temperature changes on aerial and aquatic gas exchange, pulmonary ventilation, and blood gas status in the South American lungfish, *Lepidosiren paradoxa*. *Comp. Physiol. Biochem.*, **138A**, 133–9.

Amin-Naves, J., Sanchez, A. P., Bassi, M., Giusti, H., Rantin, F. T. and Glass, M. L. (2007). Blood gases of the South American lungfish, *Lepidosiren paradoxa*: a comparison to other air-breathing fish and to amphibians. In *Fish Respiration and Environment*,

ed. M. N. Fernandes, F. T. Rantin, M. L. Glass and B. G. Kapoor. Enfield, NH: Science
Publisher, pp. 243–53.

Armbruster, J. W. (1998). Modifications of the digestive tract for holding air in
loricariid and scoloplacid catfishes. *Copeia*, 1998, 663–75.

Bassi, M., Klein, W., Fernandes, M. N., Perry, S. F. and Glass, M. L. (2005). Pulmonary
oxygen diffusing capacity of the South American lungfish *Lepidosiren paradoxa*:
physiological values by the Bohr method. *Physiol. Biochem. Zool.*, **78**, 560–9.

Benech, V. and Lek, S. (1981). Résistance á l'hypoxie et observations écologiques pour
seize espéces de poissons du Techad. *Rev. d'Hydrobiol. Trop.*, **14**, 153–68.

Berner, R. A., VandenBrooks, J. M.. and Ward, P. D. (2007). Oxygen and evolution.
Science, **316**, 557–8.

Bickler, P. E. and Buck, L. T. (2007). Hypoxia tolerance in reptiles, amphibians, and
fishes: life with variable oxygen availability. *Ann. Rev. Physiol.*, **69**, 145–70.

Brauner, C. J. and Berenbrink, M. (2007). Gas transport and exchange. In *Primitive
Fishes, Fish Physiology*, vol. 26, ed. D. J. McKenzie, A. P. Farrell, and C. J. Brauner. San
Diego: Elsevier Academic Press, pp. 213–82.

Burggren, W. W. (1982). 'Air gulping' improves blood oxygen transport during
aquatic hypoxia in the goldfish, *Carassius auratus. Physiol. Zool.*, **55**, 327–33.

Burggren, W. W. and Johansen, K. (1987). Circulation and respiration in lungfishes. In
The Biology and Evolution of Lungfishes, ed. W. E. Bemis, W. W. Burggren, and
N. E. Kemp. New York: Liss, pp. 217–36.

Carroll, R. C. (1988). *Vertebrate Paleontology and Evolution.* New York: Freeman.

Chapman, L. J. and Liem, K. F. (1995). Papyrus swamps and the respiratory ecology of
Barbus neumayeri. Env. Biol. Fishes, **44**, 183–97.

Chew, S. F., Sim, M. Y., Phua, Z. C., Wong, W. P. and Ip, Y. K. (2007). Active ammonia
excretion in the giant mudskipper, *Periophthalmodon schlosseri* (Pallas), during
emersion. *J. Exp. Zool.*, **307A**, 357–69.

Chew, S. F., Chan, N. K. Y., Loong, A. M., Hiong, K. C., Tam, W. L. and Ip, Y. K. (2004).
Nitrogen metabolism in the African lungfish (*Protopterus dolloi*) aestivating in a
mucus cocoon on land. *J. Exp. Biol.*, **207**, 777–86.

Clack, J. A. (2002). *Gaining Ground: the Origin and Early Evolution of Tetrapods.*
Bloomington: University of Indiana.

Congleton, J. L. (1980). Observations on the responses of some southern California
tidepool fishes to nocturnal hypoxic stress. *Comp. Biochem. Physiol.*, **66A**, 719–22.

Daniels, C. B., Orgeig, S., Sullivan, L. C., *et al.* (2004). The origin and evolution of the
surfactant system in fish: insights into the evolution of lungs and swim bladders.
Physiol. Biochem. Zool., **77**, 732–49.

Darwin, C. (1859). *The Origin of Species by Means of Natural Selection.* London: Murray.

de Moraes M. F. P. G., Hölle, S., da Costa, O. T. F., Glass, M. L., Fernandes, M. N.,
and Perry, S. F. (2005). Morphometric comparison of the respiratory organs of
the South American lungfish *Lepidosiren paradoxa. Physiol. Biochem. Zool.*, **78**,
546–59.

de Oliveira, C., Taboga, S. R., Smarra, A. L. S. and Bonilla-Rodriguez, G. O. (2001).
Microscopal aspects of accessory air breathing through a modified stomach in

the armoured catfish *Liposarcus anisitsi* (Siluriformes, Loricariidae). *Cytobios*, **105**, 153–62.

de Pinna, M. C. C. (1998). Phylogenetic relationships of Neotropical Siluriformes (Teleostei: Ostariophysi): historical overview and synthesis of hypotheses. In *Phylogeny and Classification of Neotropical Fishes*, ed. L. R. Malabarba, R. E. Reis, R. P. Vari, Z. M. S. Lucena, and C. A. S. Lucena. Porto Alegre, Brazil: EDIPUCRS, pp. 279–330.

Dudley, R. (1998). Atmospheric oxygen, giant Paleozoic insects and the evolution of aerial locomotor performance. *J. Exp. Biol.*, **201**, 1043–50.

Farrell, A. P. (2007). Cardiovascular systems in primitive fishes. In *Primitive Fishes, Fish Physiology*, vol. 26, ed. D. J. McKenzie, A. P. Farrell and C. J. Brauner. San Diego: Elsevier Academic Press, pp. 53–120.

Fernandes-Castilho, M., Goncalves-de-Freitas, E., Giaquinto, P. C., de Oliveira, C. P. F., de Almeida-Val, V. M. and Val, A. L. (2007). Behavior and adaptation of air-breathing fishes. In *Fish Respiration and Environment*, ed. M. N. Fernandes, F. T. Rantin, M. L. Glass and B. G. Kapoor. Enfield, NH: Science Publisher, pp. 121–45.

Fishman, A. P., Pack, A. I., DeLaney, R. G. and Galante, R. J. (1987). Estivation in Protopterus. In *The Biology and Evolution of Lungfishes*, ed. W. E. Bemis, W. W. Burggren and N. E. Kemp. New York: Liss, pp. 237–48.

Flück, M., Webster, K. A., Graham, J. B., Giomi, F., Gerlach, F. and Schmitz, A. (2007). Coping with cyclic oxygen availability: evolutionary aspects. *Int. Comp. Biol.*, **47**, 524–31.

Fritsche, R., Axelsson, M., Franklin, C. E., Grigg, G. G., Holmgren, S. and Nilsson, S. (1993). Respiratory and cardiovascular responses to hypoxia in the Australian lungfish. *Resp. Physiol.*, **94**, 173–87.

Gee, J. H. (1976). Buoyancy and aerial respiration: factors influencing the evolution of reduced swimbladder volume of some Central American catfishes (Trichomycteridae, Callichthyidae, Loricariidae, Astroblepidae). *Can. J. Zool.*, **54**, 1030–7.

Gee, J. H. and Gee, P. A. (1991). Reactions of gobiid fishes to hypoxia: buoyancy control and aquatic surface respiration. *Copeia*, **1991**, 17–28.

Gee, J. H. and Gee, P. A. (1995). Aquatic surface respiration, buoyancy control and the evolution of air-breathing in gobies (Gobiidae: Pisces). *J. Exp. Biol.*, **198**, 79–89.

Gilbert, C. R. (1993). Evolution and phylogeny. In *The Physiology of Fishes*, ed. D. H. Evans. Boca Raton: CRC Press, pp. 1–45.

Gonzales, T. T., Katoh, M. and Ishimatsu, A. (2006). Air breathing of the aquatic burrow-dwelling eel goby, *Odontamblyopus lacepedii* (Gobiidae: Amblyopinae). *J. Exp. Biol.*, **209**, 1085–92.

Gonzales, T. T., Katoh, M. and Ishimatsu, A. (2008). Respiratory vasculatures of the intertidal air-breathing eel goby, *Odontamblyopus lacepedii* (Gobiidae: Amblyopinae). *Env. Biol. Fishes*, **82**, 341–51.

Gracey, A. Y. (2008). The *Gillichthys mirabilis* Cooper array: a platform to investigate the molecular basis of phenotype plasticity. *J. Fish. Biol.*, **72**, 2118–32.

Gracey, A. Y. and Cossins, A. R. (2003). Application of microarray technology in environmental and comparative physiology. *Ann. Rev. Physiol.*, **65**, 231–59.

Gracey, A. Y., Troll, J. V. and Somero, G. N. (2001). Hypoxia-induced gene expression profiling in the euryoxic fish *Gillichthys mirabilis*. *Proc. Natl. Acad. Sci. USA*, **98**, 1993–8.

Graham, J. B. (1976). Respiratory adaptations of marine air-breathing fishes. In *Respiration of Amphibious Vertebrates*, ed. G. M. Hughes. London: Academic Press, pp. 165–87.

Graham, J. B. (1983). The transition to air breathing in fishes: II. Effects of hypoxia acclimation on the bimodal gas exchange of *Ancistrus chagresi* and *Hypostomus plecostomus* (Loricariidae). *J. Exp. Biol.*, **102**, 157–73.

Graham, J. B. (1985). Seasonal and environmental effects on the blood hemoglobin concentrations of some Panamanian air-breathing fishes. *Env. Biol. Fishes*, **12**, 291–301.

Graham, J. B. (1997). *Air Breathing Fishes: Evolution, Diversity and Adaptation*. San Diego: Academic Press.

Graham, J. B. (1999). Comparative aspects of air-breathing fish biology: an agenda for some neotropical species. In *The Biology of Tropical Fishes*, ed. A. L. Val and V. M. F. Almeida-Val. Manaus, Brazil: INPA, pp. 317–31.

Graham, J. B. (2006). Aquatic and aerial respiration. In *The Physiology of Fishes*, 3rd edn, ed. D. H. Evans and J. B. Claiborne. Boca Raton: CRC Press, pp. 85–117.

Graham, J. B. and Baird, T. A. (1982). The transition to air breathing in fishes: I. Environmental effects on the facultative air breathing of *Ancistrus chagresi* and *Hypostomus plecostomus* (Loricariidae). *J. Exp. Biol.*, **102**, 157–73.

Graham, J. B. and Lee, H. J. (2004). Breathing air in air: in what ways might extant amphibious fish biology relate to prevailing concepts about early tetrapods, the evolution of vertebrate air breathing, and the vertebrate land transition? *Physiol. Biochem. Zool.*, **77**, 720–31.

Graham, J. B., Lee, H. J. and Wegner, N. C. (2007). Transition from water to land in an extant group of fishes: air breathing and the acquisition sequence of adaptations for amphibious life in oxudercine gobies. In *Fish Respiration and Environment*, ed. M. N. Fernandes, F. T. Rantin, M. L. Glass and B. G. Kapoor. Enfield, NH: Science Publisher, pp. 255–88.

Graham J. B., Dudley, R., Aguilar, N. M. and Gans, C. (1995). Implications of the Late Palaeozoic oxygen pulse for physiology and evolution. *Nature*, **375**, 117–20.

Greenwood, P. H. (1987). The natural history of African lungfishes. In *The Biology and Evolution of Lunfishes*, ed. W. E. Bemis, W. W. Burggren and N. E. Kemp. New York: Liss, pp. 163–79.

Grigg, G. C. (1965). Studies on the Queensland lungfish, *Neoceratodus forsteri* (Krefft). III. Aerial respiration in relation to habits. *Aust. J. Zool.*, **13**, 413–21.

Halpin, P. M. and Martin, K. L. M. (1999). Aerial respiration in the salt marsh fish *Fundulus heteroclitus* (Fundulidae). *Copeia*, 1999, 743–8.

Harder, V., Souza, R. H. S., Severi, W., Rantin, F. T. and Bridges, C. R. (1999). The South American lungfish – adaptations to an extreme habitat. In *Biology of Tropical Fishes*, ed. A. L. Val and V. M. Almeida-Val. Manaus, Brazil: INPA, pp. 87–98.

Helfman, G. S., Collette, B. B. and Facey, D. E. (1997). *The Diversity of Fishes*. Malden, MA: Blackwell.

Hill, J. V., Davison, W. and Marsden, I. D. (1996). Aspects of the respiratory biology of two New Zealand intertidal fishes, *Acanthoclinus fuscus* and *Forsterygion* sp. *Env. Biol. Fishes*, **45**, 85–93.

Huey, R. B. and Ward, P. D. (2005). Hypoxia, global warming, and terrestrial Late Permian extinctions. *Science*, **308**, 398–401.

Hughes, G. M. (1976). On the respiration of *Latimeria chalumnae*. *Zool. J. Linn. Soc.*, **59**, 195–208.

Icardo, J. M., Brunelli, E., Perrotta, I., Colvee, E., Wong, W. P. and Ip, Y. K. (2005a). Ventricle and outflow tract of the African lungfish *Protopterus dolloi*. *J. Morph.*, **265**, 43–51.

Icardo, J. M., Ojeda, J. L., Colvee, E., Tota, B., Wong, W. P. and Ip, Y. K. (2005b). Heart inflow tract of the African lungfish *Protopterus dolloi*. *J. Morph.*, **263**, 30–8.

Ip, Y. K., Chew, S. F. and Randall, D. (2004). Five tropical air-breathing fishes, six different strategies to defend against ammonia toxicity on land. *Physiol. Biochem. Zool.*, **77**, 768–82.

Ip, Y. K., Lim, C. B. and Chew, S. F. (2006). Intermediary metabolism in mudskippers, *Periophthalmodon schlosseri* and *Boleophthalmus boddarti*, during immersion or emersion. *Can. J. Zool.*, **84**, 981–91.

Ishimatsu, A., Hishida, A., Takita, Y. T., *et al.* (1998). Mudskippers store air in their burrows. *Nature*, **391**, 237–8.

Ishimatsu, A., Aguilar, N. M., Ogawa, K., Hishida, Y., Takeda, T. and Khoo, K. H. (1999). Arterial blood gas levels and cardiovascular function during varying environmental conditions in a mudskipper, *Periophthalmodon schlosseri*. *J. Exp. Biol.*, **202**, 1753–62.

Ishimatsu, A., Yoshida, Y, Itoki, N., Takeda, T., Lee, H. J. and Graham, J. B. (2007). Mudskippers brood their eggs in air but submerge them for hatching. *J. Exp. Biol.*, **210**, 3946–54.

Janvier, P., (2007). Living primitive fishes and fishes from deep time. In *Primitive Fishes, Fish Physiology*, vol. 26, ed. D. J. McKenzie, A. P. Farrell and C. J. Brauner. San Diego: Elsevier Academic Press, pp. 1–51.

Kaiser, A., Klok, C. J., Socha, J. J., Lee, W-K, Quinlan, M. C. and Harrison, J. F. (2007). Increase in tracheal investment with beetle size supports hypothesis of oxygen limitation on insect gigantism. *Proc. Nat. Acad. Sci.*, **104**, 13198–203.

Kemp, A. (1987). The biology of the Australian lungfish, *Neoceratodus forsteri*. In *The Biology and Evolution of Lungfishes*, ed. W. E. Bemis, W. W. Burggren and N. E. Kemp. New York: Liss, pp. 181–98.

Kind, P. K., Grigg, G. C. and Booth, D. T. (2002). Physiological responses to prolonged hypoxia in the Queensland lungfish *Neoceratodus forsteri*. *Respir. Physiol. Neurobiol.*, **132**, 179–90.

Kok, W. K., Lim, C. B., Lim, T. J. and Ip, Y. K. (1998). The mudskipper *Periophthalmodon schlosseri* respires more efficiently on land than in water and vice versa for *Boleophthalmus boddaerti*. *J. Exp. Zool.*, **280**, 86–90.

Lee, H. J., Martinez, C. A., Hertzberg, K. J., Hamilton, A. L. and Graham, J. B. (2005). Burrow air phase maintenance and respiration by the mudskipper *Scartelaos histophorus* (Gobiidae: Oxudercinae). *J. Exp. Biol.*, **208**, 169–77.

Liem, K. L. (1988). Form and function of lungs: the evolution of air breathing mechanisms. *Amer. Zool.*, **28**, 739–59.

Liem, K. L. (1989). Respiratory gas bladders in teleosts: functional conservation and morphological diversity. *Amer. Zool.*, **29**, 333–52.

Long, J. A. (1995). *The Rise of Fishes*. Baltimore: Johns Hopkins.

Maina, J. N. (1987). The morphology of the lung of the African lungfish, *Protopterus aethiopicus*. *Cell Tiss. Res.*, **250**, 197–204.

Maina, J. N. (2002). *Functional Morphology of the Vertebrate Respiratory Systems*. Enfield, NH: Science Publishers.

Maina, J. N., Wood, C. M., Narahara, A., Bergman, H. L., Laurent, P. and Walsh, P. (1996). Morphology of the swim (air) bladder of a cichlid teleost: *Oreochromis alcalicus grahami* (Trewavas, 1983), a fish adapted to a hyperosmotic, alkaline, and hypoxic environment: a brief outline of the structure and function of the swimbladder. In *Fish Morphology: Horizon of New Research*, ed. J. S. Datta Munshi and H. N. Dutta. Lebanon: Science Publishers, pp. 179–92.

Martin, K. L. M. (1995). Time and tide wait for no fish: intertidal fishes out of water. *Env. Biol. Fishes*, **44**, 165–81.

Martin, K. L. M. and Bridges, C. R. (1999). Respiration in water and air. In *Intertidal Fishes: Life in Two Worlds*, ed. M. H. Horn, K. L. M. Martin and M. Chotowski. San Diego, Academic Press, pp. 54–78.

Mattias, A. T., Rantin, F. T. and Fernandes, M. N. (1998). Gill respiratory parameters during progressive hypoxia in the facultative air-breathing fish, *Hypostomus regaini* (Loricariidae). *Comp. Biochem. Physiol.*, **120A**, 311–15.

Mazlan, A. G., Masitah, A. and Mabani, M. C. (2006). Fine structure of the gills and skins of the amphibious mudskipper, *Periophthalmus chrysospilos* Bleeker, 1852 and a non-amphibious goby *Favonigobius reichei* (Bleeker, 1853). *Acta Ichthyol. Piscat.*, **36**, 127–133.

McKenzie, D. J., Campbell, H. A., Taylor, E. W., Micheli, M., Rantin, F. T. and Abe, A. S. (2007). The autonomic control and functional significance of the changes in heart rate associated with air breathing in the jeju, *Hoplerythrinus unitaeniatus*. *J. Exp. Biol.*, **210**, 4224–32.

Moritz, T. and Linsenmair, K. E. (2007). The air-breathing behavior of *Brevimyrus niger* (Osteoglossomorpha, Mormyridae). *J. Fish. Biol.*, **71**, 279–83.

Murdy, E. O. (1989). A taxonomic revision and cladistic analysis of the oxudercine gobies (Gobiidae: Oxudercinae). *Rec. Aust. Mus. Suppl.*, **11**, 1–93.

Nelson, J. A., Rios, F. S., Sanches, J. R., Fernandes, M. N. and Rantin, F. T. (2007). Environmental influences on the respiratory physiology and gut chemistry of a facultatively air-breathing, tropical herbivorous fish *Hypostomus regani*

(Ihering, 1905). In *Fish Respiration and Environment*, ed. M. N. Fernandes, F. T. Rantin, M. L. Glass and B. G. Kapoor. Enfield, NH: Science Publishers, pp. 191–217.

Nelson, J. S. (2006). *Fishes of the World*. Hoboken, NJ: Wiley.

Nikinmaa, M. and Rees, B. B. (2005). Oxygen-dependent gene expression in fishes. *Amer. J. Physiol. Regul. Integr. Comp. Physiol.*, **288**, R1079–90.

Nilsson, G. E., Hobbs, J-P, A., Östlund-Nilsson, S. and Munday, P. L. (2007). Hypoxia tolerance and air-breathing ability correlate with habitat preference in coral-dwelling fishes. *Coral Reefs*, **26**, 241–8.

Ong, K. J., Stevens, E. D. and Wright, P. A. (2007). Gill morphology of the mangrove killifish (*Kryptolebias marmoratus*) is plastic and changes in response to terrestrial air exposure. *J. Exp. Biol.*, **210**, 1109–15.

Palzenberger, M. and Pohla, H. (1992). Gill surface area of water-breathing freshwater fish. *Rev. Fish Biol. Fish.*, **2**, 187–216.

Park, J. Y. (2002). Structure of the skin of an air-breathing mudskipper, *Periophthalmus magnuspinnatus*. *J. Fish Biol.*, **60**, 1543–50.

Park, J. Y., Lee, Y. J., Kim, I. S. and Kim, S. Y. (2003). Morphological and cytological study of the skin of the Korean eel goby, *Odontamblyopus lacepedii* (Pisces, Gobiidae). *Korean J. Biol. Sci.*, **7**, 43–7.

Perna, S. A. and Fernandes, M. N. (1996). Gill morphometry of the facultative air-breathing loricariid fish, *Hypostomus plecostomus* (Walbaum) with special emphasis on aquatic respiration. *Fish Physiol. Biochem.*, **15**, 213–20.

Perry, S. F. (2007). Swimbladder-lung homology in basal osteichthyes revisited. In *Fish Respiration and Environment*, ed. M. N. Fernandes, F. T. Rantin, M. L. Glass and B. G. Kapoor. Enfield, NJ: Science Publishers, pp. 41–54.

Perry, S. F. and Sander, M. (2004). Reconstructing the evolution of the respiratory apparatus in tetrapods. *Resp. Physiol. Neurobiol.*, **144**, 125–39.

Perry, S. F., Gilmour, K. M., Vulesevic, B., McNeill, B., Chew, S. F. and Ip, Y. K. (2005). Circulating catecholamines and cardiorespiratory responses in hypoxic lungfish (*Protopterus dolloi*): a comparison of aquatic and aerial hypoxia. *Physiol. Biol. Zool.*, **78**, 325–34.

Podkowa, D. and Goniakowska-Witalińska, L. (1998). The structure of the air bladder of the catfish *Pangasius hypophthalmus* Roberts and Vidthayanon 1991 (previously *P. sutchi* Fowler 1937). *Folia Biol. (Kraków)*, **46**, 189–96.

Podkowa, D. and Goniakowska-Witalińska, L. (2002). Adaptations to air breathing in the posterior intestine of the catfish (*Corydoras aeneus*, Callicthyidae): a histological and ultrastructural study. *Folia Biol. (Kraków)*, **50**, 69–82.

Podkowa, D. and Goniakowska-Witalińska, L. (2003). Morphology of the air-breathing stomach of the catfish *Hypostomus plecostomus*. *J. Morph.*, **257**, 147–63.

Potter, I. C., Macey, D. J. and Roberts, A. R. (1997). Oxygen uptake and carbon dioxide excretion by the branchial and postbranchial regions of adults of the lamprey *Geotria australis* in air. *J. Comp. Physiol.*, **166B**, 331–6.

Powell, F. L. (2003). Functional genomics and the comparative physiology of hypoxia. *Ann. Rev. Physiol.*, **65**, 203–30.

Power, J. H. T., Doyle, I. R., Davidson, K. and Nicholas, T. E. (1999). Ultrastructural and protein analysis of surfactant in the Australian lungfish *Neoceratodus forsteri*: evidence for conservation of composition for 300 million years. *J. Exp. Biol.*, **202**, 2543–50.

Randall, D. J., Ip, Y. K., Chew, S. F. and Wilson, J. W. (2004). Air breathing and ammonia excretion in the mudskipper, *Periophthalmodon schlosseri*. *Physiol. Biochem. Zool.*, **77**, 783–8.

Romer, A. S. (1972). Skin breathing – primary or secondary? *Resp. Physiol.*, **14**, 183–92.

Ross, L. G. (2000). Environmental physiology and energetics. In *Tilapias: Biology and Exploitation*, ed. M. C. M. Beveridge and B. J. McAndrew. Dordrecht: Kluwer, pp. 89–128.

Sanchez, A., Soncini, R., Wang, T., Koldkjaer, P., Taylor, E. W. and Glass, M. L. (2001). The differential cardio-respiratory responses to ambient hypoxia and systemic hypoxaemia in the South American lungfish, *Lepidosiren paradoxa*. *Comp. Biochem. Physiol.*, **130A**, 677–87.

Santos, C. T. C., Fernandes, M. N. and Severi, W. (1994). Respiratory gill surface area of a facultative air-breathing loricariid fish, *Rhinelepis strigosa*. *Can. J. Zool.*, **72**, 2009–113.

Satchell, G. H. (1976). The circulatory system of air-breathing fish. In *Respiration of Amphibious Vertebrates*, ed. G. M. Hughes. London: Academic Press, pp. 105–23.

Sayer, M. D. J. (2005). Adaptations of amphibious fish for surviving life out of water. *Fish and Fish.*, **6**, 186–211.

Seymour, R. S., Wegner, N. C. and Graham, J. B. (2008). Body size and the air-breathing organ of the Atlantic tarpon *Megalops atlanticus*. *Comp. Biochem. Physiol.*, **150**A, 282–7.

Silva, J. M., Hernandez-Blanquez, F. J. and Julio, H. F. Jr. (1997). A new accessory respiratory organ in fishes: morphology of the respiratory purses of *Loricariichthys platymetopon* (Pisces, Loricariidae). *Ann. Sci. Natur. Zool. Paris*, **18**, 93–103.

Smith, H. W. (1930). Metabolism of the lungfish Protopterus aethiopicus. *J. Biol. Chem.*, **88**, 97–130.

Sturla, M., Paola, P., Carlo, G., Angela, M. M. and Maria, U. B. (2002). Effects of induced aestivation in *Protopterus annectens*: a histomorphological study. *J. Exp. Zool.*, **292**, 26–31.

Takasusuki, J., Fernandes, M. N. and Severi, W. (1998). The occurrence of aerial respiration in *Rhinelepis strigosa* during progressive hypoxia. *J. Fish. Biol.*, **52**, 369–79.

Takeda, T., Ishimatsu, A., Oikawa, S., Kanda, T., Hishida, Y. and Khoo, K.H. (1999). Mudskipper *Periophthalmodon schlosseri* can repay oxygen debts in air but not in water. *J. Exp. Zool.*, **284**, 265–70.

Taylor, D. S., Turner, B. J., Davis, W. P. and Chapman, B. B. (2008). A novel terrestrial fish habitat inside emergent logs. *Amer. Nat.*, **171**, 263–66.

Thacker, C. E. (2003). Molecular phylogeny of the gobiod fishes (Teleostei: Perciformes: Gobioidei). *Molec. Phylogen. Evol.*, **26**, 354–68.

Todd, E. S. and Ebeling, A. W. (1966). Aerial respiration in the longjaw mudsucker *Gillichthys mirabilis* (Teleostei: Gobiidae). *Biol. Bull.*, **130**, 265–88.

Torday, J. S., Rehan, V. K., Hicks, J. W., *et al.* (2007). Deconvoluting lung evolution: from phenotypes to gene regulatory networks. *Int. Comp. Biol.*, **47**, 601–9.

Urist, M. R. (1973). Testosterone-induced development of limb gills of the lungfish, *Lepidosiren paradoxa. Comp. Biochem. Physiol.*, **44A**, 131–5.

Val, A. L. and Almeida-Val., V. M. F. (1995). *Fishes of the Amazon and their Environment*. Berlin: Springer.

Val., A. L., Almeida-Val, V. M. F., and Affonso, E. G. (1990). Adaptative features of Amazon fishes: hemoglobins, hematology, and intraerythrocytic phosphates and whole blood Bohr effect of *Pterogoplichthys multilradiatus* (Siluriformes). *Comp. Biochem. Physiol.*, **97B**, 435–44.

Ward, P., Labandeira, C., Lauren, M. and Berner, R. A. (2006). Confirmation of Romer's Gap as a low oxygen interval constraining the timing of critical arthropod and vertebrate terrestrialization. *Proc. Natl. Acad. Sci.*, **103**, 16818–22.

Wells, N. A. and Dorr, J. A. (1985). Form and function of the fish *Bothriolepis* (Devonian; Placodermi, Antiarchi): the first terrestrial vertebrate? *Mich. Acad.*, **17**, 157–73.

Wells, R. M. G., Baldwin, J., Seymour, R. S., Christian, K. and Farrell, A. P. (2007). Air breathing minimizes post-exercise lactate load in the tropical Pacific tarpon, *Megalops cyprinoides* Broussonet 1782 but oxygen debt is repaid by aquatic breathing. *J. Fish Biol.*, **71**, 1649–61.

Yoshiyama, R. C., Valpey, C. J., Schalk, L. L., *et al.* (1995). Differential propensities for aerial emergence in intertidal sculpins (Teleostei: Cottidae). *J. Exp. Mar. Biol. Ecol.*, **191**, 195–207.

Zaccone, G., Fusulo, S. and Ainis, L. (1995). Gross anatomy, histology, and immunochemistry of respiratory organs of air-breathing and teleost fishes with particular reference to the neuroendocrine cells and their relationship to the lung and the gill as endocrine organs. In *Histology, Ultrastructure, and Immunochemistry of the Respiratory Organs in Non-Mammalian Vertebrates*, ed. L. M. Pastor. Murcia, Spain: Secretariado de Publicaciones de la Universidad de Murcia, pp 17–43.

Zhang, J., Taniguchi, T., Takita, T. and Ali, A. B. (2000). On the epidermal structure of *Boleophthalmus* and *Scartelaos* mudskippers with reference to their adaptation to terrestrial life. *Ichthyol. Res.*, **47**, 359–66.

Zhang, J., Taniguchi, T., Takita, T. and Ali, A. B. (2003). A study on the epidermal structure of *Periophthalmodon* and *Periophthalmus* mudskippers with reference to their terrestrial adaptation. *Ichthyol. Res.*, **50**, 310–17.

7

Air breathers under water: diving mammals and birds

LARS P. FOLKOW AND ARNOLDUS SCHYTTE BLIX

7.1 Introduction

Most people know that seals spend most of their time, and whales all of their time, in water, but research over the last few decades has shown that several species of both orders of these air-breathing mammals spend as much as 80–90% of the time *under* water. Moreover, sperm whales (*Physeter catodon*) (Watkins *et al.*, 1985) and southern elephant seals (*Mirounga leonina*) (Hindell *et al.*, 1992) normally dive to 300–600 m, but may dive to more than 1000 m and occasionally remain submerged for a staggering 2 hours. Hooded seals (*Cystophora cristata*) normally also dive to 300–600 m with dive durations of 5–25 minutes, but some individuals specialize in repetitive deep diving to more than 1000 m, with durations of up to one hour (Folkow and Blix, 1999). Even birds such as the emperor penguin (*Aptenodytes forsteri*) dive to depths of 550 m with durations of more than 15 minutes (Kooyman and Kooyman, 1995). How is this achieved? Let us look at what physiological problems life under water imposes on air-breathing animals such as whales, seals, penguins, and ducks – but before we do, we have to define 'diving,' for reasons that will be obvious as we go along. Thus, in the following, 'experimental dive' implies that the animal is held under water more or less against its own will, whereas 'voluntary dive' implies that an animal swimming freely (in a pond or in the ocean) dives of its own free will.

When mammals and birds dive, voluntarily or not, respiration has to stop immediately if drowning is to be avoided. However, because the tissues and cells continue to metabolize, and blood circulation is maintained, this results in an ever-increasing arterial hypoxia and hypercapnia, as first shown in the seal in experimental dives by Scholander (1940), and later elegantly demonstrated in voluntarily diving Weddell seals (*Leptonychotes weddellii*) by Qvist *et al.* (1986) (Fig. 7.1).

Respiratory Physiology of Vertebrates: Life with and without Oxygen, ed. Göran E. Nilsson. Published by Cambridge University Press. © Cambridge University Press 2010.

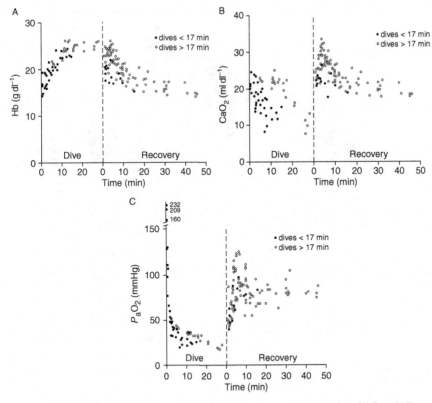

Fig. 7.1 Arterial (A) hemoglobin (Hb), (B) oxygen concentration ($[O_2]$) and (C) oxygen tension (PO_2) during diving and after resurfacing in voluntarily diving Weddell seals (*Leptonychotes weddelli*). Dives were divided into short (< 17 min) and long (> 17 min) dives (Qvist *et al.*, 1986).

There is, in principle, a straightforward solution to this problem: when you cannot renew your oxygen pool, you bring as much oxygen as you can with you, you economize with it to the best of your ability, and you do so from the very moment of submersion – that is, if you want to extend your diving capacity as much as possible. But diving animals do not always want to do that, and in the following we shall see how these animals have adapted to a variety of different ways of diving, which incur a variety of physiological challenges.

7.2   Oxygen stores

Common for all diving mammals and birds is that they have an enhanced capacity to store oxygen in blood and muscles and, in some species, also in the lungs (Fig. 7.2).

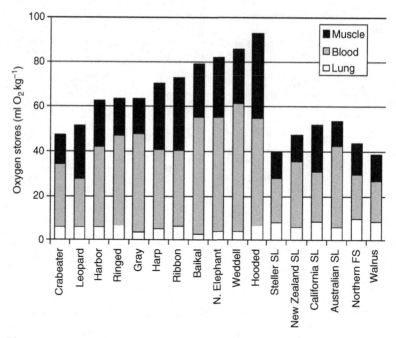

Fig. 7.2 Mass-specific total available body oxygen stores for a variety of adult seals, with relative distribution in blood, muscles, and lungs (Burns *et al.*, 2007). SL, sea lion; FS, fur seal.

7.2.1   Hemoglobin

Habitually diving animals have highly elevated hematocrit levels (Hct) and hemoglobin concentrations ([Hb]). Thus, Hct and [Hb] of deep-diving phocid seals may reach 55–60% and 20–25 g dl^{-1}, respectively (e.g. Scholander, 1940; Clausen and Ersland, 1969; Lenfant *et al.*, 1970; Burns *et al.*, 2007), whereas Hct/[Hb] levels in shallow-diving otariid seals and small cetaceans appear to be somewhat lower (40–63%/13–24 g dl^{-1}) (e.g. Ridgway and Johnston, 1966; Lenfant *et al.*, 1970; Koopman *et al.*, 1999). Hct and [Hb] values of ducks and penguins range between 45 and 53% and 11 and 20 g dl^{-1}, respectively (e.g. Milsom *et al.*, 1973; Stephenson *et al.*, 1989; Ponganis *et al.*, 1999). The red blood cell (RBC) mass of divers correlates positively with diving capacity (Mottishaw *et al.*, 1999), but a high Hct is not maintained without increased blood viscosity. This problem is temporarily overcome, at least in seals, by the sequestering of appreciable amounts of oxygenated RBCs in the spleen when the animal is not diving, to be released into the circulation when diving commences.

7.2.2   Spleen storage of RBCs

That seals have a large, and some seals a huge, spleen has been known for a long time (e.g. Bryden and Lim, 1969; Castellini and Castellini, 1993), and

Kooyman et al. (1980) reported that arterial [Hb] increased with the duration of the dive in Weddell seals. But the real significance of this was first understood by Qvist et al. (1986), who found that arterial [Hb] increased from 15 to 25 g dl^{-1} during dives, allowing the O_2 content of circulating blood to remain constant for 15–18 minutes into long dives in the Weddell seal (Fig. 7.1), and attributed this to release of oxygenated RBCs from the spleen. However, this release results in an increase in Hct from about 40% to about 60%, which will appreciably increase the viscosity of the blood (e.g. Wickham et al., 1989; Elsner and Meiselman, 1995) and thereby the peripheral vascular resistance and myocardial workload. Even so, Castellini et al. (1988) have shown that the time course for the decline in Hct after a dive is too long to return to resting levels between dives and that it remains high during a series of dives until the animal finishes diving and rests or sleeps at the surface. The main function of the spleen therefore seems to be to reduce blood viscosity between dives (Castellini and Castellini, 1993). Cabanac et al. (1997, 1999) have provided information on spleen structure and dynamics in vitro from the hooded seal.

7.2.3   Blood volume and Hb–O_2 affinity

The increased number of RBCs in breath-hold divers is naturally accompanied by a substantially elevated plasma volume, resulting in a blood volume that may amount to 100–200 ml kg^{-1} (Scholander, 1940; Lenfant et al., 1970; Stephenson et al., 1989; Burns et al., 2007), making the blood O_2 stores of the divers three to four times larger than the average for terrestrial mammals (e.g. Snyder. 1983).

The Hb–O_2 affinities of diving birds and mammals are not particularly high (Clausen and Ersland, 1969; Milsom et al., 1973; Willford et al., 1990; Snyder, 1983), which makes sense, as diving animals are not exposed to low O_2 tension (PO_2) during breathing. Deep-diving animals may instead benefit from efficient Hb–O_2 unloading during asphyxic hypoxia, because of both a low Hb temperature coefficient (e.g. Brix et al., 1990a; Willford et al., 1990) and a high Bohr coefficient (e.g. Willford et al., 1990), which facilitates O_2 unloading as acidosis develops during diving (Brix et al., 1990b).

7.2.4   Myoglobin

Diving mammals and birds also have in common a very high concentration of myoglobin ([Mb]) in their skeletal muscles (50–80 mg g^{-1}) (Robinson, 1939; Lenfant et al., 1970; Weber et al., 1974; Snyder, 1983), the highest [Mb] yet recorded (95 mg g^{-1}) being found in swimming muscles of hooded seals (Fig. 7.2) (Burns et al., 2007).

The [Mb] is lower (10–76 mg g^{-1}) in shallow-diving otariid seals (sea lions and fur seals), small cetaceans, sea otters, sirenians, and diving rodents than in the deep-diving animals (Lenfant *et al.*, 1970; Snyder, 1983; Polasek and Davis, 2001), but still higher than in most terrestrial species (e.g. Snyder, 1983). This is also the case in diving birds, which have a [Mb] of 4–64 mg g^{-1} (Weber *et al.*, 1974; Haggblom *et al.*, 1988; Stephenson *et al.*, 1989; Ponganis *et al.*, 1999). The monomeric Mb molecule, which is also found in high concentrations (28 mg g^{-1}) in the heart of diving animals (O'Brien *et al.*, 1992), has an extraordinarily high affinity for O_2 (P_{50} = 2.5 mmHg) and therefore serves as an O_2 store, but – maybe even more importantly – it also facilitates transport of O_2 from the cell membrane to the mitochondria (Scholander, 1960; Wittenberg and Wittenberg, 1989) and provides antioxidant defense (Flögel *et al.*, 2004).

7.2.5   *Lung oxygen stores*

Ventilation in diving mammals is characterized by exchange of high tidal volumes (e.g. Olsen *et al.*, 1969; Kooyman *et al.*, 1971; Wahrenbrock *et al.*, 1974; Reed *et al.*, 1994) during brief surfacing periods, enabling the animals rapidly to dispose of excess CO_2 and reload O_2 (e.g. Reed *et al.*, 1994). Lung volumes of deep-diving mammals conform to allometric relationships for terrestrial species (e.g. Lenfant *et al.*, 1970; Leith, 1976; Folkow and Blix, 1992), but such animals normally *exhale* before diving (e.g. Scholander, 1940; Kooyman *et al.*, 1971; Reed *et al.*, 1994), reflecting both a lack of reliance on lung O_2 stores and avoidance of decompression sickness during diving. Shallow-diving species, however, (e.g. the sea otter [*Enhydra lutris*] and some otariids) have relatively larger lung volumes and rely heavily on lung O_2 stores during diving (Lenfant *et al.*, 1970) (Fig. 7.2).

7.3   Respiratory sensitivity to asphyxia

Considering the extent of hypoxia and (in particular) hypercapnia encountered during prolonged dives (Qvist *et al.*, 1986), clearly, diving animals must be able to suppress breathing better than non-divers. This is achieved in part, at least in ducks, by afferent input from unspecific mechanoreceptors near the glottis and the nares (e.g. Blix *et al.*, 1976a), and in part by a decreased respiratory response to increased carbon dioxide (Andersen and Løvø, 1964). In seals, the situation appears less clear (Robin *et al.*, 1963; Skinner and Milsom, 2004), but it seems that most phocids are indeed sensitive both to hypoxia and hypercapnia, although the threshold level for eliciting a response may be higher. This at least seems to be the case while the animals are breathing air, although, more importantly, what it is when they are submerged remains to be known.

7.4   Oxygen economy during experimental and long-duration natural dives

Field research conducted during the past 30 years has shown that the majority of dives performed by diving birds and mammals are largely aerobic, and that under such conditions the cardiovascular and metabolic adjustments that were revealed in experimental dives in the laboratory take place only to a limited extent (e.g. Kooyman *et al.*, 1980; Stephenson *et al.*, 1986; for a review, see Butler and Jones, 1997). However, physiologists at least now recognize that animals that voluntarily choose to embark on a particularly long dive elicit the same responses as those seen during restrained diving in the laboratory (e.g. Kooyman *et al.*, 1980; Guppy *et al.*, 1986; Thompson and Fedak, 1993; Hochachka *et al.*, 1995; Ponganis *et al.*, 1997). As this book deals with the over-arching theme 'life with and *without* oxygen,' the present chapter therefore not only focuses on what diving birds and mammals do most of the time, but also to a large extent on what they are capable of doing in extreme situations.

Therefore, let us first see what happens when a diving mammal, such as a seal, is forced under water experimentally, as this is how we have learned most of what we know of physiological adaptations to diving, and this is when the ability to cope with hypoxia is really put to the test. It has been known for more than a century that animals respond to forced submersion with a profound, and, in the case of seals, abrupt bradycardia; however, again, it was Scholander and associates (Scholander 1940; Irving *et al.*, 1942) who were first able to put this dramatic event into perspective. In a series of elegant experiments, they provided the basis for the understanding that the bradycardia is developed in concert with a widespread *selective* peripheral arterial constriction. This was later nicely confirmed by use of angiography (Bron *et al.*, 1966) (Fig. 7.3A).

The selective peripheral vasoconstriction ensures that the now 90% reduced cardiac output (Folkow *et al.*, 1967; Sinnett *et al.*, 1978; Blix *et al.*, 1983) is almost exclusively distributed to the most hypoxia-sensitive tissues, with maintenance of systemic arterial pressure. Maintenance of arterial blood pressure under these dramatic cardiovascular changes is primarily achieved by careful balancing of cardiac output and total peripheral resistance, but particularly with regard to diastolic pressure, at least in seals, also by the presence of a huge and elastic ascending aorta. This 'windkessel' was already noted by Burow (1838) and later described in more detail by Drabek (1975). Unfortunately, the physiological responses outlined above are often referred to as 'the diving response' (singular), a term which sometimes seems to be used by ecologists and medics synonymously with the easily recorded bradycardia response. However, 'the master switch of life' (Scholander, 1963), consists of a host of

Fig. 7.3 (A) Angiograms of peripheral (abdominal) arteries of a harbor seal (*Phoca vitulina*). (1) During breathing in air at surface position, well-filled arteries of flanks (upper arrows) and hind flippers (lower arrows) are seen. (2) During experimental diving, the same arteries in the same animal are profoundly constricted and consequently poorly filled with contrast medium. bl, urinary bladder (Bron *et al.*, 1966). (B) Tissue blood flow of harbor seals in four different brain regions, determined by the use of radioactive microspheres before diving and at 2, 5, and 10 min of experimental diving as well as after 40 s of recovery from a 10 min dive. Tissue blood flow is shown as percent of pre-dive (control) values. The number of samples is indicated above the columns (Blix *et al.*, 1983). (C) Left circumflex coronary artery (LCCA) blood flow and velocity in a 3 min experimental dive of a harbor seal. Flow abruptly diminishes immediately after the beginning of the dive and is transiently restored at 30–45 s intervals. The response suggests rhythmic, neurogenic, spasm-like coronary vasoconstrictions modulated by myocardial metabolic demand (Elsner *et al.*, 1985). Arrows mark the duration of the dive.

responses and should more accurately be referred to as 'the diving responses.' For an outline of the historical development of the basal understanding of these defenses against asphyxia in diving mammals and birds, the reader should consult Andersen (1966) and Blix and Folkow (1983).

7.4.1   Organ blood flow

7.4.1.1 Brain

Of all the tissues, the brain is unquestionably the most favored with regard to blood perfusion during dives. Thus, in the voluntarily diving sea lion (*Zalophus californianus*) there is an initial 40% reduction in cerebral blood flow, followed by an almost linear increase to 123% above the pre-dive value at the end of 3-minute dives (Dormer *et al.*, 1977). In phocid seals, however, there appears to be an initial 50% reduction in cerebral blood flow, lasting for more than 5 minutes, while gradually increasing to well above pre-dive values at the end of a 10-minute experimental dive (Fig. 7.3B). In such animals it also appears that the perfusion of different parts of the brain is very different and variable over time, the cortex and the mid-brain being favored over cerebellum and pons/medulla (Blix *et al.*, 1983).

7.4.1.2 Heart

Myocardial blood flow decreases almost instantaneously upon submergence, to an average of only 10% of pre-dive values (Blix *et al.*, 1976b; Kjekshus *et al.*, 1982), and coronary flow oscillates and frequently ceases entirely for periods as long as 45 s in harbor seals (*Phoca vitulina*) during experimental dives lasting up to 15 minutes (Elsner *et al.*, 1985) (Fig. 7.3C). During such dives left ventricular volume, as well as myocyte shortening, decrease progressively, the major reduction occurring in diastole, while systolic dimensions remain relatively constant (Elsner *et al.*, 1985). These changes, which happen at an unchanged left ventricular end-diastolic pressure and at reduced left ventricular contractility, are indicative of reduced ventricular filling, wall tension, and ventricular contractility, which together with the profound bradycardia reduce myocardial workload dramatically and are therefore energetically very advantageous. Moreover, myocardial lactate and hydrogen ion production increase throughout the dive, and after surfacing there is an immediate return to myocardial uptake of lactate (Murphy *et al.*, 1980). Myocardial extraction fraction of glucose and free fatty acids decreases or remains unchanged, and the production of lactate during the dive suggests an increased reliance on anaerobic glycogenolysis/glycolysis, already from the onset of the dive when arterial O_2 tension (P_aO_2) is still high (Kjekshus *et al.*, 1982). This is supported by the finding that the heart of the harp seal (*Pagophilus groenlandicus*) is rich in glycogen, and isolated cardiomyocytes, unlike rat cardiomyocytes, are in fact able to maintain concentrations of adenosine triphosphate (ATP) throughout 1 hour of simulated ischemia (Henden *et al.*, 2004).

There is no evidence of ischemic dilation of the left ventricle, or S-T segment elevation, in the electrocardiogram in harp seals (Kjekshus *et al.*, 1982), suggesting that seals can maintain myocardial function during dives with a reduction of coronary blood flow comparable to that observed in the infarcted dog myocardium (Kjekshus *et al.*, 1972), even at very low P_aO_2. Further studies of myocardial function in diving seals may therefore have relevance for therapeutic approaches aimed at reducing myocardial ischemic injury in humans, particularly with regard to hypoxic preconditioning effects.

7.4.1.3 Kidney

Elsner *et al.* (1966) and Davis *et al.* (1983) have shown that in the Weddell seal, kidney perfusion seems to be completely shut off in both experimental dives and prolonged voluntary dives. Ronald *et al.* (1977) observed in the harp seal that the normal peristaltic motions of the ureter continued only for 10–25 s upon submergence but resumed already within 15–30 s after the end of experimental dives. Moreover, Halasz *et al.* (1974) have demonstrated that isolated harbor seal kidneys can endure 1 hour of warm (32–34°C) ischemia, and then show a prompt recovery of urine production upon reperfusion, whereas dog kidneys treated in the same way remain anuric.

7.4.1.4 Liver and intestines

Liver function in diving animals has not received the attention it deserves, but it is clear that the arterial blood supply to the liver is very low (Zapol *et al.*, 1979), or almost zero (Blix *et al.*, 1983) in experimentally diving seals. However, Davis *et al.* (1983), measuring indocyanine green (ICG) clearance during relatively short voluntary dives in Weddell seals, found that the ICG clearance rates were maintained during dives. The liver normally receives 25–30% of its blood supply from the hepatic artery and the rest from the hepatic portal vein, and as there seems to be more than usual agreement that the splanchnic circulation is shut down during prolonged dives it is possible that the duration of the dives in that study was too short to fully activate the diving responses. In fact, Sparling *et al.* (2007) have even suggested that an almost absurd post-dive increase in resting metabolic rate recorded in grey seals (*Halichoerus grypus*) represents payback of costs deferred during foraging earlier in the day due to vasoconstriction in the gut during diving. If anything, this entirely novel view on diet-induced thermogenesis seems to support the notion that the intestines become uncirculated during prolonged dives. It has also been suggested that the extremely long small intestine of the southern elephant seal is an adaptation to frequent and deep diving, in that it permits absorption from a large mucosal surface area during the brief time the animal is at the surface and the gut is fully perfused with blood (Krockenberger and Bryden, 1994). However, Mårtensson *et al.* (1998) found that this hypothesis is most likely wrong. The extreme variation in small intestinal length among pinnipeds, some of which even utilize the same type of prey, deserves further attention.

7.4.1.5 Muscle

Skeletal muscles are energetically important in diving, not because of their resting metabolic rate and not because of their metabolic scope, which both seem rather low, at least in seals (Ashwell-Erickson and Elsner, 1981), but because of their huge mass. The skeletal muscle blood flow therefore appears to be completely shut off during experimental dives, as first suggested by Scholander (1940) (Fig. 7.4), and later confirmed by Elsner *et al.* (1978), Zapol *et al.* (1979), and Blix *et al.* (1983), by use of radioactive microspheres. Moreover, although the results of Guyton *et al.* (1995) are difficult to interpret, because their method did not distinguish between Hb and Mb, their data seem to support Scholander *et al.* (1942a) in that the oxy-Mb is utilized first and is exhausted already after some 4 minutes of diving. Muscle phosphocreatine (PCr) stores represent an important source of energy for regeneration of ATP, which might further delay the onset of anaerobic metabolism. Diving mammals and birds do not seem to have larger tissue stores of creatine than non-divers (Blix, 1971), but even the levels found in

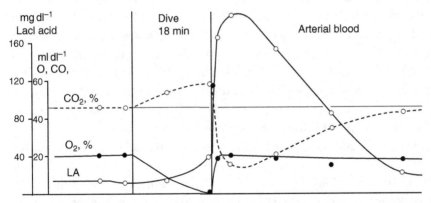

Fig. 7.4 Arterial variations in lactate (LA) concentration in a gray seal (*Halichoerus grypus*) before, during, and after an 18 min experimental dive. Also shown are concomitant changes in arterial content of oxygen (O_2) and carbon dioxide (CO_2). Redrawn from Scholander, 1940.

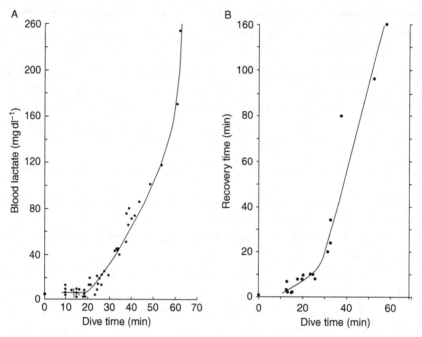

Fig. 7.5 (A) Peak arterial lactic acid concentrations obtained during recovery from various dive durations in a voluntarily diving Weddell seal (*Leptonychotes weddelli*). (B) Recovery time required for various dive durations in the same animal as in A (Kooyman *et al.*, 1980)

terrestrial mammals would provide ATP for additional minutes of diving (Butler and Jones, 1997). Eventually, however, the myocytes have to metabolize anaerobically and produce the lactate load that is evident in the blood after the dive (Scholander, 1940; Scholander *et al.*, 1942b) (Fig. 7.4), and which increases exponentially with the duration of the dive, at least in long (anaerobic) dives (Kooyman *et al.*, 1980) (Fig. 7.5A). All this is not to say that *all* muscle cells are metabolizing anaerobically during long dives. More than likely, the huge muscle mass that is not engaged in swimming, such as the muscles involved in respiration, with their very low resting metabolic rate, may rely entirely on endogenous stores of oxy-Mb, PCr, and possibly hypometabolism (see below).

In most cases, such as in Weddell seals, this lactate load has to be eliminated primarily by the liver to avoid undue pH problems before the animal can dive again, and therefore long dives imply progressively longer recovery times at the surface (Kooyman *et al.* (1980) (Fig. 7.5B). Paradoxically, this implies that if the animal indulges in very long dives, the total time the animal can spend submerged during a day will be reduced when compared with a series of short dives within its aerobic dive capacity (i.e. without accumulation of lactate, which we will address later). However, in the elephant seal (*Mirounga* sp.) this does not always appear to be the case (Le Boeuf *et al.*, 1988; Hindell *et al.*, 1992). In these extremely long-duration divers, long dives are often followed by a series of shorter dives in which one must assume that lactate is used as the substrate for aerobic metabolism (see section 5 below).

The intense vasoconstriction with its decrease in blood supply to the muscles engaged in swimming during prolonged dives is, of course, in direct conflict with the normal vasodilatation seen in response to exercise. This raises the question of how the animal manages to maintain the intense constriction of the vascular smooth muscles under the steadily increasing insult from local metabolites. Folkow *et al.* (1966) demonstrated in the Pekin duck (*Anas platyr-hynchos*) that not only were the resistance vessels *within* the muscles more densely innervated and responded much more strongly to sympathetic stimulation than in terrestrial animals, but sympathetic stimulation affected the large supplying arteries *outside* the muscles as well. Thus, the increased vascular resistance takes place upstream from the tissues to be supplied, and thereby beyond reach of the locally produced vasodilator metabolites. Similar innervation and effects were later described in the harbor seal (White *et al.*, 1973) and are illustrated in Fig. 7.3A. In this context it is worth mentioning that the adrenals are among the very few organs that receive significant blood supply during long dives (Blix *et al.*, 1983), and it is likely that the high concentrations of circulating catecholamines that are seen both in experimental (Hance *et al.*, 1982) and voluntary (Hochachka *et al.*, 1995) diving in seals contribute to the maintenance of the intense vasoconstriction.

7.4.1.6 Lung

Zapol *et al.* (1979) and Liggins *et al.* (1980) found that a very large fraction (30 and 44%, respectively) of radioactive microspheres injected into the aorta during experimental dives as found in the lungs after the dives in the Weddell seal. It was demonstrated by Blix *et al.* (1983), however, that the huge accumulation of microspheres in the lungs is caused by extensive peripheral arteriovenous shunting of the blood early in the dive, whereas bronchial arterial flow is very low (6%) during experimental dives. Sinnett *et al.* (1978) measured pulmonary arterial, right ventricular pressure, and pulmonary wedge pressure (indicative of left atrial pressure) during experimental dives in harbor seals, and found that pulmonary blood flow may cease for extended periods during the diastole when right ventricular and pulmonary wedge pressures are equal (10–16 mmHg). This elevated right ventricular pressure reflects the central pooling of the blood that follows the profound peripheral arterial constriction during diving. The pooling takes place primarily in the huge posterior caval vein and hepatic sinuses, and engorgement of the right heart is prevented by a caval sphincter of striated muscle at the level of the diaphragm (Elsner et al., 1971; Hol et al., 1975). However a dilatation of the right ventricle, unlike the left, is still conspicuous during diving (Blix and Hol, 1973).

Finally, Miller et al. (2006) have recently addressed the hitherto unappreciated problem of maintaining alveolar surfactant action at depth to ensure that inspiration is possible upon return from deep diving. Interestingly, they found very poor surface activity of surfactant in several species of seals, and they suggest that pinniped surfactant primarily has an anti-adhesive function to meet the challenges of regularly collapsing lungs at depth (Ridgway et al., 1969; Kooyman et al., 1970; Falke et al., 1985).

7.4.2   Fuel sources during diving

Fat is the main energy substrate in marine mammals at the surface and continues to be so during diving, at least as long as O_2 is available (e.g. Davis, 1983; Davis et al., 1991), with organ enzyme systems adapted accordingly (Fuson et al., 2003). As their reliance on anaerobic metabolism increases with falling P_aO_2 or ischemia, however, an adequate supply of carbohydrate is required. Under such conditions, high tissue levels of glycogen (e.g. Kerem et al., 1973) serve as an important local source of substrates, and plasma glucose is fairly well maintained throughout long dives, even as anaerobic pathways increase in importance (e.g. Robin et al., 1981; Castellini et al., 1988; Davis et al., 1991). As hepatic blood flow somehow may be maintained by way of the portal venous route even in long (anaerobic) voluntary dives (Davis et al., 1983), liver glycogen stores probably represent a major source of plasma glucose under these conditions. Moreover, the seal liver appears capable of glycogen conversion to glucose even under severely O_2-limited conditions, as shown in vitro in isolated seal liver slices (Hochachka et al., 1988). The elevated levels of catecholamines typical of long-duration dives are also likely to stimulate glycogenolysis under these conditions (Hochachka et al., 1995), and gluconeogenesis from glycerols formed during fat metabolism represents an additional source of glucose during aerobic dives (Davis, 1983). For good measure, the lungs of diving seals may also release some glucose into the circulation (Hochachka et al., 1977). During recovery, when O_2 is readily available again, a range of tissue types may utilize lactate as a substrate for energy metabolism.

7.4.3   Hypometabolism during diving

The understanding of a reduced total energy metabolism during diving was already arrived at by Richet (1899), but did not receive proper attention until the studies of Scholander (1940), who showed that the excess uptake of O_2 after a dive was much lower than expected. Later studies of freely diving animals have confirmed reduced metabolic rates during long dives (e.g. Castellini *et al.*, 1992; Reed *et al.*, 1994; Green *et al.*, 2007). Also, the diving performance of expert divers, such as the southern elephant seal, and in particular the hooded seal, which with a body mass of only 200 kg may regularly dive for about an hour and reach depths of a thousand meters (Folkow and Blix, 1999), suggests that they must have a reduced diving metabolic rate. One contributor to the reduction of metabolic rate would be to reduce body temperature, and that is exactly what seems to be the case.

7.4.4   *Body cooling during diving*

In an almost forgotten paper by Scholander *et al.* (1942b), a 2°C drop in temperature in various parts of the body, including the brain, of experimentally submerged seals was reported, but probably not taken too seriously because the temperatures were recorded with glass (mercury) thermometers, and the paper was seldom cited, even by the authors themselves. Moreover, these experiments were performed with the animal in cold water, whereas Andersen (1959) observed a general decrease in body temperature concomitant with an increase in peripheral insulation in Pekin ducks when only their head was immersed into a beaker of water. Likewise, Caputa *et al.* (1998) reported a 3–4°C reduction in brain temperature in ducks during 5–10 minute experimental dives, and Kooyman *et al.* (1980) and Hill *et al.* (1987) both recorded a 2–3°C drop in central arterial temperature in a voluntarily diving Weddell seal (Fig. 7.6A). Likewise, data logger technology has revealed that voluntary diving birds such as king penguins (*Aptenodytes patagonicus*) (Handrich *et al.*, 1997) and South Georgian shags (*Phalacrocorax georgianus*) (Bevan *et al.*, 1997) cool off appreciably during feeding. This raises the question: Do the animals cool during diving because they depress their metabolic rate, or is their metabolism reduced because they cool down? Odden *et al.* (1999) recreated the experiments of Scholander *et al.* (1942b) with modern equipment and found that brain temperature in hooded and harp seals did indeed drop 2–3°C during 10–15 minute experimental dives (Fig. 7.6B). Based on evidence of anticipatory cooling, the very rapid rate at which cooling took place and the fact that cooling does not seem to continue below a certain level, Blix *et al.* (2002) suggested that this brain cooling is the result of a physiologically regulated process, which is part of the diving responses. This begs another question: Why are the seals, like all other mammals, not shivering when their brain is cooled? Kvadsheim *et al.* (2005) have, in fact, demonstrated in the hooded seal that shivering is inhibited during diving, whereby the cooling, within limits, is not compromised, evidently as part of the diving response 'package.' It therefore seems fair to conclude that diving time is extended in expert divers by a physiologically controlled shift to a hypometabolic state, which is partially supported by

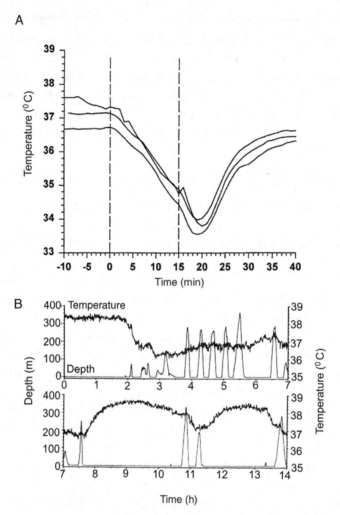

Fig. 7.6 (A) Changes in brain temperature in a hooded seal (*Cystophora cristata*) during three experimental dives each lasting 15 min (Odden *et al.*, 1999). (B) Selected period of 14 h (from a total of 118 h) of continuous recording of aortic blood temperature and dive (depth) in a voluntarily diving Weddell seal (*Leptonychotes weddelli*) (Hill *et al.*, 1987).

the Q_{10} effect of a 2–3°C cooling, which in particular will affect those tissues that are selected for perfusion during prolonged dives (Blix *et al.*, 2002).

7.4.5   *Size matters*

In this context, the high body mass of most diving mammals could be considered a specific adaptation, as an increased body mass is associated with a decreased specific basal metabolic rate (e.g. Singer *et al.*, 1993). It has been suggested that maximum diving capacity increases with body mass among pinnipeds (Ferren and Elsner, 1979). This makes sense, because the brain is the main consumer of oxygen during a dive and is smaller relative to total body mass, and hence blood volume, in larger animals. Accordingly, Irvine *et al.* (2000) found that small underyearling southern elephant seals were diving for shorter periods than were large ones.

7.4.6   Fetal hypoxia during diving

Prolonged diving during pregnancy would represent a particular challenge for the mother, not to speak of the fetus, and reduced diving performance at this time is therefore to be expected. However, Elsner *et al.* (1969) found that near-term Weddell seals were diving voluntarily for up to 60 minutes and reached depths of 310 m. Lenfant *et al.* (1969) reported both greater O_2 affinity of maternal blood and higher O_2 capacity in maternal than fetal blood of these seals, although Liggins *et al.* (1980) found that the fetus responded to 20 minute experimental dives with a prompt and almost as profound bradycardia as the mother. In that study it was also found that maternal kidneys and liver were left without perfusion, whereas the fraction of cardiac output that was directed to the placenta increased six times. Even so, as maternal P_aO_2 steadily decreased, fetal P_aO_2 dropped to less than 10 mmHg at the end of the dives.

7.4.7   Protection from respiratory acidosis

During the recovery period from long dives, accumulated lactate and metabolic H^+ in muscle and other previously ischemic tissues are washed out into the circulation by the reactive hyperemia that ensues upon the withdrawal of sympathetic nervous stimulation of the vascular beds. This inevitably is accompanied by changes in arterial pH and osmolarity, which are sometimes dramatic. Thus, Kjekshus *et al.* (1982) measured arterial pH values as low as 7.14 after 15 minute experimental dives in the harbor seal and Kooyman *et al.* (1980) recorded the extreme pH value of 6.79 after a 61 minute voluntary dive in a Weddell seal. To defend themselves against detrimental changes in pH, habitually diving species must be in the possession of powerful buffer systems. In most species Hb appears to be the most important buffering factor in the blood, whereas other factors such as plasma proteins also contribute in seals (Clausen and Ersland, 1969; Lenfant *et al.*, 1969) and penguins (Murrish, 1982), whereas in the heart and muscles PCr may counteract acidification by the binding of H^+ (Butler and Jones, 1997). In addition, Blix *et al.* (1983) demonstrated that skeletal muscles of harbor seals and gray seals were not flushed wholesale with blood upon emergence from experimental dives, but were instead perfused mosaic-wise in a very conservative way. Thus, parts of the muscles had to wait their turn, probably to mitigate the otherwise intolerable surge in H$^+$ and other leftovers from the anaerobic metabolism. This pattern was also evident in voluntarily diving Weddell seals returning from long dives, in which it took some 5 minutes for a swimming muscle (m. latissimus dorsi) to be fully resaturated with O$_2$ (Guyton et al., 1995).

7.5   Short voluntary dives

It has probably been known since *Genesis* that dabbling ducks normally 'dive' for only a few seconds, whereas it has been known for 400 years (Boyle, 1670) that these birds can endure long periods of submersion. In fact, Andersen (1959) demonstrated that even Pekin ducks can endure 15 minute submersions, given some practice. Unlike previous students of diving animals, Eliassen (1960) studied freely diving sea birds in the wild and found that they were usually, if not always, only under water for very short periods and did not show any lactate accumulation during the dive. This caused some confusion at the time, and with the beginning of studies of freely diving Weddell seals in Antarctica (Kooyman 1965), it also became clear that 97% of their dives were shorter than 26 minutes, whereas their breath-hold capacity is at least 1.2 hours (Kooyman et al., 1980). In such short dives, the dramatic cardiovascular adjustments described above may be less intense, or not expressed at all, such as in the much studied dabbling ducks (e.g. Butler and Woakes, 1975). This caused even more confusion at the time, and some even proposed that emotional stress accounted for 'a large component' of the bradycardia observed during experimental diving (Kanwisher et al., 1981), apparently because of its outward similarity to the 'freezing response.' In so doing, they overlooked several reports of 'normal' diving responses in decerebrate ducks (Andersen, 1963; Djojosugito et al., 1969; and later Gabbot and Jones, 1991), preparations where emotions normally are in short supply, and managed to sidetrack the development of our understanding of diving physiology for almost ten years. However, by now it is understood by most that expert divers, such as seals and some birds, have cortical (voluntary) control over their respiratory *and* cardiovascular system and are able to respond in accordance with the challenge of each individual dive (Kooyman and Campbell, 1972; Blix and Folkow, 1983). This was much later clearly demonstrated by Thompson and Fedak (1993) in freely diving gray seals, in which some that were in the habit of performing long-duration dives displayed a spectacular bradycardia, and those that were in the habit of performing a series of short dives did not. It is also instructive that Elsner et al. (1989) reported that blindfolding prevented the usual anticipatory increases in heart rate in ringed seals

(*Phoca hispida*) approaching an artificial breathing hole in the ice cover of a lake. Moreover, Jobsis *et al.* (2001) have nicely demonstrated that harbor seals trained for 3 minute submersions drastically reduced their heart rate and muscle blood flow when a dive unexpectedly was extended beyond the usual duration. These and many more reports (see Ramirez *et al.*, 2007) clearly reflect how higher central nervous system centers can immediately turn on the full O_2-conserving responses via powerful, modulating descending pathways (Fig. 7.7) if the animals do not know what the duration of submersion will be, or can suppress the responses if the dive is anticipated to be brief.

So what happens when animals decide to perform a series of dives that are short enough not to exceed their aerobic capacity? In the dabbling ducks, which usually stick their heads under the water for only a few seconds, the answer is nothing, short of cessation of breathing. This is because the 'dive' is too short to activate the peripheral chemoreceptors, which are known to be responsible for the activation of the spectacular cardiovascular responses expressed in these animals during experimental dives (Jones and Purves, 1970; Blix and Berg, 1974). In fact, the tufted duck (*Aythya fuligula*) increases its heart rate before 20–40 second dives and has a normal 'resting' heart rate during the dive, with tachycardia in the short recovery period after the dive (Stephenson *et al.*, 1986).

In freely diving seals and the larger penguins the cardiovascular responses are variable, depending on such factors as anticipated duration of the dive and swimming activity (Hill *et al.*, 1987): in very short dives the responses are usually not expressed, whereas in longer dives within the aerobic capacity of the animal a moderate bradycardia reflects some degree of peripheral vasoconstriction. We know, by definition, that the animal in this situation is operating fully aerobically, as no lactate is produced during the dive (Kooyman *et al.*, 1980). This raises the important questions of where the vasoconstriction occurs, and how it can occur without resulting in production of lactate. We have to assume that brain and heart are adequately supplied during such dives, as we know that the kidneys and the liver are (Davis *et al.*, 1983), and ignoring the gut for simplicity, that leaves us with the skeletal muscles. These are all very rich in Mb (Robinson, 1939; Burns *et al.*, 2007); therefore, in prolonged diving it is of paramount importance to keep the blood and muscle O_2 stores separate, because the tremendous difference in O_2 affinity between Hb and Mb (Theorell, 1934) would otherwise allow the Mb to drain the Hb for O_2. That is not so much the case in short aerobic dives, when the best O_2 economy is achieved by perfusion of the active swimming muscles, while the 'resting' muscles, such as those involved in respiration, are shut off from circulation. In that case the swimming muscles necessarily maintain their Mb fully saturated throughout the dive and compete with other perfused tissues for the blood O_2 stores. The inactive muscles, with their very low resting metabolism, however, are able to sustain themselves on their vast amounts of endogenous oxy-Mb, without the need to resort to anaerobic metabolism, and hence no lactate is produced in such dives. In this context it is worth noticing that although the Mb concentration is lower in the thoracic muscles than in the 'swimming muscles' (70 mg g^{-1} vs. 100 mg g^{-1}) (Lestyk et al., 2009) in the deep-diving hooded seal, the concentration is still very high.

This is not to say, however, that the animals are not hypoxic during such so-called aerobic dives. In fact, a P_aO_2 of <20 mmHg has been recorded (e.g. Qvist et al., 1986; Ponganis et al., 2007), and the time it takes to reach such abysmal levels is determined by the amount of exercise going into the dive (Davis et al., 1985). In this context, it is an advantage that these animals produce low drag (Williams and Kooyman, 1985) and have developed very cost-effective modes of locomotion (Williams et al., 2000). It is also significant that the swimming muscles of the harbor seal have many fewer capillaries than the same muscles in dogs (Kanatous et al., 2001; Davis et al., 2004), and both George and Ronald (1973) and Watson et al. (2003), among others, found that the swimming muscles of several species of seals are made exclusively of slow-twitch and fast-twitch oxidative fibers. This, together with the high concentrations of Mb, indicates that the seal muscles are particularly constructed for aerobic metabolism, but that does not exclude the possibility that they can also metabolize anaerobically.

Several investigators have spent vast amounts of time on attempts to calculate the aerobic dive limit (ADL) for several species of seals, using data on blood and muscle O_2 stores and the diving metabolic rate of the animals. The ADL values obtained in this way very often fall short of the diving times actually recorded in many species of both mammals and birds (e.g. Butler, 2006). The concept of ADL was first introduced by Kooyman et al. (1983) and defined as 'the dive duration of freely diving animals, at which post-dive blood lactate concentrations rise above pre-dive level.' It is good to know when this is actually *measured*, but great confusion and rather original physiological concepts are often put forward when ecologists are attempting to explain why both mammals and birds regularly exceed their *calculated* ADL. This review should have shown that it is hardly possible to accurately calculate the ADL of an animal, as this parameter is affected by a number of unknown variables, such as distribution of cardiac output and body temperature.

7.6   Central nervous integration of the physiological responses to diving

The study of the complex integration of the very many components of the diving responses, as outlined above, had its heyday in the 1970s, and the harvest from this period has been summarized in at least two comprehensive reviews (Butler and Jones, 1982; Blix and Folkow, 1983). The responses (Fig. 7.7) are evoked by stimulation of telereceptors (eyes and ears) and/or trigeminal and glossopharyngeal receptors (that are stimulated by contact with water). In some cases (notably in very short dives) initial cardiovascular responses can be occluded by cortico-hypothalamic influences. Normally, however, they are stimulated, albeit to a different extent in different species, immediately on cessation of breathing. In seals and the larger penguins initial responses are usually profound, whereas in dabbling ducks they are more modest. In prolonged dives, arterial chemoreceptors are activated and initiate secondary reinforcement of initial responses. In dabbling ducks chemoreceptors are required for full development of responses, whereas in seals they merely ensure that initial responses are maintained. Thus, the cardiovascular system of diving animals is converted by intense peripheral vasoconstriction, so that the huge blood oxygen store is delivered to the brain, heart, adrenals, and (in short dives) selected skeletal muscles. In this situation other tissues have to rely on local stores of oxygen and/or anaerobic metabolism. This dramatic redistribution of blood takes place at a largely maintained arterial blood pressure due to a well-balanced reduction of cardiac output. Arterial baroreceptors, together with cardiac volume receptors, are instrumental in the execution of this balance. (Fig. 7.7).

7.7   Hypometabolism at cellular level

7.7.1   Enhanced potential for anaerobic metabolism

Diving animals typically have not only very large local stores of glycogen, but also glycolytic enzyme systems with activity levels, isozyme distribution, and control properties that allow them to operate effectively under O_2-limited conditions (e.g. Blix and From, 1971; Messelt and Blix, 1976; Murphy et al., 1980; Fuson et al., 2003). The massive release of catecholamines that typically occurs during diving is also likely to contribute to this end, by activating glycogenolytic and glycolytic pathways (e.g. Hochachka et al., 1995). However, as we shall soon see, a high potential for anaerobic metabolism is not necessarily synonymous with maintenance of high anaerobic metabolic rates.

In hypoxia-sensitive animals, tissue demand for glucose typically increases substantially under O_2-limited conditions, in order to sustain cellular activities through glycolytic ATP production (the 'Pasteur effect'). However, this production line is very inefficient (each glucose molecule yielding only 2 ATP molecules, as opposed to, theoretically, 36 during oxidative phosphorylation) and cannot possibly supply enough energy for long-term maintenance of normal basic functions in the majority of species (Hochachka, 1986a,b; Lutz et al., 2003). This is why hypoxia typically leads to energy insufficiency in hypoxia-sensitive species, caus-ing loss of ion homeostasis with catastrophic consequences (see Chapter 1). In diving mammals and birds, and many other hypoxia-tolerant organisms, protec-tive measures are initiated before such events cause malfunction and death. In these, tissue glucose utilization generally is reduced as the availability of O_2 decreases – which is the opposite of the typical response in hypoxia-sensitive animals. It is therefore referred to as the 'reverse Pasteur effect,' and represents a depression of metabolic rate, which not only has the advantageous effect of extending the time the organism can survive on available (stored) fuels during hypoxia, but also reduces the output of potentially harmful metabolites, such as H^+ (Hochachka, 1986a,b).

7.7.2   *Metabolic depression*

Metabolic depression in the face of environmental stress is now known to be a normal part of the life cycle of many animals – e.g. hibernating mammals, dormant (and anoxic) turtles, dehydrated frogs, dormant insects, and diving mam-mals and birds (see Guppy and Withers, 1999 for a review). It includes two princi-pally different strategies: a downregulation of ATP-generating processes and a reduced rate of ATP utilization, for example, by stabilizing membrane function through reduced membrane permeability, which reduces energy use for active ion transport across membranes. These strategies of 'metabolic arrest' and 'channel arrest' were proposed by Hochachka (1986a) to represent important defense mechanisms in hypoxia-tolerant species such as diving birds and mammals (Fig. 7.8).

7.7.3   *Mechanisms of metabolic arrest*

Metabolic depression involves the reversible phosphorylation of key enzymes, particularly in the glycolytic pathway, and also changes in levels of enzymes responsible for phosphorylation/dephosphorylation (i.e. protein kinases and phosphatases), which allow enzyme activities to be decreased or increased (see Storey, 1988; Bickler and Buck, 2007; Storey and Storey, 2007). Studies of hepato-pancreatic mitochondria from both invertebrates (snails) and higher vertebrates (hibernating ground squirrels) have shown that control of metabolic depression is mainly (~75%) targeting the processes that produce the mitochondrial membrane potential, whereas only 25% of the effect is due to reduced rates of ATP turnover and demand (Bishop *et al.*, 2002; Barger *et al.*, 2003). However, Bickler and Buck (2007) pointed out that responses differ between tissues and species, and that reducing ATP demand (e.g. through chan-nel arrest) must still be a key step in metabolic depression, particularly in nervous tissue. Metabolic depression in hypoxia-tolerant organisms also involves the downregulation of energy-demanding protein synthesis (Smith

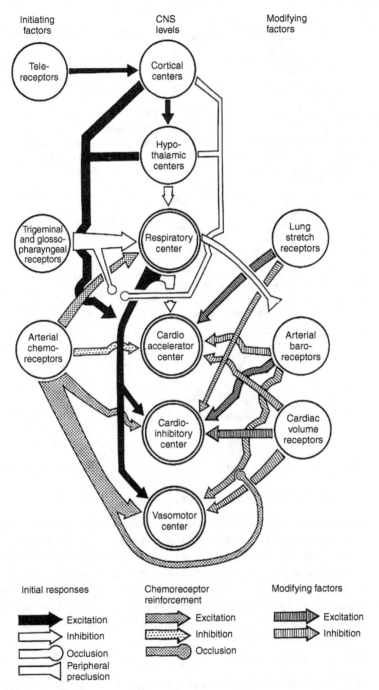

Fig. 7.7 Integration of reflexes involved in initiation and development of the defense against hypoxia in diving mammals and birds, as expressed during experimental dives in the laboratory (from Blix and Folkow 1983).

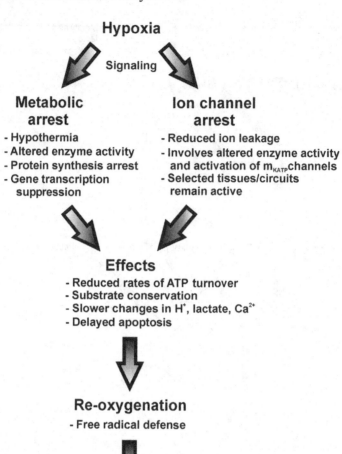

Hypoxia

Signaling

Metabolic arrest
- Hypothermia
- Altered enzyme activity
- Protein synthesis arrest
- Gene transcription suppression

Ion channel arrest
- Reduced ion leakage
- Involves altered enzyme activity and activation of m_{KATP}channels
- Selected tissues/circuits remain active

Effects
- Reduced rates of ATP turnover
- Substrate conservation
- Slower changes in H^+, lactate, Ca^{2+}
- Delayed apoptosis

Re-oxygenation
- Free radical defense

Repair and recovery

Fig. 7.8 Summary of mechanisms of hypoxia defense in hypoxia-tolerant organisms. Hypoxia induces downregulation of ATP-generating and -utilizing processes (termed metabolic arrest and channel arrest, respectively [Hochachka 1986a]), whereby the metabolic derangements and loss of ion homeostasis that typically result from lack of oxygen are delayed. As a result, cell death and apoptosis is much reduced. Upon reoxygenation, efficient defense mechanisms against free radicals further contribute to this end, thus allowing subsequent recovery and repair of damaged cells. (Modified from Bickler and Buck, 2007.)

et al., 1996; Fraser *et al.*, 2001; Pakay *et al.*, 2002) and suppression of overall rates of gene transcription (van Breukelen and Martin, 2002), although a small percentage of genes, several of which are involved in protective mechanisms, show specific upregulation (Storey and Storey, 2007).

7.7.4   Channel arrest

With regard to control mechanisms of ATP consumption, particular attention has been given to cell membrane functions, as active membrane ion transport mechanisms are main consumers of energy (Hochachka, 1986b). In fact, 40–60% of the resting energy metabolism of the vertebrate brain is used for such ion transport in order to maintain the transmembrane ion gradients that form the basis of its electrical activity (Erecińska *et al.*, 2004). In hypoxia-sensitive species, insufficient rates of ATP production in hypoxia therefore perturb transmembrane ion differences. This first causes membrane depolarization, which, particularly in excitable cells (myocytes and neurons), initiates a cascade of detrimental effects, primarily caused by the uncontrolled influx of Ca^{2+} (Hochachka, 1986b) (see also Chapter 1).

By contrast, some anoxia-tolerant species demonstrate channel arrest that stabilizes membrane function at a low energetic cost by reducing membrane ion leakage. The first evidence of ion-specific channel arrest mechanisms as part of a general metabolic depression was obtained from studies of the turtle brain (Bickler, 1992; Pérez-Pinzón *et al.*, 1992; Doll *et al.*, 1993). A detailed discussion of the mechanisms allowing long-term anoxia tolerance in turtles is presented in Chapter 9.

7.7.5   Mechanisms of channel arrest

From studies of anoxia-tolerant vertebrates such as turtles, we know that reversible phosphorylation of key proteins is an important feature also in the modification of ion channels and membrane receptors in connection with channel arrest (see reviews by Bickler and Buck, 2007; Storey and Storey, 2007), and mitochondria appear to play a central role in controlling these events. Thus, in anoxia-tolerant turtle neurons, opening of ATP-sensitive mitochondrial K^+ channels (mK_{ATP}) in anoxia may depolarize the mitochondrial membrane and cause efflux of mitochondrial Ca^{2+} into the cytoplasm. This in turn produces channel arrest of cell membrane N-methyl-D-aspartate receptors (NMDAR) which gate Ca^{2+} influx, via activation of dephosphorylating phosphatases (Bickler and Buck, 2007). Additionally, changes in ATP concentrations that result from anoxia cause levels of adenosine to rise and adenosine receptors to be activated, thereby increasing cytosolic $[Ca^{2+}]$ and producing channel arrest (Buck and Bickler, 1998; Bickler and Buck, 2007). Adenosine is also known to have several additional physiological effects that may be beneficial in conjunction with hypoxia – e.g. vasodilation, promotion of glycolysis and glycogenolysis, reduction of neuronal excitability and of neurotransmitter release – and plays a vital role in anoxia adaptation in hypoxia-tolerant turtles (e.g. Nilsson and Lutz, 1992; Buck and Bickler, 1998; Bickler and Buck, 2007).

In addition to adenosine, other candidate substances that may have signaling functions in hypoxia/ischemia include reactive oxygen species (ROS) and hydrogen sulfide (H_2S). Decreased O_2 availability decreases the rate of mitochondrial oxidative phosphorylation, and thereby alters the rate of mitochondrial production of ROS. Evidence from hypoxic cardiac preconditioning studies imply that ROS may be involved in the activation of mK_{ATP} channels and thereby invoke channel arrest (Pain *et al.*, 2000). Along similar lines, endogenous H_2S has been shown to contribute to cardioprotection by metabolic inhibition preconditioning in rats (Pan *et al.*, 2006). These possible avenues are likely to be further pursued in future research.

7.7.6   *Evidence of metabolic depression in divers*

Apart from the lower-than-expected diving metabolic rates in both diving birds and mammals mentioned above, evidence of metabolic arrest in diving animals has emerged from in vitro experiments with liver tissue from Weddell seals, in which lactate production in hypoxia was found to be much lower than in hypoxia-sensitive tissues, and also lower than expected based on normoxic ATP production rates (i.e. the reverse Pasteur effect) (Hochachka *et al.*, 1988). Also, K^+ efflux and Ca^{2+} influx rates in Weddell seal liver slices subjected to chemical anoxia (antimycin A) were much lower than typically found in hypoxia-sensitive tissues under similar conditions, suggesting ion-specific channel arrest (Hochachka *et al.*, 1988). Possible signs of channel arrest have also been obtained in studies of isolated kidney slices, in which intracellular $[K^+]$ decreased and intracellular $[Na^+]$ increased much less in the harbor seal kidney than in the rat kidney (Hong *et al.*, 1982). Finally, recent studies of electrophysiological responses of isolated brain slices from hooded seals and eider ducks (*Somateria mollissima*) also imply neuronal hypometabolism under severely hypoxic conditions (Folkow *et al.*, 2008; Ludvigsen, and Folkow, 2009).

7.7.7   *Tolerating the consequences of hypoxia: antioxidant defense*

Hypoxia in itself presents a critical challenge, but the subsequent oxidative stress caused by the release of ROS in the re-oxygenation phase after a dive may well represent as great a challenge. Reperfusion after ischemia provides O_2 as a substrate for numerous enzyme oxidation reactions that produce free radicals to such an extent that antioxidant systems may be overwhelmed and cause oxidative damage, such as lipid peroxidation, protein oxidation, and DNA damage (e.g. Zheng *et al.*, 2003). The antioxidant system includes enzymes such as catalase (CAT), superoxide dismutases (SOD), glutathione peroxidase (GPX), glutathione-S-transferase (GST), and low-molecular weight scavengers, such as melatonin, the water-soluble glutathione, urate and ascorbate, and lipid-soluble scavengers such as α-tocopherol (vitamin E).

Enhancement of antioxidant defenses is widely seen in hypometabolic states of a range of species (see reviews by Bickler and Buck, 2007; Storey and Storey, 2007). Even though this aspect has received limited attention in diving animals, their extraordinary tolerance for episodic regional ischemia and abrupt reperfusion suggests that post-ischemic ROS generation and oxidative stress is well taken care of. Thus, diving mammals have an inherently higher antioxidant capacity than non-diving mammals (Wilhelm Filho *et al.*, 2002; Zenteno-Savín *et al.*, 2002). For example, the total SOD activity in ringed seal hearts is higher than in pig hearts (Elsner *et al.*, 1998), and the total antioxidant capacity of seal hearts and kidneys (i.e. anti-peroxidative capacity of tissue homogenates) is also higher than in those of pigs (Zenteno-Savín *et al.*, 2002). In addition, a recent study shows that not only SOD, but also GPX and GST, activities protect the ringed seal heart from deleterious ROS effects, while CAT activity is high in their liver, GPX-activity in their muscles, and SOD and GPX in their lungs (Vázquez-Medina *et al.*, 2006). Possibly, the typically high skeletal muscle Mb levels of divers may also contribute to this end, as Mb has been shown to have an antioxidant function (Flögel *et al.*, 2004). Another important component, not to be forgotten in this context, is that the tissue cooling that occurs during diving, in both mammals and birds, may provide additional protection from oxidative stress, as repeatedly demonstrated (Liu and Yenari, 2007). However, more information on the functions and importance of antioxidant systems in diving animals is badly needed.

7.8   Brain function during diving-induced hypoxia

We have outlined above that the brain, unlike most other organs, experiences no or only limited ischemia even during long dives, as cerebral blood flow largely remains unchanged or, in the end, even increases compared with pre-dive levels. Therefore, brain neurons of diving animals are probably never substrate limited, while they will experience severe hypoxia as P_aO_2 drops. In fact, as mentioned above, their P_aO_2 may fall below 20 mmHg, toward the end of dives even in freely diving Weddell seals (Qvist *et al.*, 1986) (see Fig. 7.1), during natural sleep apnea in northern elephant seals (*Mirounga angustirostris*) (Stockard *et al.*, 2007) and in freely diving emperor penguins (Ponganis *et al.*, 1999). At such low P_aO_2, the brain of most mammals, which typically has a very limited tolerance to acute hypoxia, displays several signs of malfunction: loss of consciousness, purposeful movement, and normal electroencephalographic (EEG) activity occurs within seconds (Siesjö, 1978; Lipton, 1999), and within 2 minutes of stroke onset neurons and glia undergo a sudden and profound loss of membrane potential (Anderson *et al.*, 2005) (see Chapter 1). In fact, even intermittent exposure to relatively mild hypoxia (10% O_2 in 90 second episodes) causes increased apoptosis of hippocampal neurons in rats (Gozal et al., 2001).

By contrast, habitually diving animals do not show signs of brain malfunction despite sometimes being repeatedly exposed to severely hypoxic conditions in their everyday life. Elsner and associates characterized the changes in the EEG of seals that were subjected to experimental diving for extended durations (Elsner et al., 1970; Kerem and Elsner, 1973). They found EEG signal changes characteristic of metabolic impairment (changes from resting alpha and low-voltage fast activity to high-voltage slow waves) only when P_aO_2 fell below 8–10 mmHg.

Kerem and Elsner (1973) also noted that capillary density was higher, and that mean capillary distance was lower, in the brain of the yearling northern elephant seal than in mouse, cat, and human. A high neocortical capillary density was also reported for striped dolphins (*Stenella coeruleoalba*) by Glezer et al. (1987). These findings imply that the enhanced brain hypoxia tolerance of diving mammals in part may be due to a more efficient use of bloodborne O_2, as a result of shorter diffusion distances from capillaries to neurons. In addition, the diving-associated brain cooling of both birds (Caputa et al., 1998) and mammals (e.g. Scholander et al., 1942b; Odden et al., 1999) not only has a hypometabolic effect but is also likely to confer neuroprotection during the hypoxic event, as well as in recovery when oxidative stress ensues as ROS are liberated (e.g. Globus et al., 1995; Liu and Yenari, 2007).

However, recent studies in our laboratory show that seal neurons also survive hypoxia by virtue of a high *intrinsic* hypoxia tolerance: in vitro intracellular and extracellular recordings from isolated neocortical slices of adult hooded seals showed maintenance of a near-normal membrane potential and preservation of the ability to generate action potentials for up to 1 hour in severe hypoxia (slice perfusate PO_2 = 15–30 mmHg; tissue PO_2 = not measurable), whereas mouse neurons depolarized and went silent within 5–10 minutes (Folkow et al., 2008). In birds, Bryan and Jones (1980) concluded that the increased cerebral tolerance to apneic asphyxia in ducks vs. fowl was entirely due to the enhanced O_2 stores and O_2-conserving cardiovascular adjustments of the former, as cerebral NADH: NAD^+ ratios (an index of the oxidative state) were similar at similar brain PO_2 in both species. However, Hochachka (1979), in citing an early manuscript of that publication, argued that their technique would not give insight into cytoplasmic (e.g. glycolytic) events, and recent electrophysiological studies of cerebellar slices from eider ducks and fowl show significant differences in hypoxia tolerance that in part seem to depend on glycolytic mechanisms (Ludvigsen and Folkow, 2009). Brain creatine levels are not higher in diving than in non-diving mammals (Blix, 1971), and ATP regeneration from PCr stores therefore cannot explain these observations. Thus, cellular mechanisms that may explain the enhanced intrinsic hypoxia tolerance of the brain of diving mammals and birds include: (1) a high cerebral potential for anaerobic metabolism, allowing ATP production to persist (albeit at lower levels) even in severe hypoxia; (2) metabolic arrest, coupled to channel arrest, causing reduced neuronal requirements for ATP; (3) enhanced cellular O_2 transport, e.g. through facilitated O_2 diffusion, which would enable neural tissue better to exploit the minute amounts of O_2 that are present even under conditions of severe hypoxia for oxidative ATP production.

7.8.1   *The capacity for cerebral anaerobic metabolism in diving animals*

Murphy *et al.* (1980) found that cerebral dependence on anaerobic metabolism in Weddell seals did not increase much, even as P_aO_2 dropped to levels considered critical to non-diving mammals (~25 mmHg). However, even under resting conditions the proportion of blood glucose utilized by the brain that was released as lactate was higher in seals (20–25%) than in rats (5–15%), which in itself may reflect an adaptation toward enhanced hypoxia tolerance (Murphy *et al.*, 1980). Kerem and Elsner (1973), however, noted that blood lactate content in what they considered to be the cerebral venous effluent of harbor seals on average increased by close to 700% during 'endpoint' dives (i.e. dives that were extended until reversible EEG anomalies were observed). Most of this increase seems to have taken place during the final 4–5 minutes of dives with an average endpoint time of ~18.5 minutes. These data suggest that the seal brain has a fairly high basal contribution from anaerobic metabolism, which is further increased as blood O_2 stores are drained. Thus, lactate dehydrogenase (LDH) isozyme patterns and activities in the brains of divers vs. non-divers are suggestive of biochemical adaptations to hypoxic conditions that may be induced by prolonged diving (Blix and From, 1971; Messelt and Blix, 1976; Murphy *et al.*, 1980). However, although seal brain glycogen stores are larger than in most terrestrial mammals, they are still small compared with those of other tissues, such as skeletal muscle and heart (Kerem *et al.*, 1973), and the role of resident brain glycogen is not obvious. In terrestrial mammals, brain glycogen is stored predominantly in astrocytes (glial cells) and is mobilized to supply substrate to neurons during hypoglycemia, which is an unlikely state in diving seals (Robin *et al.*, 1981; Castellini *et al.*, 1988; Davis *et al.*, 1991). However, astrocyte glycogenolysis is also activated in situations of intense brain activation, because neural energy demand may then temporarily exceed glucose supply (Brown and Ransom, 2007). Under anaerobic conditions this temporary deficit between demand and supply is presumably much more pronounced, and the elevated brain glycogen stores of divers may provide the emergency supply of substrate they will need to maintain sufficient brain function. Anyway, any sustained cerebral anaerobic glycolytic activity during diving would mainly depend on an adequate supply of bloodborne glucose. Upon recovery, however, when lactate washout peaks are high, the seal brain, along with some other tissues (heart and lung) utilizes lactate as substrate (Murphy *et al.*, 1980), which is not surprising given that lactate, rather than glucose, seems to be the preferred oxidative energy substrate of neurons (Pellerin *et al.*, 2007).

7.8.2   Cerebral metabolic depression in diving?

Regardless of adaptations for cerebral anaerobic metabolism in seals, it is, as pointed out by Lutz *et al.* (2003), difficult to envisage a brain with its glycolytic capacity enhanced to the extent that anaerobic ATP production can keep pace with the high rate at which ATP is expended in the fully active brain. Adaptations that promote cerebral glycolytic capacity in diving mammals therefore probably must be coupled to a metabolic depression in order to enable neurons to survive under severely hypoxic conditions. Estimates of brain metabolism in diving animals are few, and results are equivocal. Based on changes in arterio-venous differences in glucose and lactate concentrations, Murphy *et al.* (1980) concluded that brain metabolism in Weddell seals was not O_2 limited in experimental dives lasting for up to 30 minutes, and consequently largely remained unchanged at P_aO_2 down to 25 mmHg. Based on arterio-venous differences in blood O_2 content, however, Kerem and Elsner (1973) estimated that cerebral O_2 consumption rate dropped by up to 50% during long simulated dives in harbor seals, which may be indicative of depressed cerebral metabolism. Such a depression may in part be ascribed to the Q_{10} effect of brain cooling, which occurs during diving in both birds and mammals (Scholander *et al.*, 1942b; Caputa *et al.*, 1998; Odden *et al.*, 1999; Blix *et al.*, 2002). However, results from in vitro studies of seal and duck brain slices suggest that additional mechanisms of metabolic depression are involved.

Thus, we (Folkow *et al.*, 2008; Ludvigsen, and Folkow, 2009) have recently found in isolated cortical and cerebellar slices from eider ducks (*Somateria mollissima*) and hooded seals that different neuronal populations may display two distinctly different responses to severe hypoxia (slice perfusate $PO_2 = 15$–30 mmHg; tissue PO_2 = not measurable): while a majority of spontaneously active neurons went silent within 3–5 minutes of hypoxia exposure, and resumed activity upon re-oxygenation 60 minutes later, some neurons maintained persistent activity, sometimes for a full 60 minutes of hypoxia exposure (Fig. 7.9). The seal and duck neurons that went silent appear to have survived the hypoxic challenge by adopting a suspended, metabolically depressed state, presumably fuelled by anaerobic metabolism.

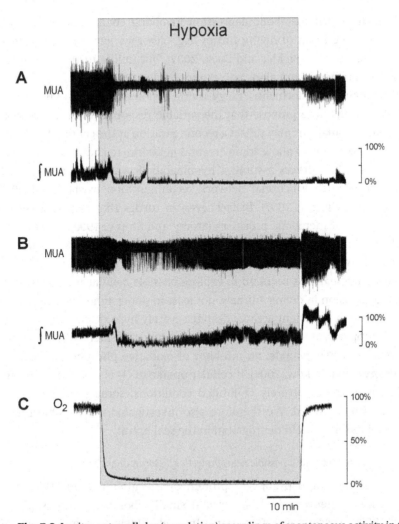

Fig. 7.9 In vitro extracellular (population) recordings of spontaneous activity in the Purkinje cell layer of isolated (400 μm thick) eider duck cerebellar slices before, during, and after a 60 minute exposure to severe hypoxia, reflecting the capacity of eider duck neurons not only to survive, but to remain active despite severe hypoxia. Recordings are shown both as filtered (high-pass 100 Hz; low-pass 3 kHz) multi-unit activity (MUA) and as integrated activity ($\int$MUA; time constant 50 ms), where 100% refers to peak activity level in the pre-exposure (control) period. (A) Typical response involving cessation of activity within ~5 minutes of exposure to hypoxia, followed by partial recovery upon reoxygenation after 60 minutes in severe hypoxia. (B) Hypoxic response involving reduced but maintained activity throughout 60 minutes of hypoxia exposure, as displayed by about 40% of the studied slices. (C) Changes in oxygen content (O_2) of the artificial cerebrospinal fluid superfusing the cerebellar slices, in response to switching from gas bubbling with 95% O_2/5% CO_2 (normoxia) to 95% N_2/5% CO_2 (hypoxia) and back. (data from Ludvigsen, and Folkow, 2009).

We know that metabolic depression involving channel arrest is pivotal in protecting the brain of diving turtles (e.g. *Chrysemys picta*), which may survive anoxia for months (Bickler and Buck, 2007), but unlike turtles, the diving seal must remain active and alert, and cannot escape the effects of hypoxia by altogether assuming a dormant, energy-saving hypometabolic state (Ramirez *et al.*, 2007). We therefore propose that the variable responses of both seal and duck neuronal populations may reflect a reconfiguration at the cellular level, which in the intact organ may allow some cerebral networks to continue to control vital functions, while others assume a hypometabolic state (Ramirez *et al.*, 2007). Evidence of such functional network reconfiguration exists in other species (Marder and Bucher, 2007). In fact, even in turtles that display a very strong suppression of metabolic processes (Storey and Storey, 1990), entry into hypometabolism is not simply a general shutdown, but seems to be differentially regulated, within and between cells as well as organs (Hochachka *et al.*, 1996).

Further research is required to explore possible cellular mechanisms of neuronal depression in diving animals, not least to understand how some networks may maintain persistent activity despite severely hypoxic conditions. The latter activity appears, at least in part, to depend on oxidative metabolism, as it tends to be reduced when cyanide, an inhibitor of oxidative phosphorylation, is added (Ludvigsen and Folkow, 2009). If cellular uptake of O_2 is possible to any appreciable extent under severely O_2-limited conditions, some adaptive mechanism seems to be required. We therefore also investigated the possible role of the oxygen-binding protein neuroglobin in the seal brain.

7.8.3   *Neuroglobin as a possible neuroprotective factor in diving mammals*

Neuroglobin (Ngb) is a protein of the globin family that is widely expressed in neurons in humans and small laboratory rodents (Burmester *et al.*, 2000; Pesce *et al.*, 2002; Hankeln *et al.*, 2004). Like myoglobin (Mb), it binds O_2 with a very high affinity, but its intracellular levels are much lower than those of Mb (Burmester *et al.*, 2000, Pesce *et al.*, 2002). Ngb may serve to facilitate O_2 uptake into, and transport within, neurons (Burmester *et al.*, 2000; Bentmann *et al.*, 2005), and in mouse cortical neuron cultures Ngb has been reported to be upregulated in long-term hypoxia, and also to protect neurons from ischemia/reperfusion injury (Sun *et al.*, 2001).

In a recent study, Williams *et al.* (2007) attempted to determine levels of Hb and resident globins (i.e. Ngb and cytoglobin [Cygb], another intracellularly based globin [Burmester *et al.*, 2002]) in the brains of diving and non-diving mammals, using spectrophotometric techniques). However, their work is likely to be flawed owing to the difficulties in separating the absorption spectra of Hb and Ngb/Cygb, on the one hand, and other cellular heme proteins, such as cytochromes, on the other. Also, no respiratory function has yet been documented for Cygb (Hankeln *et al.*, 2004).

However, our immunohistochemical studies have shown that the distribution of Ngb in the hooded seal brain is quite unusual compared with that in mice and other non-divers: Mitz *et al.* (2009) found that while in the mouse brain Ngb was as expected primarily localized in neurons (Laufs *et al.*, 2004), seal brain glial cells contained more Ngb than seal neurons. As amounts of Ngb correlate with regional rates of O_2 consumption (Bentmann *et al.*, 2005), these findings indicate that glial cells play a more prominent role in the oxidative metabolism of the seal brain than neurons, which may rely on anaerobic metabolism to survive severe hypoxia. In this context it is interesting that neurons in the striped dolphin cerebral cortex are surrounded by glia in unusual abundance (Glezer *et al.*, 1987). There is increasing evidence of an important role of glial cells in regulating neuronal oxidative metabolism and activity in mammals, among other things as substrate suppliers (e.g. Brown and Ransom, 2007; Pellerin *et al.*, 2007). Glial cells are also known to clear extracellular fluid of excess K^+ (e.g. Walz, 2000) and glutamate (Danbolt, 2001), thereby possibly slowing the detrimental hypoxia-induced rise in extracellular K^+ and glutamate caused by a loss of membrane ion balance as a result of impaired ion pumping (see Chapter 1). Thus, it is conceivable that glia may play a particularly prominent role in maintaining neocortical activity under extreme hypoxia in diving animals.

References

Andersen, H. T. (1959). Depression of metabolism in the duck during diving. *Acta Physiol. Scand.* **46**, 234–9.

Andersen, H. T. (1963). The reflex nature of the physiological adjustments to diving, and their afferent pathway. *Acta Physiol. Scand.* **58**, 263–73.

Andersen, H. T. (1966). Physiological adaptation in diving vertebrates. *Physiol. Rev.* **46**, 212–43.

Andersen, H. T. and Løvø, A. (1964). The effect of carbon dioxide on the respiration of avian divers (ducks). *Comp. Biochem. Physiol.* **12**, 451–6.

Anderson, T. R., Jarvis, C. R., Biedermann, A. J., Molnar, C. and Andrew, R. D. (2005). Blocking the anoxic depolarization protects without functional compromise following simulated stroke in cortical brain slices. *J. Neurophysiol.* **93**, 963–79.

Ashwell-Erickson, S. and Elsner, R. (1981). The energy cost of free existence for Bering Sea harbour and spotted seals. In: *The Eastern Bering Sea Shelf: Oceanography and Resources*, ed. D. W. Hood and J. A. Calder. Washington, DC: US Dept. of Commerce 2, pp. 879–99.

Barger, J. L., Brand, M. D., Barnes, B. M. and Boyer, B. B. (2003). Tissue-specific depression of mitochondrial proton leak and substrate oxidation in hibernating arctic ground squirrels. *Am. J. Physiol.* **284**, R1306–13.

Bentmann, A., Schmidt, M., Reuss, S., Wolfrum, U., Hankeln, T. and Burmester, T. (2005). Divergent distribution in vascular and avascular mammalian retinae links neuroglobin to cellular respiration. *J. Biol. Chem.* **280**, 20660–5.

Bevan, R. M., Boyd, I. L., Butler, P. J., Reid, K. R., Woakes, A. J. and Croxall, J. (1997). Heart rates and abdominal temperatures of free-ranging South Georgian shags. *J. Exp. Biol.* **200**, 661–75.

Bickler, P. E. (1992). Cerebral anoxia tolerance in turtles: regulation of intracellular calcium and pH. *Am. J. Physiol.* **263**, R1298–302.

Bickler, P. E. and Buck, L. T. (2007) Hypoxia tolerance in reptiles, amphibians and fishes: life with variable oxygen availability. *Annu. Rev. Physiol.* **69**, 145–70.

Bishop, T., St-Pierre, J. and Brand, M. D. (2002). Primary causes of decreased mitochondrial oxygen consumption during metabolic depression in snail cells. *Am. J. Physiol.* **282**, R372–82.

Blix, A. S. (1971). Creatine in diving animals – a comparative study. *Comp. Biochem. Physiol. A* **40**, 805–7.

Blix, A. S. and Berg, T. (1974). Arterial hypoxia and the diving responses of ducks. *Acta Physiol. Scand.* **92**, 566–8.

Blix, A. S., and Folkow, B. (1983). Cardiovascular adjustments to diving in mammals and birds. In: *Handbook of Physiology. The Cardiovascular System III. Peripheral Circulation and Organ Blood Flow*, ed. J. T. Shepherd and F. M. Abboud. Bethesda: American Physiological Society, pp. 917–45.

Blix, A. S. and From, S. H. (1971). Lactate dehydrogenase in diving animals – a comparative study with special reference to the eider (*Somateria mollissima*). *Comp. Biochem. Physiol. B* **40**, 579–84.

Blix, A. S. and Hol, R. (1973). Ventricular dilatation in the diving seal. *Acta Physiol. Scand.* **87**, 431–2.

Blix, A. S., Elsner, R. and Kjekshus, J. K. (1983). Cardiac output and its distribution through A-V shunts and capillaries during and after diving in seals. *Acta Physiol. Scand.* **118**, 109–16.

Blix, A. S., Folkow, L. P. and Walloe, L. (2002). How seals may cool their brains during prolonged diving. *J. Physiol.* **543**.P, 7P.

Blix, A. S., Rettedal, A. and Stokkan, K. -A. (1976a). On the elicitation of the diving responses in ducks. *Acta Physiol. Scand.* **98**, 478–83.

Blix, A. S., Kjekshus, J. K., Enge, I. and Bergan, A. (1976b). Myocardial blood flow in the diving seal. *Acta Physiol. Scand.* **96**, 227–8.

Boyle, R. (1670). New pneumatical experiments about respiration. *Philos. Trans. R. Soc. Lond.* **5**, 2011–31.

Brix, O., Condo, S. G., Bargard, A., Tavazzi, B. and Giardina, B. (1990a). Temperature modulation of oxygen transport in a diving mammal (*Balaenoptera acutorostrata*). *Biochem. J.* **271**, 509–13.

Brix, O., Ekker, M., Condo, S. G., Scatena, R., Clementi, M. E. and Giardina, B. (1990b). Lactate does facilitate oxygen unloading from the haemoglobin of the whale *Balaenoptera acutorostrata*, after diving. *Arct. Med. Res.* **49**, 39–42.

Bron, K. M., Jr. Murdaugh, H. V., Millen, J. E., Lenthall, R., Raskin, P. and Robin, E. D. (1966). Arterial constrictor response in a diving mammal. *Science* **152**, 540–3.

Brown, A. M.and Ransom, B. R. (2007). Astrocyte glycogen and brain metabolism. *Glia* **55**, 1263–71.

Bryan, R. M.and Jones, D. R. (1980). Cerebral energy metabolism in diving and non-diving birds during hypoxia and apnoeic asphyxia. *J. Physiol.* **299**, 323–36.

Bryden, M. M.and Lim, G. H. K. (1969). Blood parameters of the southern elephant seal (*Mirounga leonina*) in relation to diving. *Comp. Biochem. Physiol.* **28**, 139–48.

Buck, L. T.and Bickler, P. E. 1998. Adenosine and anoxia reduce N-methyl-D-aspartate receptor open probability in turtle cerebrocortex. *J. Exp. Biol.* **210**, 289–97.

Burmester, T., Ebner, B., Weich, B. and Hankeln, T. (2002). Cytoglobin: a novel globin type ubiquitously expressed in vertebrate tissue. *Mol. Biol. Evol.* **19**, 416–21.

Burmester, T., Weich, B., Reinhardt, S. and Hankeln, T. (2000). A vertebrate globin expressed in the brain. *Nature* **407**, 520–3.

Burns, J. M., Lestyk, K. C., Folkow, L. P., Hammill, M. O. and Blix, A. S. (2007). Size and distribution of oxygen stores in harp and hooded seals from birth to maturity. *J. Comp. Physiol. B* **177**, 687–700.

Burow, A. (1838). Ueber des Gefässystem der Robben. *Arch. Anat. Physiol. Leipzig* **5**, 230–58.

Butler, P. J. (2006). Aerobic dive limit. What is it and is it always used appropriately? *Comp. Biochem. Physiol. A* **145**, 1–6.

Butler, P. J. and Jones, D. R. (1982). The comparative physiology of diving vertebrates. In *Advances in Comparative Physiology and Biochemistry, Vol. 8*, ed. O. E. Lowenstein. New York: Academic Press, pp. 179–364.

Butler, P. J. and Jones, D. R. (1997). Physiology of diving of birds and mammals. *Physiol. Rev.* **77**, 837–99.

Butler, P. J. and Woakes, A. J. (1975). Changes in heart rate and respiratory frequency associated with natural submersion of ducks. *J. Physiol.* **256**, 73–4.

Cabanac, A., Folkow, L. P. and Blix, A. S. (1997). Volume capacity and contraction control of the seal spleen. *J. Appl. Physiol.* **82**, 1989–94.

Cabanac, A. J., Messelt, E. B., Folkow, L. P. and Blix, A. S. (1999). The structure and blood-storing function of the spleen of the hooded seal. *J. Zool.* **248**, 75–81.

Caputa, M., Folkow, L. and Blix, A. S. (1998). Rapid brain cooling in diving ducks. *Am. J. Physiol.* **275**, R363–71.

Castellini, J. M. and Castellini, M. A. (1993). Estimation of splenic volume and its relationship to long-duration apnea in seals. *Physiol. Zool.* **66**, 619–27.

Castellini, M. A., Davis, R. W. and Kooyman, G. L. (1988). Blood chemistry regulation during repetitive diving in Weddell seals. *Physiol. Zool.* **61**, 379–86.

Castellini, M. A., Kooyman, G. L. and Ponganis, P. J. (1992). Metabolic rates of freely diving Weddell seals: correlations with oxygen stores, swim velocity and diving duration. *J. Exp. Biol.* **165**, 181–94.

Clausen, G. and Ersland, A. (1969). The respiratory properties of the blood of the bladdernose seal (*Cystophora cristata*). *Respir. Physiol.* **7**, 1–6.

Danbolt, N. C. (2001). Glutamate uptake. *Prog. Neurobiol.* **65**, 1–105.

Davis, R. W. (1983). Lactate and glucose metabolism in the resting and diving harbour seal (*Phoca vitulina*). *J. Comp. Physiol.* **153**, 275–88.

Davis, R. W., Williams, T. M. and Kooyman, G. L. (1985). Swimming metabolism of yearling and adult harbour seals. *Physiol. Zool.* **58**, 590–6.

Davis, R. W., Castellini, M. A., Kooyman, G. L. and Maue, R. (1983). Renal glomerular filtration rate and hepatic blood flow during voluntary diving in Weddell seals. *Am. J. Physiol.* **245**, R743–8.

Davis, R. W., Castellini, M. A., Williams, T. M. and Kooyman, G. L. (1991). Fuel homeostasis in the harbour seal during submerged swimming. *J. Comp. Physiol. B* **160**, 627–35.

Davis, R. W., Polasek, L., Watson, R., Fuson, A., Williams, T. M. and Kanatous, S. B. (2004). The diving paradox: new insight into the role of the dive response in air-breathing vertebrates. *Comp. Biochem. Physiol. A* **138**, 263–8.

Djojosugito, A. M., Folkow, B. and Yonce, L. R. (1969). Neurogenic adjustments of muscle blood flow, cutaneous A-V shunt flow and of venous tone during 'diving' in ducks. *Acta Physiol. Scand.* **75**, 377–86.

Doll, C. J., Hochachka, P. W. and Reiner, P. B. (1993). Reduced ionic conductances in turtle brain. *Am. J. Physiol.* **265**, R929–33.

Dormer, K. J., Denn, M. J. and Stone, H. L. (1977). Cerebral blood flow in the sea lion (*Zalophus californianus*) during voluntary dives. *Comp. Biochem. Physiol. A* **58**, 11–8.

Drabek, C. M. (1975). Some anatomical aspects of the cardiovascular system of Antarctic seals and their possible functional significance in diving. *J. Morphol.* **145**, 85–92.

Eliassen, E. (1960). Cardiovascular responses to submersion asphyxia in avian divers. *Årbok Univ. Bergen, Mat.-Nat.*, ser. no. **2**, pp. 1–76.

Elsner, R. and Meiselman, H. J. (1995). Splenic oxygen storage and blood viscosity in seals. *Mar. Mamm. Sci.* **11**, 93–6.

Elsner, R., Blix, A. S. and Kjekshus, J. K. (1978). Tissue perfusion and ischemia in diving seals. *Physiologist* **21**, 33.

Elsner, R., Hanafee, W. N. and Hammond D. D. (1971). Angiography of the inferior vena cava of the harbour seal during simulated diving. *Am. J. Physiol.* **220**, 1155–7.

Elsner, R., Kooyman, G. L. and Drabek, C. M. (1969). Diving durations in pregnant Weddell seals. In *Antrctic Ecology*, ed. M. Holdgate. New York: Academic Press, pp. 477–82.

Elsner, R., Franklin, D. L., van Citters, R. L. and Kenney, D. W. (1966). Cardiovascular defence against asphyxia. *Science* **153**: 941–9.

Elsner, R., Øyasæter, S., Almaas, R. and Saugstad, O. D. (1998). Diving seals, ischemia-reperfusion and oxygen radicals. *Comp. Biochem. Physiol. A* **119**, 975–80.

Elsner, R, Shurley, J. T., Hammond, D. D. and Brooks, R. E. (1970). Cerebral tolerance to hypoxemia in asphyxiated Weddell seals. *Resp. Physiol.* **9**, 287–97.

Elsner, R., Wartzok, D., Sonfrank, N. B. and Kelly, B. P. (1989). Behavioral and physiological reactions of arctic seals during under-ice pilotage. *Can. J. Zool.* **67**, 2506–13.

Elsner, R., Millard, R. W., Kjekshus, J. K., White, F., Blix, A. S. and Kemper, W. S. (1985). Coronary blood flow and myocardial segment dimensions during simulated dives in seals. *Am. J. Physiol.* **249**, H1119–26.

Erecińska, M., Cherian, S. and Silver, I. A. (2004). Energy metabolism in mammalian brain during development. *Prog. Neurobiol.* **73**, 397–445.

Falke, K. J., Hill, R. D., Qvist, J., *et al.* (1985). Seal lung collapse during free diving: evidence from arterial nitrogen tensions. *Science* **229**, 556–8.

Ferren, H. and Elsner, R. (1979). Diving physiology of the ringed seal: adaptations and implications. *Proc. 29th Alaska Sci. Conf.*, pp. 379–87.

Flögel, U., Godecke, A., Klotz, L. O. and Schrader, J. (2004). Role of myoglobin in the antioxidant defense of the heart. *FASEB J.* **18**, 1156–8.

Folkow, L. P. and Blix, A. S. (1992). Metabolic rates of minke whales (*Balaenoptera acutorostrata*) in cold water. *Acta Physiol. Scand.* **146**, 141–50.

Folkow, L. P. and Blix, A. S. (1999). Diving behaviour of hooded seals (*Cystophora cristata*) in the Greenland and Norwegian Seas. *Polar Biol.* **22**, 61–74.

Folkow, B., Fuxe, K. and Sonnenschein, R. R. (1966). Responses of skeletal musculature in its vasculature during 'diving' in the duck: peculiarities of the adrenergic vasoconstrictor innervation. *Acta Physiol. Scand.* **67**, 327–42.

Folkow, B., Nilsson, N. J. and Yonce, L. R. (1967). Effects of 'diving' on cardiac output in ducks. *Acta Physiol. Scand.* **70**, 347–61.

Folkow, L. P., Ramirez, J. M., Ludvigsen, S., Ramirez, N. and Blix, A. S. (2008). Remarkable neuronal hypoxia tolerance in the deep-diving adult hooded seal (*Cystophora cristata*). *Neurosci. Lett.* **446**, 147–150.

Fraser, K. P. P., Houlihan, D. F., Lutz, P. L., Leone-Kabler S., Manuel, L. and Brechin, J. G. (2001) Complete suppression of protein synthesis during anoxia with no post-anoxia protein synthesis debt in the red-eared slider turtle *Trachemys scripta elegans*. *J. Exp. Biol.* **204**, 4353–60.

Fuson, A. L., Cowan, D. F., Kanatous, S. B., Polasek, L. K. and Davis, R. W. (2003). Adaptations to diving hypoxia in the heart, kidneys and splanchnic organs of harbour seals (*Phoca vitulina*). *J. Exp. Biol.* **206**, 4139–54.

Gabbot, G. R. J. and Jones, D. R. (1991). The effect of brain transection on the response to forced submergence in ducks. *J. Auton. Nerv. Syst.* **36**, 65–74.

George, J. C. and Ronald K. (1973). The harp seal. XXV. Ultrastructure and metabolic adaptation of skeletal muscle. *Can J. Zool.* **51**, 833–40.

Glezer, I. I., Jacobs, M. S. and Morgane, P. J. (1987). Ultrastructure of the blood brain barrier in the dolphin (*Stenella coeruleoalba*). *Brain Res.* **414**, 205–18.

Globus, M. Y., Alonso, O., Dietrich, W. D., Busto, R. and Ginsberg, M. D. (1995). Glutamate release and free radical production following brain injury: effects of posttraumatic hypothermia. *J. Neurochem.* **65**, 1704–11.

Gozal, D., Daniel, J. M. and Dohanich, G. P. (2001). Behavioral and anatomical correlates of chronic episodic hypoxia during sleep in the rat. *J. Neurosci.* **21**, 2442–50.

Green, J. A., Halsey, L. G., Butler, P. J. and Holder, R. L. (2007). Estimating the rate of oxygen consumption during submersion from the heart rate of diving animals. *Am. J. Physiol.* **292**, R2028–38.

Guppy, M. and Withers, P. (1999). Metabolic depression in animals: physiological perspectives and biochemical generalizations. *Biol. Rev.* **74**, 1–40.

Guppy, M., Hill, R. D., Schneider, R. C., Qvist, J., Liggins, G. C., Zapol, W. M. and Hochachka, P. W. (1986). Microcomputer assisted metabolic studies of voluntary diving of Weddell seals. *Am. J. Physiol.* **250**, R175–87.

Guyton, G. P., Stanek, K. S., Sneider, R. C., *et al.* (1995). Myoglobin saturation in free-diving Weddell seals. *J. Appl. Physiol.* **79**, 1148–55.

Haggblom, L., Terwilliger, R. C. and Terwilliger, N. B. (1988). Changes in myoglobin and lactate dehydrogenase in muscle tissues of a diving bird, the pigeon guillemot, during maturation. *Comp. Biochem. Physiol. B* **91**, 273–77.

Halasz, N. A., Elsner, R., Garvie, R. S. and Grotke, G. T. (1974). Renal recovery from ischemia: a comparative study of harbour seal and dog kidneys. *Am. J. Physiol.* **227**, 1331–5.

Hance, A. J., Robin, E. D., Halter, J. B., *et al.* (1982). Hormonal changes and enforced diving in the harbour seal. II. Plasma catecholamines. *Am. J. Physiol.* **242**, R528–32.

Handrich, Y. R., Bevan, R., Charrassin, J.-B., *et al.* (1997). Hypothermia in foraging king penguins. *Nature* **388**, 64–7.

Hankeln, T., Wystub, S., Laufs, T., *et al.* (2004). The cellular and subcellular localization of neuroglobin and cytoglobin – a clue to their function? *IUBMB Life* **56**, 671–9.

Henden, T., Aasum, E., Folkow, L., Mjøs, O. D., Lathrop, D. A. and Larsen, T. S. (2004). Endogenous glycogen prevents calcium overload and hypercontracture in harp seal myocardial cells during simulated ischemia. *J. Mol. Cell. Cardiol.* **37**, 43–50.

Hill, R. D., Schneider, R. C., Liggins, G. C. *et al.* (1987). Heart rate and body temperature during free diving of Weddell seals. *Am. J. Physiol.* **253**, R344–51.

Hindell, M. A., Slip, D. J., Burton, H. R. and Bryden, M. M. (1992). Physiological implications of continuous, prolonged and deep dives of the southern elephant seal (*Mirounga leonina*). *Can. J. Zool.* **70**, 370–9.

Hochachka, P. W. (1979). Metabolic status during diving and recovery in marine mammals. *Int. Rev. Physiol.* **20**, 253–87.

Hochachka, P. W. (1986a). Metabolic arrest. *Intensive Care Med.* **12**, 127–33.

Hochachka, P. W. (1986b). Defense strategies against hypoxia and hypothermia. *Science* **231**, 234–41.

Hochachka, P. W., Castellini, J. M., Hill, R. D., *et al.* (1988). Protective metabolic mechanisms during liver ischemia: transferable lessons from long-diving animals. *Mol. Cell. Biochem.* **84**, 77–85.

Hochachka, P. W., Liggins, G. C., Guyton, G. P., *et al.* (1995). Hormonal regulatory adjustments during voluntary diving in Weddell seals. *Comp. Biochem. Physiol. B* **112**, 361–75.

Hochachka, P. W., Liggins, G. C., Qvist, J., *et al.* (1977). Pulmonary metabolism during diving: conditioning blood for the brain. *Science* **198**, 831–3.

Hochachka, P. W., Buck, L. T., Doll, C. and Land, S. C. (1996). Unifying theory of hypoxia tolerance: molecular/metabolic defense and rescue mechanisms for surviving oxygen lack. *Proc. Natl. Acad. Sci. USA* **93**, 9493–9.

Hol, R., Blix, A. S. and Myhre, H. O. (1975). Selective redistribution of the blood volume in the diving seal (*Pagophilus groenlandicus*). *Rapp. P-v. Reun. Cons. Int. Explor. Mer.* **169**, 423–32.

Hong, S. K., Ashwell-Erickson, S., Gigliotti, P. and Elsner, R. (1982). Effects of anoxia and low pH on organic ion transport and electrolyte distribution in harbour seal (*Phoca vitulina*) kidney slices. *J. Comp. Physiol. B* **149**, 19–24.

Irvine, L. G., Hindell, M. A., van den Hoff, J. and Burton, H. R. (2000). The influence of body size on dive duration of underyearling southern elephant seals. *J. Zool.* **251**, 463–71.

Irving, L., Scholander, P. F. and Grinnell, S. W. (1942). The regulation of arterial blood pressure in the seal during diving. *Am. J. Physiol.* **135**, 557–66.

Jobsis, P. D., Ponganis, P. J. and Kooyman, G. L. (2001). Effects of training on forced submersion responses in harbour seals. *J. Exp. Biol.* **204**, 3877–85.

Jones, D. R. and Purves, M. J. (1970). The carotid body in the duck and the consequences of its denervation upon the cardiac responses to immersion. *J. Physiol.* **211**, 279–94.

Kanatous, S. B., Elsner, R. and Mathieu-Costello, O. (2001). Muscle capillary supply in harbour seals. *J. Appl. Physiol.* **90**, 1919–26.

Kanwisher, J., Gabrielsen, G. and Kanwisher, N. (1981). Free and forced diving in birds. *Science* **211**, 717–19.

Kerem, D. and Elsner, R. (1973). Cerebral tolerance to asphyxial hypoxia in the harbour seal. *Resp. Physiol.* **19**, 188–200.

Kerem, D., Hammond, D. D. and Elsner, R. (1973). Tissue glycogen levels in the Weddell seal: a possible adaptation to asphyxial hypoxia. *Comp. Biochem. Physiol.* **45**, 731–6.

Kjekshus, J. K., Maroko, P. K. and Sobel, B. E. (1972). Distribution of myocardial injury and its relation to epicardial ST-segment changes after coronary artery occlusion in the dog. *Cardiovasc. Res.* **6**, 490–9.

Kjekshus, J. K., Blix, A. S., Hol, R., Elsner, R. and Amundsen, E. (1982). Myocardial blood flow and metabolism in the diving seal. *Am. J. Physiol.* **242**, R97–104.

Koopman, H. N., Westgate, A. J. and Read, A. J. (1999). Hematology values of wild harbour porpoises (*Phocoena phocoena*) from the Bay of Fundy, Canada. *Mar. Mamm. Sci.* **15**, 52–64.

Kooyman, G. L. (1965). Techniques used in measuring diving capacities of Weddell seals. *Polar Rec.* **12**, 391–4.

Kooyman, G. L. and Campbell, W. B. (1972). Heart rate in freely diving Weddell seals (*Leptonychotes weddellii*). *Comp. Biochem. Physiol. A* **43**, 31–6.

Kooyman, G. L. and Kooyman, T. G. (1995). Diving behaviour of emperor penguins nuturing chicks at Coulman Island, Antarctica. *Condor* **97**, 536–49.

Kooyman, G. L., Hammond, D. D. and Schroeder, J. P. (1970). Bronchograms and tracheograms of seals under pressure. *Science* **169**, 82–4.

Kooyman, G. L., Castellini, M. A., Davis, R. W. and Maue, R. A. (1983). Aerobic diving limits of immature Weddell seals. *J. Comp. Physiol.* **151**, 171–4.

Kooyman, G. L., Kerem, D. H., Campbell, W. B. and Wright, J. J. (1971). Pulmonary function in freely diving Weddell seals, *Leptonychotes weddelli*. *Respir. Physiol.* **12**, 271–82.

Kooyman, G. L., Wahrenbrock, E. A., Castellini, M. A., Davis, R. A. and Sinnett, E. E. (1980). Aerobic and anaerobic metabolism during diving in Weddell seals: evidence of preferred pathways from blood chemistry and behavior. *J. Comp. Physiol.* **138**, 335–46.

Krockenberger, M. B. and Bryden, M. M. (1994). Rate of passage of digesta through the alimentary tract of southern elephant seals (*Mirounga leonina*) (Carnivora: Phocidae). *J. Zool.* **234**, 229–37.

Kvadsheim, P. H., Folkow, L. P and Blix, A. S. (2005). Inhibition of shivering in hypothermic seals during diving. *Am. J. Physiol.* **289**, R326–31.

Laufs, T. L., Wystub, S., Reuss, S., Burmester, T., Saaler-Reinhardt, S. and Hankeln, T. (2004). Neuron-specific expression of neuroglobin in mammals. *Neurosci. Lett.* **362**, 83–6.

Le Boeuf, B. J., Costa, D. P., Huntley, A. C. and Feldkamp, S. D. (1988). Continuous, deep diving in female northern elephant seals. *Can. J. Zool.* **66**, 446–58.

Leith, D. E. (1976). Comparative mammalian respiratory mechanics. *Physiologist* **19**, 485–510.

Lenfant, C., Johansen, K. and Torrance, J. D. (1970). Gas transport and oxygen storage capacity in some pinnipeds and the sea otter. *Respir. Physiol.* **9**, 277–86.

Lenfant, C., Elsner, R., Kooyman, G. L. and Drabek, C. M. (1969). Respiratory function of the blood of the adult and fetal Weddell seal. *Am J. Physiol.* **216**, 1595–7.

Lestyk, K. C., Folkow, L. P., Blix, A. S., Hammill, M. D. and Burns, J. M. Development of myoglobin concentration and acid buffering capacity in harp (*Pagophilus greenlandicus*) and hooded (*Cystophora cristat*) seals from birth to maturity. *J.Comp. Physiol.B.doi* 10.1007/s 00360-009-0378-9.

Liggins, G. C., Qvist, J., Hochachka, P. W., *et al.* (1980). Fetal cardiovascular and metabolic responses to simulated diving in the Weddell seal. *J. Appl. Physiol.* **49**, 424–30.

Lipton, P. (1999). Ischemic cell death in brain neurons. *Physiol. Rev.* **79**, 1431–568.

Liu, L. and Yenari, M. A. (2007). Therapeutic hypothermia: neuroprotective mechanisms. *Front. Biosci.* **12**, 816–25.

Ludvigsen, S. and Folkow, L. P. (2009). Differences in in-vitro cerebellar neuronal responses to hypoxia in eider ducks, chicken and rats. *J.Comp.Physiol. A. doi* 10.1007/s 00359-009-0476-X.

Lutz, P. L., Nilsson, G. E. and Prentice, H. M. (2003). *The Brain without Oxygen*, 3rd edn. Dordrecht: Kluwer Academic Publishers, pp. 62–3.

Marder, E. and Bucher, D. (2007). Understanding circuit dynamics using the stomatogastric nervous system of lobsters and crabs. *Annu. Rev. Physiol.* **69**, 291–316.

Mårtensson, P. -E., Nordøy, E. S., Messelt, E. B. and Blix, A. S. (1998). Gut length, food transit time and diving habit in phocid seals. *Polar Biol.* **20**, 213–17.

Messelt, E. B and Blix, A. S. (1976). The LDH of the frequently asphyxiated beaver (*Castor fiber*). *Comp. Biochem. Physiol.* B **53**, 77–80.

Miller, N. J., Postle, A. D., Orgeig, S., Koster, G. and Daniels, C. B. (2006). The composition of pulmonary surfactants from diving mammals. *Resp. Physiol. Neurobiol.* **152**, 152–68.

Milsom, W. K., Johansen, K. and Millard, R. W. (1973). Blood respiratory properties in some Antarctic birds. *Condor* **75**, 472–4.

Mitz, S. A., Reuss, S., Folkow, L. P., *et al.* (2009). When the brain goes diving: glial oxidative metabolism may confer hypoxia tolerance to the seal brain. *Neuroscience*, **163**, 552–560.

Mottishaw, P. D., Thornton, S. J. and Hochachka, P. W. (1999). The diving response mechanism and its surprising evolutionary path in seals and sea lions. *Am. Zool.* **39**, 434–50.

Murphy, B., Zapol, W. M. and Hochachka, P. W. (1980). Metabolic activities of heart, lung and brain during diving and recovery in the Weddell seal. *J. Appl. Physiol.* **48**, 596–605.

Murrish, D. E. (1982). Acid-base balance in three species of Antarctic penguins exposed to thermal stress. *Physiol. Zool.* **55**, 137–43.

Nilsson, G. E. and Lutz, P. L. (1992). Adenosine release in anoxic turtle brain as a mechanism for anoxic survival. *J. Exp. Biol.* **162**, 345–51.

O'Brien, P. J., Shen, H., McCutcheon, L. J., *et al.* (1992). Rapid, simple and sensitive microassay for skeletal and cardiac muscle myoglobin and hemoglobin: use in various animals indicates functional role of myohemoproteins. *Mol. Cell. Biochem.* **112**, 42–52.

Odden, Å., Folkow, L. P., Caputa, M., Hotvedt, R. and Blix, A. S. (1999). Brain cooling in diving seals. *Acta Physiol. Scand.* **166**, 77–8.

Olsen, C. R., Elsner, R., Hale, F. C. and Kenney, D. W. (1969). Blow of the pilot whale. *Science* **163**, 953–5.

Pain, T., Yang, X. M., Critz, S. D., *et al.* (2000). Opening of mitochondrial K_{ATP} channels triggers the preconditioned state by generating free radicals. *Circ. Res.* **87**, 460–6.

Pakay, J. L., Withers, P. C., Hobbs, A. A. and Guppy, M. (2002). The in vivo down-regulation of protein synthesis in the snail *Helix aspera* during estivation. *Am. J. Physiol.* **283**, R197–204.

Pan, T. -T., Feng, Z. -N., Lee, S. W., Moore, P. K. and Bian, J. -S. (2006). Endogenous hydrogen sulfide contributes to the cardioprotection by metabolic inhibition preconditioning in the rat ventricular myocytes. *J. Mol. Cell. Cardiol.* **40**, 119–30.

Pellerin, L., Bouzier-Sore, A.-K., Aubert, A., *et al.* (2007). Activity-dependent regulation of energy metabolism by astrocytes: an update. *Glia* **55**, 1251–62.

Pérez-Pinzón, M. A., Rosenthal, T., Sick, T. J., Lutz, P. L., Pablo, J. and Mash, D. (1992). Downregulation of sodium channels during anoxia: a putative survival strategy of turtle brain. *Am. J. Physiol.* **262**, R712–15.

Pesce, A., Bolognesi, M., Ascenzi, P., *et al.* (2002). Neuroglobin and cytoglobin: fresh blood for the vertebrate globin family. *EMBO Rep.* **3**, 1146–51.

Polasek, L. K. and Davis, R. W. (2001). Heterogeneity of myoglobin distribution in the locomotory muscles of five cetacean species. *J. Exp. Biol.* **204**, 209–15.

Ponganis, P. J., Stockard, T. K., Meir, J. U., *et al.* (2007). Returning on empty: extreme blood O_2 depletion underlies dive capacity of emperor penguins. *J. Exp. Biol.* **210**, 4279–85.

Ponganis, P. J., Kooyman, G. L., Starke, L. N., Kooyman, C. A. and Kooyman, T. G. (1997). Post-dive blood lactate concentrations in emperor penguins, *Aptenodytes forsteri*. *J. Exp. Biol.* **200**, 1623–6.

Ponganis, P. J., Starke, L. N., Horning, M. and Kooyman, G. L. (1999). Development of diving capacity in emperor penguins. *J. Exp. Biol.* **202**, 781–6.

Qvist, J., Hill, R. D., Schneider, R. C., *et al.* (1986). Hemoglobin concentrations and blood gas tensions of free-diving Weddell seals. *J. Appl. Physiol.* **61**, 1560–9.

Ramirez, J. -M., Folkow, L. P. and Blix, A. S. (2007) Hypoxia tolerance in mammals and birds: from the wilderness to the clinic. *Annu. Rev. Physiol.* **69**, 113–43.

Reed, J. Z., Chambers, C., Fedak, M. A. and Butler, P. B. 1994. Gas exchange of captive freely diving grey seals (*Halichoerus grypus*). *J. Exp. Biol.* **191**, 1–18.

Richet, C. (1899). De la résistance des canards à l'asphyxie. *J. Physiol. Pathol. Gén.* **1**, 641–50.

Ridgway, S. H. and Johnston, D. G. (1966). Blood oxygen and ecology of porpoises of three genera. *Science* **151**, 456–8.

Ridgway, S. H., Scronce, B. L. and Kanwisher, N. (1969). Respiration and deep diving in the bottlenose porpoise. *Science* **166** 1651–54.

Robin, E. D., Ensick, J., Hance, A. J., *et al.* (1981). Glucoregulation and simulated diving in the harbor seal *Phoca vitulina*. *Am. J. Physiol.* **241**, R293–300.

Robin, E. D., Jr, Murdaugh, H. V., Pyron, W., Weiss, E. and Soteres, P. (1963). Adaptations to diving in the harbour seal – gas exchange and ventilatory responses to CO_2. *Am. J. Physiol.* **205**, 1175–7.

Robinson, D. (1939). The muscle haemoglobin of seals as an oxygen store in diving. *Science* **90**, 276–7.

Ronald, K., McCarter, R. and Selley, L. J. (1977). Venous circulation in the harp seal. In *Functional Anatomy of Marine Mammals, Vol. 3*, ed. R. J. Harrison. London: Academic Press, pp. 235–70.

Scholander, P. F. (1940). Experimental investigations on the respiratory function in diving mammals and birds. *Hvalrådets Skr.* **22**, 1–131.

Scholander, P. F. (1960). Oxygen transport through haemoglobin solutions. *Science* **131**, 585–90.

Scholander, P. F. (1963). The master switch of life. *Sci. Am.* **209**, 92–106.

Scholander, P. F., Irving, L. and Grinnell, S. W. (1942a). Aerobic and anaerobic changes in seal muscles during diving. *J. Biol. Chem.* **142**, 431–40.

Scholander, P. F., Irving, L. and Grinnell, S. W. (1942b). On the temperature and metabolism of the seal during diving. *J. Cell. Comp. Physiol.* **19**, 67–78.

Siesjö, B. K. (1978). *Brain Energy Metabolism.* New York: John Wiley.

Singer, D., Bach, F., Bretschneider, H. J. and Kuhn H. J. (1993). Metabolic size allometry and the limits to beneficial metabolic reduction: hypothesis of a uniform specific minimal metabolic rate. In *Surviving Hypoxia: Mechanisms of Control and Adaptation*, ed. P. W. Hochachka, P. L. Lutz, T. Sick, M. Rosenthal and G. Thillart. London: CRC Press, pp. 447–58.

Sinnett, E. E., Kooyman, G. L. and Wahrenbrock, E. A. (1978). Pulmonary circulation of the harbour seal. *J. Appl. Physiol.* **45**, 718–27.

Skinner, L. A. and Milsom, W. K. (2004). Respiratory chemosensitivity during wake and sleep in harbour seal pups (*Phoca vitulina richardsii*). *Physiol. Biochem. Zool.* **77**, 847–63.

Smith, R. W., Houlihan, D. F., Nilsson, G. E. and Brechin, J. G. (1996). Tissue-specific changes in protein synthesis rates in vivo during anoxia in crucian carp. *Am. J. Physiol.* **271**, R897–904.

Snyder, G. K. (1983). Respiratory adaptations in diving mammals. *Respir. Physiol.* **54**, 269–94.

Sparling, C. E., Fedak, M. A. and Thompson, D. (2007). Eat now, pay later? Evidence of deferred food-processing costs in diving seals. *Biol. Lett.* **3**, 94–8.

Stephenson, R., Butler, P. J. and Woakes, A. J. (1986). Diving behaviour and heart rate in tufted ducks (*Aythya fuligula*). *J. Exp. Biol.* **126**, 341–59.

Stephenson, R., Turner, D. L. and Butler, P. J. (1989). The relationship between diving activity and oxygen storage capacity in the tufted duck (*Aythya fuligula*). *J. Exp. Biol.* **141**, 265–75.

Stockard, T. K., Levenson, D. H., Berg, L., Fransioli, J. R., Baranov, E. A. and Ponganis, P. J. 2007. Blood oxygen depletion during rest-associated apneas of northern elephant seals (*Mirounga angustirostris*). *J. Exp. Biol.* **210**, 2607–17.

Storey, K. B. (1988). Suspended animation: the molecular basis of metabolic depression. *Can. J. Zool.* **66**, 124–32.

Storey, K. B. and Storey, J. M. (1990). Metabolic rate depression and biochemical adaptation in anaerobiosis, hibernation and estivation. *Q. Rev. Biol.* **65**, 145–74.

Storey, K. B. and Storey, J. M. (2007). Tribute to P. L. Lutz: putting life on 'pause' – molecular regulation of hypometabolism. *J. Exp. Biol.* **210**, 1700–14.

Sun, Y., Jin, K., Mao, X. O., Zhu, Y. and Greenberg, D. A. (2001). Neuroglobin is up-regulated by and protects neurons from hypoxic-ischemic injury. *Proc. Natl. Acad. Sci. USA* **98**, 15306–11.

Theorell, H. (1934). Kristallinisches Myoglobin: I. Mitteilung: Kristallisieren und Reinigung des Myoglobins sowie Vorläufige Mitteilung über sein Molekulargewicht. *Biochem. Z.* **252**, 1–7.

Thompson, D. and Fedak, M. A. (1993). Cardiac responses of gray seals during diving at sea. *J. Exp. Biol.* **174**, 139–64.

van Breukelen, F. and Martin, S. L. (2002). Reversible depression of transcription during hibernation. *J. Comp. Physiol. B* **172**, 355–61.

Vázquez-Medina, J. P., Zenteno-Savín, T. and Elsner, R. (2006). Antioxidant enzymes in ringed seal tissues: potential protection against dive-associated ischemia/reperfusion. *Comp. Biochem. Physiol. C.* **142**, 198–204.

Wahrenbrock, E. A., Maruschak, G. F., Elsner, R. and Kenney, D. W. (1974). Respiration and metabolism in two baleen whale calves. *Marine Fish. Rev.* **36**, 3–8.

Walz, W. (2000). Role of astrocytes in the clearance of excess extracellular potassium. *Neurochem. Int.* **36**, 291–300.

Watkins, W. A., Moore, K. E. and Tyack, P. (1985). Investigations of sperm whale acoustic behaviours in the southeast Carribean. *Cetology* **49**, 1–15.

Watson, R. R., Miller, T. A. and Davis, R. A. (2003). Immunohistochemical fiber typing of harbour seal skeletal muscle. *J. Exp. Biol.* **206**, 4105–11.

Weber, R. E., Hemmingsen, E. A. and Johansen, K. (1974). Functional and biochemical studies of penguin myoglobin. *Comp. Biochem. Physiol. B* **49**, 197–214.

White, F. N., Ideda, M. and Elsner, R. (1973). Adrenergic innervation of large arteries in the seal. *Comp. Gen. Pharmacol.* **4**, 271–6.

Wickham, L. L., Elsner, R., White, F. C. and Cornell, L. H. (1989). Blood viscosity in phocid seals: possible adaptations to diving. *J. Comp. Physiol. B* **159**, 153–8.

Wilhelm Filho, D., Sell, F., Ribeiro, L., *et al.* (2002). Comparison between the antioxidant status of terrestrial and diving mammals. *Comp. Biochem. Physiol. A* **133**, 885–92.

Willford, D. C., Gray, A. T., Hempleman, S. C., Davis, R. W. and Hill, E. P. (1990). Temperature and the oxygen-hemoglobin dissociation curve of the harbour seal, *Phoca vitulina. Respir. Physiol.* **79**, 137–44.

Williams, T. M. and Kooyman, G. L. (1985). Swimming performance and hydrodynamic characteristics of the harbour seal. *Physiol. Zool.* **58**, 576–89.

Williams, T. M., Davis, R. W., Fuiman, L. A., *et al.* (2000). Sink or swim: strategies for cost-efficient diving by marine mammals. *Science* **288**, 133–5.

Williams, T. M., Zavanelli, M., Miller, M. A., *et al.* (2008). Running, swimming and diving modifies neuroprotecting globins in the mammalian brain. *Proc. Biol. Sci.* **275**, 751–8.

Wittenberg, B. A. and Wittenberg, J. B. (1989). Transport of oxygen in muscle. *Annu. Rev. Physiol.* **51**, 857–78.

Zapol, W. M., Liggins, G. C., Schneider, R. C., *et al.* 1979. Regional blood flow during simulated diving in the conscious Weddell seal. *J. Appl. Physiol.* **47**, 968–73.

Zenteno-Savín, T., Clayton-Hernández, E. and Elsner, R. (2002). Diving seals: are they a model for coping with oxidative stress? *Comp. Biochem. Physiol. C* **133**, 527–36.

Zheng, Z., Lee, J. E. and Yenari, M. A. (2003). Stroke: molecular mechanisms and potential targets for treatment. *Curr. Mol. Med.* **3**, 361–72.

8

Vertebrate life at high altitude

FRANK L. POWELL AND SUSAN R. HOPKINS

8.1 Introduction

The physiological stresses and limited resources at high altitude pose limits on vertebrate life in this environment. Primary stresses include low oxygen pressure, temperatures, and humidity, and increased radiation. High-altitude ecosystems are characterized by less diversity, rugged topography, and marginal availability of certain nutrients. However, given the amazing physiological abilities to cope with low oxygen described in this book, it is not surprising that there are numerous examples of life at high altitude. Representatives from every class of vertebrates are found living at altitudes of 4000 m above sea level, where the PO_2 is less than 100 Torr, including fish (trout) in Andean lakes and rivers (Bouverot *et al.*, 1985). (A pressure of 1 Torr = 1/760 atmosphere = 1 mmHg). The primary focus of this chapter on species that are native to high altitudes is to emphasize adaptations to life with limited oxygen instead of reviewing physiological acclimatization to high altitude. Adaptations to hypoxia in fishes are covered in Chapter 5, so here we focus on air-breathing vertebrates.

8.2 The high-altitude environment

Paul Bert first demonstrated that the primary physiological challenge at high altitude is reduced oxygen partial pressure (PO_2) as a result of reduced barometric pressure (Bouverot *et al.*, 1985). Various algorithms have been devised to estimate the fall in barometric pressure with altitude, such as the International Civil Aviation Organization (1964) or National Oceanic and Atmospheric Administration (1976) standard atmospheres. However, these

Respiratory Physiology of Vertebrates: Life with and without Oxygen, ed. Göran E. Nilsson. Published by Cambridge University Press. © Cambridge University Press 2010.

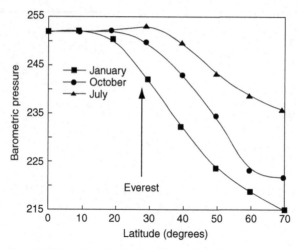

Fig. 8.1 The effect of season and latitude on barometric pressure. For a given elevation, barometric pressure is highest at the equator and lowest at the poles. At the equator any seasonal variation is minimal, but at 70° latitude barometric pressure varies by almost 5%. Although this change is small, it potentially results in large differences in oxygen availability and exercise performance. Redrawn from West *et al.*, 1983b.

may be in considerable error, particularly as a function of season and latitude (West *et al.*, 1983b) (Fig. 8.1), and, for example, predict a barometric pressure 17 Torr lower than the measured value for the summit of Mount Everest. Hence, West (West, 1996) developed an equation that predicts barometric pressure within 1% of actual measured values for many locations of interest at high altitude within latitudes of 15° (in all seasons) and 30° (in the summer):

$$PB\ (Torr) = \exp\left(6.63268 - 0.1112\,h - 0.00149\,h^2\right) \tag{8.1}$$

where h is the altitude in kilometers.

Although the primary effect of hypoxia at altitude is the reduction in inspired PO_2, there are a few reports of specific effects of decreased barometric pressure on respiration. Ventilation is greater in ducks exposed to the same degree or hypobaric, rather than normobaric, hypoxia (Shams *et al.*, 1990; Shams and Scheid, 1993). This appears to be a result of increased lactic acid in hypobaria, which stimulates ventilation. However, the mechanism of lactic acidosis in hypobaria is unknown. The opposite effect has been observed in humans at simulated altitude, who had decreased ventilation compared with the same inspired air PO_2 (P_IO_2) at sea level (Loeppky *et al.*, 1997). Hence, the effects of hypobaria at altitude are variable and less robust than the response to hypoxia per se; however, in the majority of situations the effects are very similar on most physiological variables.

In addition to decreased PO_2 (i.e. hypoxia), the high-altitude environment is also cold. This has a considerable impact on all vertebrates. For homeothermic birds and mammals, it imposes an additional energetic requirement to maintain body temperature. For ectothermic amphibians and reptiles, it limits activities that depend on thermal warming from the environment. There are several algorithms for cooling with altitude, but in general ambient temperature falls about 6°C for every 1000 m altitude (Bouverot *et al.*, 1985). Solar and ionizing radiation are also increased at altitude. At 4000 m altitude, solar radiation is increased 100% compared with that at sea level because of the reduced air density (Ward *et al.*, 2000). At 3000 m altitude, increased cosmic radiation yields a dose of about 0.7 mGy per year for a person, which can be compared to a normal annual dose of 0.5–20 mGy from all sources. There are no documented effects of these increased levels of radiation exposure, but they may pose particular risks to ectotherms basking in the sun and be a further limitation to the distribution of amphibians and reptiles at high altitude.

The high-altitude environment is typically very dry, and this also has an impact on an animal's physiology. Evaporative water loss is proportional to the difference between inhaled and exhaled relative humidity and temperature. Homeotherms exhale saturated gas at relatively high body temperatures, so they experience increased evaporative water loss with increasing altitude. A further problem is that the vapor pressure for water reduces the available partial pressure for oxygen in the lungs, and this is fixed by body temperature, in contrast to decreasing barometric pressure with altitude. Hence, the relative impact of humidification increases with altitude; and in the extreme case of 19 215 m altitude, the vapor pressure of water equals the barometric pressure, leaving no room for oxygen (Luft, 1965)!

8.3   Vertebrate diversity at high altitude

Multiple physical and biological factors, as discussed above, determine the distributions of animals in nature. Any of these factors may play a role in determining which species are pre-adapted to life at high altitude, and which species are able to adapt genetically to high-altitude environments. Ultimately, multidisciplinary approaches involving biogeography and comparative physiological genomics will be necessary to understand evolutionary adaptations to specific variables in the environment, such as oxygen level (Powell, 2003). However, examining the respiratory physiology of vertebrates native to high-altitude environments reveals both diverse and common solutions to the problem of life without oxygen. This chapter focuses on the vertebrate species that have been studied most extensively in terms of high-altitude respiratory physiology and does not attempt to catalog the entire diversity of vertebrate life at high altitudes.

8.3.1   Mammals

Mammals are the most widely studied class of vertebrates at high altitude. It is notable that some of the species indigenous to the highest altitudes, including humans, are generalists that live at altitudes from below sea level to almost 6000 m above sea level. The highest altitude documented for human habitation in modern times is the Aucanquilcha mine in the Andes of Chile at 5950 m (West, 1986). These miners descend to lower altitudes on the weekend, and their families are born and raised at lower altitudes. However, there are several human populations reproducing above 4000 m in the Andes and Himalayan regions (Vitzthum and Wiley, 2003). Other well-studied mammalian species at high altitude include the camelids of the Andes, which live at over 5000 m on the altiplano (Bouverot et al., 1985), the domestic llama (Lama glama) and alpaca (Lama pacos), and the wild vicuña (Vicugna vicugna) and guanaco (Lama guanicoe). The deer mouse (Peromyscus maniculatus), which has been described as the most widely distributed mammal in North America, occurs at altitudes below sea level to over 4000 m in the Rocky Mountains and White Mountains of the Great Basin (Dunmire, 1960).

8.3.2   Birds

Birds are an interesting case, because they can readily ascend to very high altitudes directly from sea level by flying. The avian altitude record comes from a collision of an Old World vulture (Rüppell's griffon, Gyps ruepellii) with a commercial aircraft at 11278 m over the Ivory Coast (Laybourne, 1974). In general, birds appear pre-adapted to high altitude, so that even species that are normally found only at low altitudes (e.g. house sparrows) can survive simulated altitudes much better than comparably sized mammals (Tucker, 1968). Consequently, there are several studies of domestic species such as chickens, ducks, and pigeons that have provided valuable information about respiratory physiology in birds at high altitudes (Powell and Whittow, 2000). However, the bar-headed goose (Anser indicus), which migrates directly over the highest peaks in the Himalayas (Swan, 1970), has provided some of the most interesting data about high-altitude adaptations in birds (Scott and Milsom, 2006). Adaptations in avian embryos and eggs at altitude have been reviewed elsewhere (Monge and Leon-Velarde, 1991; Leon-Velarde and Monge, 2004).

8.3.3   Ectotherms

Although more limited, amphibians and reptiles occur at altitudes comparable to those described for homeothermic vertebrates. Lizards occur at 5500 m in the Himalayas and at 4900 m in the Andes (Bouverot *et al.*, 1985). All the lizards and snakes found above 3000 m are ovoviviparous (Bouverot *et al.*, 1985). This is interesting, because birds show special adaptations in eggs laid at high altitude (Leon-Velarde and Monge, 2004) that may not be possible in reptiles. The distribution of amphibians at altitude is further limited by the distribution of aquatic environments. Salamanders (Mount Lyell salamander, *Eurycea platycephala*) have been documented at 3292 m in the Sierra Nevada (Grinnell and Storer, 1924). The purely aquatic Lake Titicaca toad (*Telmatobius culeus*) is native to 3812 m, but less aquatic species of anurans (e.g. *Bufo spinulosus*) reach elevations near 4500 m in the Andes (Navas and Chaui-Berlinck, 2007).

8.4   The oxygen cascade at high altitude

Unfortunately, there have been no systematic studies of adaptations to high altitude at every step of the oxygen cascade in non-human vertebrates. Different groups of investigators have focused on specific aspects of the oxygen cascade, with comparative studies between low- and high-altitude species. However, it is difficult to make many generalizations from these studies because of limitations inherent to comparative designs. For example, it is difficult to ascribe differences between two species from low and high altitude to natural selection for high altitude unless the phylogenetic relationship between both species is carefully considered (Garland and Adolph, 1994; Garland, 2001). Similarly, it is difficult to infer adaptations from comparisons between different populations of the same species that are native to different altitudes because of the complications from the original genetic stock of the populations and the effects of development (Brutsaert, 2001). Hence, we briefly review the comparative studies here but focus on the most complete data sets available for high-altitude vertebrates to emphasize a strong foundation for future comparative studies. Human physiology at high altitude is considered separately, in the last section on climbing Mount Everest.

8.4.1   Mammals

Considering the first step in the O_2 cascade of ventilation, high-altitude mammals tend to have a normal (not blunted) hypoxic ventilatory response. This has been observed in yaks and llamas, as well as in sheep and dogs born and raised at high altitude (Bouverot *et al.*, 1985; Weil *et al.*, 1986). The mechanisms responsible for the hypoxic desensitization and a blunted hypoxic ventilatory response that is observed in human natives of the Andes, or animals such as steers, ponies, and cats during acclimatization to high altitude are not known (Weil et al., 1986). It has been hypothesized that a blunted hypoxic ventilatory response could be adaptive by decreasing oxygen demand for breathing when other changes in more distal steps of the O_2 cascade have adapted to increase O_2 delivery. However, the relatively low O_2 cost of breathing and increased ventilation in species that are successful in the high-altitude ecological niche suggests that increased ventilatory supply of O_2 is important in both acute and chronic exposures.

No differences in pulmonary diffusing capacity for O_2 are reported for the few high-altitude mammalian species that have been studied (Monge and Leon-Velarde, 1991). However, in humans, individuals born and raised at high altitude and genetically high-altitude natives have increased lung volumes and an increased diffusing capacity for carbon monoxide (for example, see Wu et al., 2005). Bouverot (Bouverot et al., 1985) pointed out that the inverse relationship between red blood cell volume and the velocity of red cell oxygenation would favor diffusion equilibrium for O_2 in high-altitude species with very small red cell volumes, such as llama and vicuña. Mechanisms involved in hypoxic pulmonary vasoconstriction are discussed in Chapter 4 (section 3.9), and the maladaptive effects of hypoxic pulmonary vasoconstriction and pulmonary hypertension on the development of high-altitude pulmonary edema (HAPE) and other diseases affecting pulmonary gas exchange are discussed further below (section 6). One might expect reduced hypoxic pulmonary vasoconstriction in high-altitude mammals, but this does not appear to be a general observation. A notable exception is the yak, however, which shows no increased pulmonary artery pressure at 4800 m altitude, in sharp contrast to the pulmonary hypertension observed in cattle (Anand et al.,1986). Studies of crossbreeds between yak and cattle led the authors to suggest that the control of pulmonary artery pressure in these species may be under the control of a single autosomal dominant gene.

For cardiovascular transport of O_2, adaptations in hemoglobin and blood oxygen affinity appear more important than changes in cardiovascular function for maintaining O_2 delivery (Bouverot et al., 1985; Monge and Leon-Velarde, 1991). For example, cardiac output is essentially constant in llamas exposed to simulated altitudes between 1600 m and 6400 m, but is increased in sheep (Banchero and Grover, 1972) and alpacas at sea level or 3300 m (Sillau et al., 1976). Surprisingly, the mixed-venous PO_2 was also higher in llamas than sheep (decreasing only 8 Torr vs. a 26 Torr decrease in sheep between 1600 and 6400 m). The llamas need to extract less O_2 despite the lower cardiac output because of their higher O_2 capacity and O_2 affinity of the llama blood: P_{50} = 23 Torr vs.

40 Torr in sheep (Bouverot *et al.*, 1985). Hence, the major adaptation to high altitude in cardiovascular O_2 delivery for non-human mammals appears to be in hemoglobin.

8.4.1.1 Deer mice

The deer mouse (*Peromyscus maniculatus*) is the most widely distributed mammal in North America and is found over an extremely wide range of altitudes. For example, the subspecies *P. maniulatus sonoriensis* ranges from below sea level in Death Valley to over 4000 m in the Sierra Nevada and White Mountains of California (Dunmire, 1960). Evolutionary adaptations to high altitude in the O_2 cascade of deer mice are supported by evidence for selection for maximum aerobic performance in free-living deer mice at altitude (Hayes and O'Connor, 1999). However, it should be considered that the success of deer mice at high altitude may result from their extreme phenotypic plasticity, which has allowed them to thrive in such a wide range of habitats (MacMillen and Garland, 1989).

Two fundamental observations about O_2 transport in *P. maniculatus* have made it an attractive animal model for studying high-altitude adaptations. First, they have extensive complex hemoglobin polymorphisms that are related to both O_2 transport and their native altitude. The a^0c^0/a^0c^0 haplotype has a significantly greater O_2 affinity ($P_{50} \approx 32$ Torr) compared with the a^1c^1/a^1c^1 haplotype ($P_{50} \approx 36$ Torr) (Chappell and Snyder, 1984). These haplotypes are significantly correlated with the average regional altitude of different populations, so a^0c^0/a^0c^0 is more common at high altitude (Snyder *et al.*, 1998). Second, maximal oxygen consumption ($\dot{V}O_{2max}$) determined during exercise or cold exposure is greater in a^0c^0/a^0c^0 than in a^1c^1/a^1c^1 at high altitude; conversely, $\dot{V}O_{2max}$ in a^1c^1/a^1c^1 is greater than a^0c^0/a^0c^0 at low altitude (Chappell and Snyder, 1984). Considering the evidence for selection for maximum aerobic performance in free-living deer mice at altitude (Hayes and O'Connor, 1999), these findings support natural selection for increased hemoglobin–O_2 affinity at high altitude.

More recent experiments have elucidated the molecular genetics behind these differences in low- and high-altitude populations (Storz *et al.*, 2007). Differences in P_{50} are due to independent or combined effects of five amino acid substitutions on the alpha globin molecule that affects O_2 binding to hemoglobin. These functionally distinct protein alleles are maintained as long-term balanced polymorphisms, which is consistent with natural selection favoring the different genotypes at different native altitudes. However, questions remain about the role of the genetically based differences in P_{50} for explaining the differences observed in $\dot{V}O_{2max}$ in different populations at different altitudes. Theoretical models of integrated O_2 transport (Wagner, 1997) predict no increases in $\dot{V}O_{2max}$ for the decrease in P_{50} observed in the high-altitude *P. maniculatus*. One possible explanation is that the O_2 transport models do not account for blood with different O_2 affinities. The erythrocytes in *P. maniculatus* contain a heterogeneous mixture of hemoglobin isoforms with different P_{50} that are hypothesized to provide a mechanism for fine-tuning blood–O_2 affinity in response to variation in metabolic demands (Storz *et al.*, 2007).

Other aspects of physiological O_2 transport have not been investigated in detail in *P. maniculatus*. Phenotypic plasticity in lung mass and hematocrit are determined primarily by oxygen level at altitude, whereas heart mass depends more on temperature (i.e. cold increases O_2 demand) than oxygen level (Hammond *et al.*, 2001). This is consistent with the increase in lung diffusing capacity with no change in cardiac performance described above for other high-altitude mammals. Differences between left and right ventricular masses were not measured, so it is not known whether hypoxic pulmonary vasoconstriction (and the associated right-ventricular hypertrophy) is reduced in *P. maniculatus* or not. The increase in hematocrit demonstrates that erythropoietic responses can be retained in animals after natural selection for hemoglobin–O_2 affinity too. Also consistent with other mammalian studies, there is no effect of altitude on capillaries for tissue O_2 exchange in *P. maniculatus* when corrections are made for sarcomere length (Mathieu-Costello, 1989). Potential differences in ventilatory control, pulmonary gas exchange, and O_2 extraction remain to be investigated in *P. maniculatus*.

8.4.1.2 Fetal llamas

Studies comparing oxygen transport in fetal llamas with domestic sheep found important differences that benefit the llama at high altitude (Llanos *et al.*, 2003; Llanos *et al.*, 2007). Even at sea level, the hypoxic stress on the fetus is comparable to a climber on the summit of Mount Everest. The fetus copes with this primarily with high O_2-affinity fetal hemoglobin (Longo, 1987) but also has the capacity for physiological responses to acute hypoxia. When pregnant sheep are exposed to hypoxia at sea level, the fetus responds with bradycardia and systemic and pulmonary vasoconstriction that redistributes blood flow to the heart, brain, and adrenals to sustain O_2 consumption in these organs (Llanos *et al.*, 2003). By contrast, the fetus of the domestic llama of the Andean altiplano (*Lama glama*) shows a much stronger peripheral vasoconstriction and does not show an increase in cerebral blood flow during hypoxia, and brain O_2 consumption decreases (Llanos *et al.*, 2007). The vasoconstriction depends on α-adrenergic mechanisms, as well as arginine vasopressin and endothelin (Llanos *et al.*, 2007). Cerebral hypometabolism in the hypoxic llama fetus is achieved in part by decreased sodium/potassium-ATPase activity (Llanos *et al.*, 2007), which is similar to the decreased O_2 demand strategy used by hypoxic turtles (Hochachka and Somero, 1984) and avoids seizure and neuronal death.

8.4.2   Birds

In general, birds tolerate altitude much better than do mammals. This section reviews the physiological basis for a general avian advantage at altitude, and the specific adaptations in birds that are native to high altitude are covered in a section on flying over Mount Everest later in this chapter.

The most obvious difference for birds at altitude is their respiratory system, which is unique among vertebrates in separating the functions of ventilation and gas exchange between air sacs and a parabronchial lung, respectively (Powell, 2000). This allows flow-through ventilation of the parabronchi and a cross-current model of gas exchange that is theoretically more efficient than alveolar gas exchange in mammals (Powell and Scheid, 1989). However, limitations to O_2 transport such as ventilation-perfusion mismatching and post-pulmonary shunts can reduce arterial PO_2 in birds to levels similar to those in mammals in normoxia (Powell, 1993). The effects of ventilation-perfusion mismatching and post-pulmonary shunting are reduced in hypoxia in birds, however, so a cross-current advantage is revealed for arterial PO_2 at high altitude (Fig. 8.2). By contrast, CO_2 exchange is affected less by pulmonary limitations, so the 'cross-current' advantage is revealed under both normoxic and hypoxic conditions, and arterial PCO_2 is always lower in birds than in mammals (Fig. 8.2).

The actual advantage of cross-current gas exchange for O_2 at extreme altitudes – especially with the high levels of O_2 consumption that would be necessary for a bird to fly to high altitude – has not been resolved (Shams and Scheid, 1989; Powell, 1993). However, compared with mammals, diffusion limitations are predicted to be less during hypoxic exercise (Maina *et al.*, 1989; Powell and Scheid, 1989), and ventilation-perfusion mismatching does not worsen with exercise in birds as it does in mammals (Schmitt *et al.*, 2002) (Fig. 8.3). The low CO_2 levels associated with cross-current exchange may reflect the most important avian adaptation to altitude. The ability of birds to tolerate an extreme respiratory alkalosis allows them to hyperventilate much more than mammals at high altitude to preserve O_2 delivery. The cerebral circulation in birds is less sensitive to hypocapnia, so hypoxic vasodilation is more effective at increasing cerebral blood flow at altitude in birds than in mammals (Grubb *et al.*, 1977; Grubb *et al.*, 1978; Faraci, 1991; Schmitt *et al.*, 2002). Also, intracellular pH

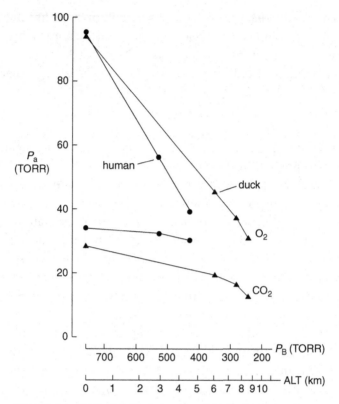

Fig. 8.2 Arterial O_2 (P_aO_2) in humans and ducks during acute exposure to simulated high altitude (Black and Tenney, 1980a; Wagner *et al.*, 1986). At sea level, measured P_aO_2 is similar in birds and mammals despite predictions that cross-current gas exchange in birds should produce greater P_aO_2 than alveolar exchange in mammals. This is because ventilation-perfusion mismatching and pulmonary shunts have a relatively greater impact on cross-current gas exchange in birds during normoxia (Powell, 1993). In hypoxia, the amount of ventilation-perfusion mismatching is not altered in birds, but the impact on cross-current gas exchange is relatively less, so P_aO_2 in birds exceeds that in mammals (Powell, 1993). Arterial CO_2 (P_aCO_2) is affected less by these limitations under all conditions, so P_aCO_2 is less in birds than in mammals at all altitudes, despite similar levels of CO_2 production and ventilation.

is constant in pigeons during large changes in blood pH with exposures to simulated altitudes to 9000 m (Weinstein *et al.*, 1985). These results suggest that tighter regulation of the intracellular milieu may provide much of the avian advantage at altitude, but this remains to be investigated in detail.

8.4.3   *Ectotherms*

There are relatively few systematic studies of the oxygen cascade in amphibians and reptiles native to high altitudes. Ventilatory responses to acute

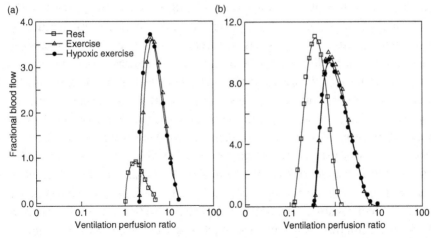

Fig. 8.3 The effect of exercise and hypoxia on ventilation-perfusion inequality in a representative normal healthy human subject (a) and an emu (b). In the human, the log standard deviation of the perfusion distribution is increased from 0.45 at rest to 0.53 during normoxic heavy exercise and increases further to 0.55 during heavy hypoxic exercise ($F_IO_2 = 0.125$). Data based on Bebout *et al.*, 1989; Hopkins *et al.*, 1994; Podolsky *et al.*, 1996. In the bird, the log standard deviation of the perfusion distribution does not change under the three conditions and remains at ~ 0.6 throughout. Data from Schmitt *et al.*, 2002.

hypoxia are relatively weak in most ectotherms (Shelton *et al.*, 1986) and it is unknown how they change with chronic exposure. Generally, ectotherms respond to decreased oxygen availability with decreased oxygen demand. For example, several species of lizard (*Varanus exanthematicus, Iguana iguana*, and *Ctenosaura pectinata*) have been shown to select lower basking temperatures to drop body temperature and metabolic rate when exposed to 7% inspired O_2 (Hicks and Wood, 1985). However, this may not be a common adaptation for ectotherms native to high altitudes. The Lake Titicaca toad (*Telmatobius culeus*) is reported to have a low metabolic rate compared with other anurans (Hutchison *et al.*, 1976). However, this is not a general characteristic of anurans that are native to high altitudes (Navas and Chaui-Berlinck, 2007). For example, Packard (Packard, 1971) found that boreal chorus frogs (*Pseudacris triseriata*) collected at different altitudes had a similar oxygen consumption.

Other adaptations in oxygen transport described for high-altitude ectotherms include enhanced tissue gas exchange and blood properties in the Lake Titicaca toad (Hutchison *et al.*, 1976). These toads have pronounced folds in their skin with extensive capillary networks which they 'ventilate' by vigorous bobbing in the water when they are prevented from surfacing. Lake Titicaca toads also have very small erythrocyte volume, high hematocrit, and a hemoglobin with among the highest O_2 affinity of all amphibians (P_{50} = 15.6 Torr at pH = 7.6 and 10°C). High hematocrit and hemoglobin concentration is not a general feature of high-altitude anurans but a low P_{50} is, as revealed in a study of three subspecies of toads indigenous to elevations from sea level to 4100 m (*Bufo spinulosus limensis, B. f. tirolium*, and *B. f. favolineatus*) (Ostojic et al., 2000). Hence, amphibians fit the general model of high-altitude species having adaptations in hemoglobin to increase O_2 affinity. By contrast, reptiles do not appear to show any correlations between altitude-based hematological parameters (Weber, 2007).

One of the most important factors determining oxygen delivery in amphibians and reptiles is central cardiac shunting (Wang and Hicks, 1996). *Bufo marinus* has been showed to decrease right-to-left shunting in acute hypoxia (Navas and Chaui-Berlinck, 2007), but comparative studies have not been done on animals acclimatized to, or native to, high altitude. The decrease in temperature expected at altitude decreases intracardiac mixing, but this is associated with changes in activity (Hedrick et al., 1999) that are not expected to be depressed in high-altitude natives.

8.5   Performance at extreme altitude

It is an interesting coincidence that the highest point on Earth is very near the absolute limits of human aerobic performance. West pointed out that it took over 50 years for men to climb the last 300 m of Mount Everest and it was only accomplished with supplemental oxygen. Physiological models predicted that Mount Everest could not be climbed by humans without supplemental O_2 (West, 1983), but the models were proved wrong when Habeler and Messner reached the summit without supplemental oxygen in 1978. The physiological basis for this performance, as well as the more amazing performance of birds flying over Mount Everest, is considered next.

8.5.1   *Humans climbing to extreme altitude*

What makes an elite high-altitude climber? The ultimate test of an elite climber is the ability to attain the summit and return alive, a test that becomes more difficult as the difficulty and the elevation of the mountain increase (Fig. 8.4) (Huey et al., 2001). Climbers of either sex appear equally likely to attain the summit of very high mountains such as Mount Everest and survive the return from the summit (Huey et al., 2007). Up until about age 40 there is no relationship between age and chance of success; however, after age 40 the probability of success is reduced and after age 60 the probability of death is markedly increased with increasing age (Huey et al., 2007). However, one of the most notable characteristics of the elite climber is the distinct lack of many

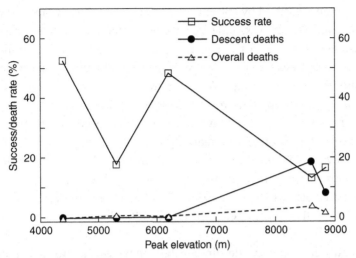

Fig. 8.4 Success and death rates on peaks of various elevations: Rainier (4392 m), Foraker (5306 m), Denali (6193 m), K2 (8616 m), and Everest (8850 m). On average, success rates decrease with increasing elevation, with lower success rates for peaks of greater difficulty (Foraker and K2). In addition, both overall death rates and death rates after successfully reaching the summit increase with increasing elevation. On K2, a very high and technically difficult mountain, less than 13% of climbers reach the summit, and 19% of individuals who do die on the descent. Data from Huey *et al.*, 2007.

physiological characteristics that one might think would be associated with elite performance at high altitude. For example, although it might be predicted that a high $\dot{V}O_{2max}$ might be a predictor of success, there has been little correlation between this characteristic and performance at very high altitudes. Although it is true that elite climbers as a whole tend to have a higher degree of aerobic fitness than sedentary individuals, their $\dot{V}O_{2max}$ is substantially less than similarly elite distance runners (Oelz *et al.*, 1986) and some have values that are not substantially different from untrained individuals. Similarly, tests of anaerobic power do not distinguish elite climbers from untrained individuals (Oelz *et al.*, 1986). The measures of muscle fiber type give similar results: elite climbers have an intermediate pattern between elite runners and untrained sedentary subjects. Capillary-to-fiber ratio in leg muscle is increased (Oelz *et al.*, 1986; Hoppeler *et al.*, 1990), a characteristic that is also found in some high-altitude birds (Mathieu-Costello *et al.*, 1998), and high-altitude sherpas (Kayser *et al.*, 1991). However, the ultimate determinant of muscle capillarity appears to be a complex interaction between hypoxia, diet, cold exposure, and exercise, and the results of such studies need to be interpreted in this context (see Mathieu-Costello, 2001 for a review).

Lung volume measured by spirometry is on average slightly larger than that predicted for age, height, sex, and race (Oelz et al., 1986), a trait that is associated with resistance to HAPE in the general climbing population (Cremona et al., 2002). However, values are not different from other athletic populations and are still within the normal range. Additionally, many successful high-altitude climbers have had a previous history of HAPE (Wiseman et al., 2006). High-altitude climbers exhibit a brisk hypoxic ventilatory response when compared with both sedentary controls and marathon runners (Schoene, 1982). This may confer an advantage, both in terms of elevating alveolar PO_2 and in leftward shifting of the oxygen hemoglobin dissociation curve because of profound respiratory alkalosis (West et al., 1983a). In addition, resting ventilation while at sea level may also be elevated (Oelz et al., 1986), the significance of which is unclear.

In general, elite high-altitude climbers tend to be resistant to acute mountain sickness (AMS) and high-altitude cerebral edema (HACE), possibly related to their brisk hypoxic ventilatory response and greater ventilation when exposed to hypoxia (Schoene, 1982; Oelz et al., 1986). Climbers of peaks in the 5000–8000 m elevation range, particularly those who climb without supplemental oxygen, have been shown to suffer subtle but measurable cognitive impairment post-climb, and structural brain abnormalities on magnetic resonance imaging (MRI) (Hornbein et al., 1989; Garrido et al., 1993; Fayed et al., 2006). In addition, some of the cognitive deficits were correlated with a greater hypoxic ventilatory response. It has therefore been suggested that, paradoxically, climbers with a brisk hypoxic ventilatory response with resulting hypocapnia, who feel well at altitude and are able to climb higher faster, are vulnerable to hypoxia-induced-brain injury because hypocapnia causes cerebral vasoconstriction, which may affect oxygen delivery (Hornbein et al., 1989).

Although it has not been directly studied, it seems logical to think that limited weight loss while at altitude might be a determinant of success at extreme altitude. In keeping with this idea, Tibetan Sherpas, who are noted for their abilities at high altitude, have been reported to lose little weight at high altitude (Boyer and Blume, 1984). Anorexia or loss of appetite is commonly associated with AMS; however, appetite returns once acclimatization is complete unless the altitude is very high. Above ~5000 m energy balance cannot be maintained, and even acclimatized individuals will lose weight (Ward et al., 2000). This is true even in chamber studies of simulated ascents, when many of the obstacles to eating at high altitude, such as cold and limited food supplies, are removed (Rose et al., 1988). The reasons for weight loss are multifactorial and include diminished appetite (possibly due to increased in leptin, a hormone that controls satiety) (Tschop et al., 1998), increased basal metabolic rate (Brooks and

Butterfield, 2001), and increased energy expenditure. In addition, there are alterations in body composition, with loss of muscle mass in addition to loss of fat, for reasons that are not clear.

As mentioned above, Sherpas are high-altitude people, largely Tibetan, that are particularly renowned for their abilities at high altitude. During his climb of Mount Everest, Edmund Hillary was accompanied on the summit by the Sherpa Tensing Norgay. Sherpa climbers are part of most North American and European expeditions to this peak, and form a significant portion of those who are successful. Lung volumes and diffusing capacity are high in Sherpas (Havryk et al., 2002), as is the case for any individual who is born and raised at high altitude, and resting arterial oxygen saturations tend to be higher and alveolar–arterial differences lower (Zhuang et al., 1996). In part, this can be explained on the basis of a greater hypoxic ventilatory response (Beall et al., 1997) than other high-altitude natives or other acclimatized lowlanders. In addition, Sherpas have a greater peak oxygen consumption, ventilation (Sun et al., 1990), heart rate, and cardiac output (Chen et al., 1997) during acclimatized exercise at altitude than acclimatized lowlanders. MRI studies of brain structure show no findings in the majority of Sherpas studied, in contrast to lowland elite climbers (Garrido et al., 1996), suggesting that the brain may be less vulnerable in this population. Taken together, these findings suggest the ability to maintain cerebral blood flow despite brisk hyperventilation.

8.5.2   Birds flying over Mount Everest

The bar-headed goose (*Anser indicus*) has been extensively studied for its impressive ability to live and exercise at extremely high altitudes (Black and Tenney, 1980a; Black and Tenney, 1980b; Faraci et al., 1984a; Faraci et al., 1984b; Faraci et al., 1985; Faraci and Fedde, 1986; Fedde et al., 1989; Weber, 2007). Fortunately, most of these data have been synthesized in a recent theoretical analysis of the factors limiting exercise performance in birds at altitude (Scott and Milsom, 2006). Using an integrative model of avian O_2 exchange, Scott and Milsom (2006) determined that at extreme altitude, high total ventilation ($\dot{V}I$), high hemoglobin–O_2 affinity (low P_{50}), and high tissue diffusion capacity for O_2 (D_tO_2) should produce the greatest increase in O_2 consumption ($\dot{V}O_2$) and have the greatest benefit for flight at high altitude. There were additive interactions between the variables too, such that increasing D_tO_2 had a greater effect on $\dot{V}O_2$ when P_{50} was low.

New experiments support adaptations to increase $\dot{V}_I$ in bar-headed geese (Scott and Milsom, 2007). The hypoxic ventilatory response is greater in the bar-headed goose than in low-altitude birds such as the domestic Pekin duck (*Anas platyrhynchos*) (Scott and Milsom, 2007). This contrasts with earlier comparative studies that showed that the ventilatory response to moderate hypoxia was blunted in bar-headed geese compared with ducks in moderate hypoxia, although the ventilatory response was much stronger in bar-headed geese in deep hypoxia (Black and Tenney, 1980a). Tidal volume is also greater in hypoxia for bar-headed geese compared with ducks (Scott and Milsom, 2007), which increases effective parabronchial ventilation more than breathing frequency and would be predicted to increase $\dot{V}O_{2max}$ at a given level of O_2 (Scott and Milsom, 2006). Comparisons with the greylag goose (*Anser anser*), which is more closely related to the bar-headed goose than a duck, suggest that adaptations in respiratory mechanics instead of differences in O_2 sensitivity may allow higher levels of $\dot{V}_I$ in bar-headed geese (Scott and Milsom, 2007). However, the results could not rule out reduced sensitivity to hypocapnia in bar-headed geese, which would contribute to an increased hypoxic ventilatory response (Scott and Milsom, 2007).

The prediction of benefits from high hemoglobin–O_2 affinity is consistent with experimental results in bar-headed geese. The molecular basis for a low P_{50} in bar-headed goose hemoglobin has been established and appears to be an evolutionary adaptation to high altitude (Weber, 2007). The prediction of a high D_tO_2 has not been confirmed with measurements in bar-headed geese yet. However, D_tO_2 has been shown to be a major determinant of $\dot{V}O_{2max}$ in mammals (Wagner *et al.*, 1986) and is expected to be important in birds too. Also, birds show unique transverse anastomoses in muscle capillaries that increase D_tO_2 compared with mammals (Mathieu-Costello, 1991). Hence, an important question for future experiments is to determine whether an increased D_tO_2 in birds provides an avian advantage at altitude compared with mammals.

8.6   Maladaptations to high altitude

Not all physiological responses to acute hypoxia are adaptive for chronic hypoxia at high altitude. High-altitude illnesses are experienced by most people as the result of acute rapid exposure to high altitude. Chronic exposure to altitude is also associated with some maladaptative syndromes in both people and domestic animals. Domestic animals have been studied at altitude, in part because of agricultural economics, but in the future they may offer good models for studying the genetics of adaptation to high altitude. In general, the organs systems most commonly affected in high-altitude maladaptations are the lungs and brain.

8.6.1   *Cerebral manifestation of maladaptations*

8.6.1.1   Acute mountain sickness

Acute mountain sickness (AMS) is the most common of the human high-altitude illnesses and is characterized by headache, nausea, vomiting, anorexia, dizziness, lethargy, fatigue, and sleep disturbance experienced during rapid ascent to a higher altitude. The approximate incidence of AMS is about 50% of all individuals traveling to altitude, and individual susceptibility appears to play a significant role (Hackett and Roach, 2001; Basnyat and Murdoch, 2003). Symptoms typically appear within 6–12 hours of the initial exposure to high altitude, and the development of symptoms is modulated by rate of ascent, altitude reached, sleeping altitude, and previous acclimatization history. Acute mountain sickness is usually benign and self-limited, and most individuals recover within a few days. However, a minority of individuals may progress to HACE, which is fatal if untreated. It has been suggested that AMS and HACE form a spectrum of cerebral maladaptive responses to high altitude and thus share a common pathophysiology (Singh *et al.*, 1969; Hansen and Evans, 1970). Multiple mechanisms appear to be involved in AMS development, among them increased cerebral blood flow during hypoxia (Hackett, 1999a; Muza *et al.*, 1999), resulting in a vascular-type headache with possible subclinical cerebral edema.

A leading hypothesis for the development of AMS is known as the 'tight-brain hypothesis' (Hackett, 1999a; Hackett, 1999b). This hypothesis suggests that increases in cerebral blood flow, blood–brain barrier permeability, and intracellular fluid are initial maladaptive responses to hypoxic exposure in all subjects that ultimately lead to brain swelling. Anatomically, the intracranial space has a fixed volume consisting of parenchyma, blood, and cerebral spinal fluid, so this swelling results in increased intracranial pressure and symptom development unless compensated for by decreasing blood and cerebral spinal fluid volumes. If the brain is not able to accomplish this shift of cerebrospinal fluid out of the intracranial space, pressure will rise, producing symptoms of AMS.

8.6.1.2 High-altitude cerebral edema

High-altitude cerebral edema is much less common than AMS (Hackett and Roach, 2001). Whereas AMS is relatively benign and self-limited, HACE is progressive and ultimately fatal. HACE is more common in individuals who are already ill with HAPE (Hackett and Roach, 2001). The typical sign that suggests that HACE is developing in a person with either AMS or HAPE is the development of ataxia and altered consciousness in addition to the nausea, malaise, and headache characteristic of AMS. Magnetic resonance imaging studies in individuals who are ill with HACE show that the edema predominately affects the white matter (Matsuzawa *et al.*, 1992; Hackett *et al.*, 1998), although this has not been a consistent finding, and some find a more diffuse swelling (Kobayashi *et al.*, 1987; Icenogle *et al.*, 1999; Muza *et al.*, 1999).

The primary treatment of AMS includes descent to a lower altitude, with more gradual ascent and acclimatization after recovery (Hackett and Roach,

2001). Acetazolamide, a carbonic anhydrase inhibitor that creates a metabolic acidosis, has been shown to be beneficial both in the prevention and the treatment of AMS, probably because of its effect in stimulating ventilation and increasing arterial oxygenation. The corticosteroid dexamethasone, a powerful anti-inflammatory, has also been shown to be of benefit in both AMS and HACE, although because of side effects some believe that the use of dexamethasone should be confined to people suffering from HACE. In HACE the treatment is similar to that for AMS but of much greater urgency, and immediate descent, oxygen, dexamethasone, and oxygen are all important in treating HACE (Hackett and Roach, 2001).

8.6.2   *Pulmonary manifestations of maladaptation*

8.6.2.1 **High-altitude pulmonary edema**

High-altitude pulmonary edema typically develops after 2–4 days at altitude in unacclimatized individuals, but re-entry HAPE, in which high-altitude residents develop HAPE when they return to high altitude after a brief sojourn at lower elevation (Schoene *et al.*, 2001), has also been described. Dyspnea, cough, exercise intolerance, and cyanosis characterize HAPE in the early stages, with frothy, pink sputum developing as the disease progresses. Approximately 2–15% of individuals exposed to high altitude are affected severely enough to seek treatment (Hackett and Rennie, 1976; Schoene *et al.*, 2001), although the incidence of subclinical fluid accumulation may be higher (Cremona *et al.*, 2002).

The strongest predicting factor for developing HAPE is a previous history of HAPE. It is associated with rapid ascent and strenuous exercise, and HAPE-susceptible individuals have also been shown to have greater resting pulmonary vascular resistance (Eldridge *et al.*, 1996), higher pulmonary artery pressures, and higher pulmonary capillary wedge pressures (Eldridge *et al.*, 1996) during normoxic exercise. In addition, HAPE-susceptible individuals have augmented hypoxic pulmonary vasoconstriction and increased pulmonary vascular pressures when exposed to hypoxia (Viswanathan *et al.*, 1969; Hultgren *et al.*, 1971; Kawashima *et al.*, 1989; Yagi *et al.*, 1990; Grunig *et al.*, 2000). Mechanisms involved in hypoxic pulmonary vasoconstriction are discussed in Chapter 4, section 3.9). These increased pulmonary vascular pressures are likely to be important, as mechanical stress-induced damage to the pulmonary capillaries has been implicated as the inciting mechanism for HAPE (Hopkins *et al.*, 1997; Swenson *et al.*, 2002; Hopkins *et al.*, 2005). Alterations in nitric oxide regulation (Duplain *et al.*, 2000; Busch *et al.*, 2001), may also affect pulmonary vasoreactivity and susceptibility to HAPE, and especially when combined with a blunted ventilatory response to hypoxia (Hackett *et al.*, 1988; Matsuzawa *et al.*, 1989) may in part explain why pulmonary pressures are increased in these individuals. In addition, a defect of transepithelial sodium transport in HAPE-susceptible individuals (Sartori *et al.*, 1999; Sartori *et al.*, 2002; Mairbaurl *et al.*, 2003) impairs alveolar fluid clearance.

HAPE is easily treated by descent, and recovery is remarkably rapid at lower elevation, but untreated HAPE can be fatal. Pulmonary vasodilators such as sildenafil and nifedipine reduce high pressure in the pulmonary circulation resulting from hypoxic pulmonary vasoconstriction and, combined with oxygen, reduce the severity of HAPE (Oelz *et al.*, 1989; Scherrer *et al.*, 1996). Interestingly, a recent study (Maggiorini *et al.*, 2006) suggests that prophylaxis with dexamethasone prevents the rise in pulmonary artery pressure and the development of HAPE to the same extent as a pulmonary vasodilator, possible because of an effect on pulmonary arterial endothelial function.

8.6.2.2 Brisket disease in cattle

Brisket disease is characterized by edema of the chest and right-heart failure from severe pulmonary hypertension. It was first described by veterinarians working in Colorado on cattle grazing at over 2400 m (Glover and Newsom, 1915). Cattle from certain lineages develop extreme hypoxic pulmonary vasoconstriction, leading to the disease, which is observed to be more common in animals brought from low to high altitudes than in cattle born at high altitude (Bisgard, 1977). Breeding experiments and embryo transfers with susceptible and resistant cattle indicate a genetic basis for the disease that is likely to involve an autosomal-dominant gene (Weir *et al.*, 1974; Will *et al.*, 1975; Cruz *et al.*, 1980; Holt and Ramirez, 1998). Yaks (*Bos grunniens*), which are the only bovine native to high altitude, are known to have blunted hypoxic pulmonary vasoconstriction (Durmowicz *et al.*, 1993). It would be interesting to determine whether there is any relationship between the genetic determinants of brisket disease and reduced hypoxic pulmonary vasoconstriction in the yak.

8.6.2.3 Ascites in domestic chickens

Domestic chickens (*Gallus domesticus*) raised at altitude also show problems with pulmonary circulation. Ascites, or accumulation of fluid in the peritoneal cavities, is recognized as a major cause of illness and death in chickens reared for meat at altitudes above 3500 m (Julian, 1993). This edema is caused by portal hypertension from pulmonary hypertension and right-ventricular hypertrophy (Julian, 1993). Although birds show hypoxic pulmonary vasoconstriction similar to mammals (Holle *et al.*, 1978; Black and Tenney, 1980b), pulmonary hypertension is not the result of an excessive increase in

pulmonary vascular resistance as in brisket disease in cattle. The parabronchial lungs of birds are relatively non-compliant, in contrast to mammalian
alveolar lungs, so avian pulmonary capillaries show limited recruitment and
distension with increased blood flow (Powell *et al.*, 1985; West *et al.*, 2007).
Hypoxia stimulates increased blood flow through a fixed resistance and pulmonary vascular pressure rises (Powell and Whittow, 2000). In the 1980s,
ascites also became common at low altitude in chickens being bred for very
fast growth (Julian, 1993). The etiology is the same as ascites at high altitude,
except blood flow increases to support increased O_2 demand, associated with
fast growth, instead of compensating for decreased O_2 supply (Julian, 1993;
Powell, 2000).

A reasonable hypothesis is that the ultimate cause of pulmonary hypertension and ascites in chickens is the inability of the lungs to evolve increased
vascular capacity in proportion to increases in O_2 demand. The pulmonary
capillary volume of chickens is significantly smaller than the allometric prediction for a bird of comparable size (Maina *et al.*, 1989), and this may have
resulted from intense artificial selection for increased body mass (i.e. meat
production) during the domestication of chickens (Powell, 2000). Dissociation
between selection of increased O_2 demand (body mass and growth) and O_2
supply (pulmonary capillaries) suggests that the genomics of O_2 supply and
demand are organized independently and may be differentially sensitive to
natural selection (Powell, 2003). Some breeds of chickens tolerate high altitude
better than others (Mejia *et al.*, 1994), and breeding experiments have shown
adaptations to high altitude in chickens (Smith *et al.*, 1959). It would be interesting to determine whether genetic control of O_2 demand vs. O_2 supply (body
size vs. pulmonary capillary volume) could explain some breeds being more
successful at high altitude.

8.6.3   *Multi-organ maladaptation*

8.6.3.1 Chronic mountain sickness

Chronic mountain sickness, also known as Monge's disease, is a syn-
drome of polycythemia, hypoxemia, and, in severe cases, pulmonary hypertension leading to right-heart failure, in very long-term residents of high altitude
(Monge, 1942). The hallmark of chronic mountain sickness is hemoglobin that is
elevated more than normal for a particular elevation. Since the range accepted
for a normal hemoglobin value rises with increasing elevation, considerable
effort has gone into defining both normal and optimal hemoglobin levels of
high-altitude residents (Leon-Velarde *et al.*, 2005). It has been suggested that the
primary trigger of chronic mountain sickness is a blunted hypoxic ventilatory

response (Leon-Velarde and Richalet, 2006); however, this is frequently a characteristic of high-altitude residents (Weil *et al.*, 1971). Sleep-disordered breathing has also been suggested as a possible cause (Sun *et al.*, 1996). Recent work (see Moore *et al.*, 2007 for a review) also suggests a perinatal origin: men with early chronic mountain sickness are much more likely to have had mothers who suffered from pre-eclampsia during their pregnancy, leading these researchers to hypothesize that this acts as a trigger. In addition to polycythemia, cyanosis, venous dilation, paresthesia, headache, and tinnitus are also found. The treatments (aside from moving to lower elevation, which resolves the symptoms) are directed toward relieving hypoxemia (oxygen, ventilatory stimulants) and the high hematocrit (phlebotomy) (Leon-Velarde *et al.*, 2005).

8.7   Common themes of hypoxia tolerance

Reviewing the respiratory physiology of successful strategies for animals living at high altitude reveals some common themes, considered below.

8.7.1   Oxygen conservation

In the face of low oxygen supply, many high-altitude species employ techniques to reduce the demand for oxygen, including behavioral strategies and alterations in metabolism. One example of a conserving strategy is seen in high-altitude birds. Flapping (active) fight is energetically costly in absolute terms but not in terms of cost per unit distance (Maina, 2000). However, many birds exploit air currents for gliding or use 'flap-gliding' gaits to conserve energy during migration (Butler and Bishop, 2000). It is noteworthy that the high-altitude record for a bird is for a soaring vulture, Rüppell's griffon (Laybourne, 1974), mentioned previously. Hovering flight is very energetically expensive and more than double the cost of forward flight (Maina, 2000); therefore it is surprising that hummingbirds occupy an altitude niche as high as 5000 m (Altshuler *et al.*, 2004; Altshuler, 2006). Although oxygen availability is reduced and the energetic costs of flight are increased because of decreased air density, increased wing size appears to compensate for this (Altshuler *et al.*, 2004).

There is evidence that substrate utilization may be altered under conditions of hypoxia to favor maximal ATP generation in many species (for a review see Hochachka, 1998; Hochachka and Monge, 2000). In addition, hypoxia-induced metabolic depression is well documented in reptiles, small mammals, and neonates. Newborn mammals when faced with environmental hypoxia respond with a reduction in metabolic rate, resulting in relative hyperventilation as ventilation is maintained or even increased (Mortola, 1999). This effect occurs regardless of ambient temperature, although hypoxia alone will decrease thermogenesis and lower the preferred ambient temperature (Mortola, 1999). This is also seen in adult animals with small body mass, and the magnitude of the hypoxic depression of metabolism is approximately proportional to the mass specific $\dot{V}O_2$ of the animal. In addition, many heterothermic mammals, such as ground squirrels and marmots, which lower their metabolic rate by a factor of more than ten during hibernation, are very hypoxia tolerant, particularly during hibernation but also during periods of arousal (Drew et al., 2004).

Adult humans do not appear to exhibit hypoxia-induced hypometabolism, and in lowlanders, basal oxygen consumption at altitude is elevated above sea level values even several days after arrival at altitude (Brooks and Butterfield, 2001). However, in both Tibetan and Andean high-altitude native populations, basal metabolic rates do not differ from lowlander norms (Beall, 2007); thus a lack of elevation may reflect a certain extent of oxygen conservation. Interestingly, recent data show that after exposure to intermittent hypoxia, such as with using hypoxic tents at sea level, an improvement in running economy has been reported in elite runners (Saunders et al., 2004; Neya et al., 2007); thus for a given oxygen consumption, running velocity (and presumably performance) is increased.

8.7.2   Efficient gas exchange in the lung and tissues

Besides a strategy of minimizing oxygen use, another feature of high-altitude-adapted species relates to efficient pulmonary gas exchange. The potential advantage of cross-current gas exchange in birds, which is theoretically more efficient than alveolar gas exchange in mammals, was considered above. The high levels of O_2 uptake necessary for flapping flight in the bar-headed goose migrating over the highest peaks in the Himalayas support the effectiveness of the avian respiratory system at altitude. Another impressive example of avian respiration at altitude is the hummingbird, which maintains constant O_2 uptake levels while hovering at altitudes from sea level to 6000 m (Berger, 1974), especially considering that these O_2 uptake levels exceed the $\dot{V}O_{2max}$ of a comparably sized mammal at sea level.

Many animal species and humans (Dempsey et al., 1984; Hopkins et al., 1994; Hopkins et al., 1998) show less efficient gas exchange during exercise, particularly during hypoxia, and this results from the combined effects of more ventilation-perfusion mismatching and diffusion limitation of oxygen transport. By contrast, in the only bird studied during hypoxic exercise to date, ventilation-perfusion inequality decreased with hypoxic exercise, and the overall increase of the ventilation relative to perfusion minimizes the effect of any heterogeneity on gas exchange (Schmitt et al., 2002). Comparative studies of ventilation-perfusion matching between high- and low-altitude species have not been done.

A large tissue diffusing capacity for oxygen would be ideal to facilitate oxygen delivery during conditions of reduced availability. It was long thought that adaptation to hypoxia resulted in an increased capillary-to-fiber ratio, which might facilitate oxygen delivery to tissues. When high-altitude finches are compared with the same species living at low altitude they show (1) an increase in capillary fiber number associated with an increase in tortuosity and branching, without a decrease in fiber cross-sectional area; and (2) an increase in mitochondrial volume (Hepple et al., 1998). This effectively increases the diffusing capacity for tissue oxygenation in these high-altitude native animals. However, it is now recognized that in many mammals, including humans, an apparent increase in capillarity results from a decrease in fiber surface area when exposed to hypoxia (MacDougall et al., 1991). As mentioned earlier, the precise nature of the response depends on the integration of several factors acting on the muscle, including hypoxia, cold, and exercise among others (see Mathieu-Costello, 2001 for a review), and the generalized response is not clearly defined.

8.7.3   High hemoglobin-oxygen affinity

Perhaps one of the most robust features observed across widely divergent species that are expected to be adapted to hypoxia is a high affinity of hemoglobin for oxygen (i.e. a low P_{50}), and this has been covered in several recent reviews (Tenney, 1995; Weber, 1995; Weber, 2007). These observations are consistent with gas-exchange models that predict that a high hemoglobin–O_2 affinity will increase O_2 uptake at extremely high altitudes, where pulmonary diffusion is limiting (Bencowitz et al., 1982). The affinity of hemoglobin for oxygen varies inversely with body size (Lahiri, 1975), so it is difficult to compare directly across species – for example from a mouse to a human. However, this pattern can be observed by comparing a high-altitude or hypoxia-tolerant species to a related species or subspecies that is not particularly noted for hypoxia tolerance. Figure 8.5 shows the P_{50} for seven different animal types, illustrating the robustness of this finding.

Remarkably, this trait is independent of the absolute values of P_{50} in the low-altitude species, which is demonstrated most markedly by comparing birds and mammals in Fig. 8.5. Similarly, although all camelids have a relatively low P_{50} (17 to 22 Torr), those indigenous to high altitude (guanacos, llamas, alpacas, and vicuñas, which live at 2000–5000 m in the Andean altiplano) have a lower P_{50} than the two species living in the low-altitude deserts of Africa and Asia (Hochachka and Somero, 2002). In humans, Sherpas have a lower P_{50} than do acclimatized lowlanders, although this relationship does not hold true for Andean natives (Monge and Whittembury, 1974).

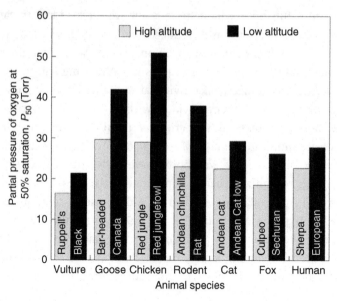

Fig. 8.5 Comparision of blood oxygen affinity between high- and low-altitude-adapted animals P$_{50}$ for hypoxia-tolerant (gray) and similar species resident at low (or lower) elevations (black). Of the birds, Rüppell's griffon (*Gyps rueppellii*) and the European black vulture (*Aegypius monachus*, Aegypiinae) have a higher oxygen affinity than the geese or the chickens. The European black vulture is found at altitudes above 4500 m and therefore is not a lowland species, but has not been recorded at the extreme elevations of the Rüppell's griffon (Weber, 2007). For the chicken, a red junglefowl (*Gallus gallus*), data presented are for members of the same species at different elevations. The bar-headed goose (*Anser indicus*) has a higher affinity than the Canada goose (*Branta canadensis*), which flies at relatively modest elevations (Saunders and Fedde, 1994). The data for the Andean cat (*Felis jacobida*) are for members of the same species found at different elevations (León-Velarde *et al.*, 1996). The fox data compare the South American red fox or culpeo fox (*Dusicyon culpaeus*) with a habitat above 4000 m to the Sechuran desert fox (*Dusicyon sechurae*) found at sea level (León-Velarde *et al.*, 1996). The human data compare a high-altitude people, Sherpas, to people of lowland European origin. In all cases cited, the hypoxia-tolerant animal has a greater affinity of hemoglobin for oxygen (lower P$_{50}$).

Relatively few amino acid substitutions can explain the differences in oxygen affinity between the species in mammals, birds, and amphibians: for example, a histidine-to-asparagine substitution in the beta chains that suppresses two 2,3-diphosphoglycerate-binding sites per tetramer in the Andean camelids (Poyart *et al.*, 1992). Differences in hemoglobin between closely related low- and high-altitude species commonly comprise substitutions of amino acids at binding sites for modulators (e.g. phosphates in homeotherms and chloride in amphibians), and at contact points between subunits that stabilize hemoglobin in either the low-affinity (tense) or high-affinity (relaxed) conformations (Weber, 2007). Recent evidence from yaks, deer mice, and Rüppell's griffon also indicates that multiple isoforms of hemoglobins with different O_2-binding properties may be a common adaptation in high-altitude species (Storz et al., 2007; Weber, 2007). The fact that these changes require relatively few amino acid substitutions supports natural selection for altitude as the mechanism causing differences in P_{50}.

It remains to be determined whether hemoglobin adaptations appear to be the most common high-altitude adaptation in vertebrates because they involve relatively few mutations or because they are easier to isolate and quantify compared with other complex physiological traits. Differences in P_{50} can clearly occur without other changes in the O_2 cascade. This is well illustrated in a comparison of house finches (*Carpodacus mexicanus*) at sea level and rosy finches (*Leucosticte arctoa*) that breed at altitudes of 3500 m in eastern California. P_{50} is lower in the rosy finch (31.0 vs. 37.4 Torr), but ventilation and oxygen consumption are similar in both species when measured at a given altitude (Clemens, 1988; Clemens, 1990). However, a strong correlation has been observed between P_{50} and the PO_2 threshold for a hypoxic ventilatory response across species (Boggs, 1995), suggesting there may be common adaptations in heme proteins for O_2 binding and those hypothesized to be involved in O_2 sensing (Fidone et al., 1986).

8.8   Conclusions

Despite the harsh environment, diverse species live at high altitude. Many species of birds are notably well adapted, although there are also many high-altitude mammals and reptiles, as well as high-altitude humans. Common traits of high-altitude species include enhanced oxygen delivery from efficient pulmonary gas exchange, optimal oxygen-carrying capacity, oxygen-conserving strategies, and resistance to cerebral and pulmonary manifestations of high-altitude illness.

References

Altshuler, D. L. (2006). Flight performance and competitive displacement of hummingbirds across elevational gradients. *Am. Nat.*, **167**, 216–29.

Altshuler, D. L., Dudley, R. and McGuire, J. A. (2004). Resolution of a paradox: hummingbird flight at high elevation does not come without a cost. *Proc. Natl. Acad. Sci. USA*, **101**, 17731–6.

Anand, I. S., Harris, E., Ferrari, R., Pearce, P. and Harris, P. (1986). Pulmonary hemodynamics of the yak, cattle and cross breeds at high altitude. *Thorax*, **41**, 696–700.

Banchero, N. and Grover, R. F. (1972). Effect of different simulated altitude on O_2 transport in llama and sheep. *Am. J. Physiol.*, **222**, 1239–45.

Basnyat, B. and Murdoch, D. R. (2003). High-altitude illness. *Lancet*, **361**, 1967–74.

Beall, C. M. (2007). Two routes to functional adaptation: Tibetan and Andean high-altitude natives. *Proc. Natl. Acad. Sci. USA*, **104** (Suppl. 1), 8655–60.

Beall, C. M., Strohl, K. P., Blangero, J., *et al.* (1997). Ventilation and hypoxic ventilatory response of Tibetan and Aymara high altitude natives. *Am. J. Phys. Anthropol.*, **104**, 427–47.

Bebout, D. E., Storey, D., Roca, J., *et al.* (1989). Effects of altitude acclimatization on pulmonary gas exchange during exercise. *J. Appl. Physiol.*, **67**, 2286–95.

Bencowitz, H. Z., Wagner, P. D. and West, J. B. (1982). Effect of change in P_{50} on exercise tolerance at high altitude: a theoretical study. *J. Appl. Physiol.*, **53**, 1487–95.

Berger, M. (1974). *Energiewechsel von Kolibris beim Schwirrflug unter Höhenbedingungen,* **115**, 273–88.

Bisgard, G. E. (1977). Pulmonary hypertension in cattle. *Adv. Vet. Sci. Comp. Med.* **21**, 151–72.

Black, C. P. and Tenney, S. M. (1980a). Oxygen transport during progressive hypoxia in high-altitude and sea-level waterfowl. *Respir. Physiol.*, **39**, 217–39.

Black, C. P. and Tenney, S. M. (1980b). Pulmonary hemodynamic responses to acute and chronic hypoxia in two waterfowl species. *Comp. Biochem. Physiol.*, **67A**, 291–3.

Boggs, D. F. (1995). Hypoxic ventilatory control and hemoglobin oxygen affinity. In *Hypoxia and the Brain*, ed. J. R. Sutton, C. S. Houston and G. Coates. Burlington: Queen City Printers, pp. 69–88.

Bouverot, P., Farner, D. S., Heinrich, B., *et al.* (1985). *Adaptation to Altitude-hypoxia in Vertebrates.* Berlin: Springer-Verlag.

Boyer, S. J. and Blume, F. D. (1984). Weight loss and changes in body composition at high altitude. *J. Appl. Physiol.*, **57**, 1580–5.

Brooks, G. A. and Butterfield, G. (2001). Metabolic responses of lowlanders to high altitude exposure: malnutrition versus the effects of hypoxia. In *Lung Biology in Health and Disease, High Altitude*, ed. T. F. Hornbein and R. B. Schoene. New York: Marcel Dekker, pp. 569–99.

Brutsaert, T. D. (2001). Limits on inferring genetic adaptation to high altitude in Himalayan and Andean populations. *High Altitude Med. Biol.*, **2**, 211–25.

Busch, T., Bartsch, P., Pappert, D., *et al.* (2001). Hypoxia decreases exhaled nitric oxide in mountaineers susceptible to high-altitude pulmonary edema. *Am. J. Respir. Crit. Care Med.*, **163**, 368–73.

Butler, P. J. and Bishop, C. M. (2000). Flight. In *Sturkie's Avian Physiology*, ed. G. C. Whittow. San Diego: Academic Press, pp. 391–435.

Chappell, M. A. and Snyder, L. R. (1984). Biochemical and physiological correlates of deer mouse alpha-chain hemoglobin polymorphisms. *Proc. Natl. Acad. Sci. USA*, **81**, 5484–8.

Chen, Q. H., Ge, R. L., Wang, X. Z., *et al.* (1997). Exercise performance of Tibetan and Han adolescents at altitudes of 3,417 and 4,300 m. *J. Appl. Physiol.*, **83**, 661–7.

Clemens, D. T. (1988). Ventilation and oxygen consumption in rosy finches and house finches at sea level and high altitude. *J. Comp. Physiol. B*, **158**, 547–66.

Clemens, D. T. (1990). Interspecific variation and effects of altitude on blood properties of rosy finches (*Leucosticte arctoa*) and house finches (*Carpodacus mexicanus*). *Physiol. Zool.*, **63**, 288–307.

Cremona, G., Asnaghi, R., Baderna, P., *et al.* (2002). Pulmonary extravascular fluid accumulation in recreational climbers: a prospective study. *Lancet*, **359**, 303–9.

Cruz, J. C., Reeves, J. T., Russell, B. E., Alexander, A. F. and Will, D. H. (1980). Embryo transplanted calves: the pulmonary hypertensive trait is genetically transmitted. *Proc. Soc. Exp. Biol. Med.* **164**, 142–5.

Dempsey, J. A., Hanson, P. G. and Henderson, K. S. (1984). Exercised-induced arterial hypoxaemia in healthy human subjects at sea level. *J. Physiol. (Lond)*, **355**, 161–75.

Drew, K. L., Harris, M. B., Lamanna, J. C., *et al.* (2004). Hypoxia tolerance in mammalian heterotherms. *J. Exp. Biol.*, **207**, 3155–62.

Dunmire, W. W. (1960). An altitudinal survey of reproduction in Peromyscus maniculatus. *Ecology*, **41**, 174–82.

Duplain, H., Sartori, C., Lepori, M., *et al.* (2000). Exhaled nitric oxide in high-altitude pulmonary edema: role in the regulation of pulmonary vascular tone and evidence for a role against inflammation. *Am. J. Respir. Crit. Care Med.*, **162**, 221–4.

Durmowicz, A. G., Hofmeister, S., Kadyraliev, T. K., Aldashev, A. A. and Stenmark, K. R. (1993). Functional and structural adaptation of the yak pulmonary circulation to residence at high altitude. *J. Appl. Physiol.*, **74**, 2276–85.

Eldridge, M. W., Podolsky, A., Richardson, R. S., *et al.* (1996). Pulmonary hemodynamic response to exercise in subjects with prior high-altitude pulmonary edema. *J. Appl. Physiol.*, **81**, 911–21.

Faraci, F. M. (1991). Adaptations to hypoxia in birds: how to fly high. *Ann. Rev. Physiol.*, **53**, 59–70.

Faraci, F. M. and Fedde, M. R. (1986). Regional circulatory responses to hypocapnia and hypercapnia in bar-headed geese. *Am. J. Physiol.*, **250**, R499–504.

Faraci, F. M., Kilgore, D. L. and Fedde, M. R. (1984a). Attenuated pulmonary pressor response to hypoxia in bar-headed geese. *Am. J. Physiol.*, **247**, R402–3.

Faraci, F. M., Kilgore, D. L. and Fedde, M. R. (1984b). Oxygen delivery to the heart and brain during hypoxia: pekin duck vs. bar-headed goose. *Am. J. Physiol.*, **16**, R69–75.

Faraci, F. M., Kilgore, D. L., Jr. and Fedde, M. R. (1985). Blood flow distribution during hypocapnic hypoxia in Pekin ducks and bar-headed geese. *Respir. Physiol.*, **61**, 21–30.

Fayed, N., Modrego, P. J. and Morales, H. (2006). Evidence of brain damage after high-altitude climbing by means of magnetic resonance imaging. *Am. J. Med.*, **119**, 168e1–6.

Fedde, M. R., Orr, J. A., Shams, H. and Scheid, P. (1989). Cardiopulmonary function in exercising bar-headed geese during normoxia and hypoxia. *Respir. Physiol.*, **77**, 239–62.

Fidone, S. J., Gonzalez, C., Cherniack, N. S. and Widdicombe, J. G. (1986). Initiation and control of chemoreceptor activity in the carotid body. In *Handbook of Physiology: the Respiratory System – Control of Breathing*, ed. A. P. Fisherman. Baltimore, MD: Waverly Press, pp. 247–312.

Garland, T., Jr. (2001). Phylogenetic comparison and artificial selection: two approaches in evolutionary physiology. *Adv. Exp. Med. Biol.*, **502**, 107–32.

Garland, T., Jr. and Adolph, S. C. (1994). Why not to do two species comparisons – limitations on inferring adaptation. *Physiol. Zool.*, **67**, 797–828.

Garrido, E., Castello, A., Ventura, J. L., Capdevila, A. and Rodriguez, F. A. (1993). Cortical atrophy and other brain magnetic resonance imaging (MRI). changes after extremely high-altitude climbs without oxygen. *Int. J. Sports Med.*, **14**, 232–4.

Garrido, E., Segura, R., Capdevila, A., *et al.* (1996). Are Himalayan Sherpas better protected against brain damage associated with extreme altitude climbs? *Clin. Sci. (Lond)*, **90**, 81–5.

Glover, G. and Newsom, I. (1915). Dropsy of high altitudes. *Colo. Agric. Exp. Sta. Bull.* **204**, 3–24.

Grinnell, J. and Storer, T. I. (1924). *Animal Life in the Yosemite*, Berkeley: University of California.

Grubb, B., Colacino, J. M. and Schmidt-Nielsen, K. (1978). Cerebral blood flow in birds: effect of hypoxia. *Am. J. Physiol.*, **234**, H230–4.

Grubb, B., Mills, C. D., Colacino, J. M. and Schmidt-Nielsen, K. (1977). Effect of arterial carbon dioxide on cerebral blood flow in ducks. *Am. J. Physiol.*, **232**, H596–601.

Grunig, E., Mereles, D., Hildebrandt, W., *et al.* (2000). Stress Doppler echocardiography for identification of susceptibility to high altitude pulmonary edema. *J. Am. Coll. Cardiol.*, **35**, 980–7.

Hackett, P. H. (1999a). High altitude cerebral edema and acute mountain sickness: a pathophysiology update. *Adv. Exp. Med. Biol.*, **474**, 23–45.

Hackett, P. H. (1999b). The cerebral etiology of high-altitude cerebral edema and acute mountain sickness. *Wilderness Environ. Med.*, **10**, 97–109.

Hackett, P. H. and Rennie, D. (1976). The incidence, importance, and prophylaxis of acute mountain sickness. *Lancet*, **2**, 1149–55.

Hackett, P. H. and Roach, R. C. (2001). High-altitude illness. *New Engl. J. Med.*, **345**, 107–14.

Hackett, P. H., Roach, R. C., Schoene, R. B., Harrison, G. L. and Mills, W. J. (1988). Abnormal control of ventilation in high-altitude pulmonary edema. *J. Appl. Physiol.*, **64**, 1268–72.

Hackett, P. H., Yarnell, P. R., Hill, R., *et al.* (1998). High-altitude cerebral edema evaluated with magnetic resonance imaging: clinical correlation and pathophysiology. *JAMA*, **280**, 1920–5.

Hammond, K. A., Szewczak, J. M. and Krol, E. (2001). Effects of altitude and temperature on organ phenotypic plasticity along an altitudinal gradient. *J. Exp. Biol.*, **204**, 1991–2000.

Hansen, J. E. and Evans, W. O. (1970). A hypothesis regarding the pathophysiology of acute mountain sickness. *Arch. Environ. Health*, **21**, 666–9.

Havryk, A. P., Gilbert, M. and Burgess, K. R. (2002). Spirometry values in Himalayan high altitude residents (Sherpas). *Respir. Physiol. Neurobiol.*, **132**, 223–32.

Hayes, J. P. and O'Connor, C. S. (1999). Natural selection on thermogenic capacity of high-altitude deer mice. *Evolution*, **53**, 1280–7.

Hedrick, M. S., Palioca, W. B. and Hillman, S. S. (1999). Effects of temperature and physical activity on blood flow shunts and intracardiac mixing in the toad *Bufo marinus*. *Physiol. Biochem. Zool.*, **72**, 509–19.

Hepple, R. T., Agey, P. J., Hazelwood, L., *et al.* (1998). Increased capillarity in leg muscle of finches living at altitude. *J. Appl. Physiol.*, **85**, 1871–6.

Hicks, J. W. and Wood, S. C. (1985). Temperature regulation in lizards: effects of hypoxia. *Am. J. Physiol.*, **248**, R595–600.

Hochachka, P. W. (1998). Mechanism and evolution of hypoxia-tolerance in humans. *J. Exp. Biol.*, **201**, 1243–54.

Hochachka, P. W. and Monge, C. (2000). Evolution of human hypoxia tolerance physiology. *Adv. Exp. Med. Biol.*, **475**, 25–43.

Hochachka, P. W. and Somero, G. N. (1984). *Biochemical Adaptation*. Princeton, New Jersey: Princeton University Press.

Hochachka, P. W. and Somero, G. N. (2002). *Biochemical Adaptation: Mechanism and Process in Physiological Evolution*, Oxford: Oxford University Press.

Holle, J. P., Heisler, N. and Scheid, P. (1978). Blood flow distribution in the duck lung and its control by respiratory gases. *Am. J. Physiol.*, **234**, R146–54.

Holt, T. N. and Ramirez, G. (1998). Genetic adaptation of cattle to high altitude. *Am. Zool.*, **38**, 10A.

Hopkins, S. R., Bayly, W. M., Slocombe, R. F., Wagner, H. and Wagner, P. D. (1998). Effect of prolonged heavy exercise on pulmonary gas exchange in horses. *J. Appl. Physiol.*, **84**, 1723–30.

Hopkins, S. R., Garg, J., Bolar, D. S., Balouch, J. and Levin, D. L. (2005). Pulmonary blood flow heterogeneity during hypoxia and high-altitude pulmonary edema. *Am. J. Respir. Crit. Care Med.*, **171**, 83–7.

Hopkins, S. R., McKenzie, D. C., Schoene, R. B., Glenny, R. and Robertson, H. T. (1994). Pulmonary gas exchange during exercise in athletes I: ventilation-perfusion mismatch and diffusion limitation. *J. Appl. Physiol.*, **77**, 912–17.

Hopkins, S. R., Schoene, R. B., Henderson, W. R., *et al.* (1997). Intense exercise impairs the integrity of the pulmonary blood-gas barrier in elite athletes. *Am. J. Respir. Crit. Care Med.*, **155**, 1090–4.

Hoppeler, H., Howald, H. and Cerretelli, P. (1990). Human muscle structure after exposure to extreme altitude. *Experientia*, **46**, 1185–7.

Hornbein, T. F., Townes, B. D., Schoene, R. B., Sutton, J. R. and Houston, C. S. (1989). The cost to the central nervous system of climbing to extremely high altitude. *New Engl. J. Med.*, **321**, 1714–19.

Huey, R. B., Eguskitza, X. and Dillon, M. (2001). Mountaineering in thin air. Patterns of death and of weather at high altitude. *Adv. Exp. Med. Biol.*, **502**, 225–36.

Huey, R. B., Salisbury, R., Wang, J. L. and Mao, M. (2007). Effects of age and gender on success and death of mountaineers on Mount Everest. *Biol. Lett.*, **3**, 498–500.

Hultgren, H. N., Grover, R. F. and Hartley, L. H. (1971). Abnormal circulatory responses to high altitude in subjects with a previous history of high-altitude pulmonary edema. *Circulation*, **44**, 759–70.

Hutchison, V. H., Haines, H. B. and Engbretson, G. (1976). Aquatic life at high altitude: respiratory adaptations in the Lake Titicaca frog, *Telmatobius culeus*. *Respir. Physiol.* **27**, 115–29.

Icenogle, M., Kilgore, D., Sanders, J., Caprihan, A. and Roach, R. C. (1999). Cranial CSF volume (cCSF) is reduced by altitude exposure but is not related to early acute mountain sickness (AMS) (Abstract). In *Hypoxia: into the Next Millennium*, ed. R. C. Roach, P. D. Wagner, and P. H. Hackett. New York: Plenum/Kluwer Academic Publishing, p. 392.

Julian, R. J. (1993). Ascites in poultry. *Avian Pathol.*, **23**, 419–54.

Kawashima, A., Kubo, K., Kobayashi, T. and Sekiguchi, M. (1989). Hemodynamic responses to acute hypoxia, hypobaria, and exercise in subjects susceptible to high-altitude pulmonary edema. *J. Appl. Physiol.*, **67**, 1982–9.

Kayser, B., Hoppeler, H., Claassen, H. and Cerretelli, P. (1991). Muscle structure and performance capacity of Himalayan Sherpas. *J. Appl. Physiol.*, **70**, 1938–42.

Kobayashi, T., Koyama, S., Kubo, K., Fukushima, M. and Kusama, S. (1987). Clinical features of patients with high-altitude pulmonary edema in Japan. *Chest*, **92**, 814–21.

Lahiri, S. (1975). Blood oxygen affinity and alveolar ventilation in relation in body weight in mammals. *Am. J. Physiol.*, **229**, 529–36.

Laybourne, R. C. (1974). Collison between a vulture and an aircraft at an altitude of 37,000 ft. *Wilson Bull*, **86**, 461–2.

Leon-Velarde, F. and Monge, C. C. (2004). Avian embryos in hypoxic environments. *Respir. Physiol. Neurobiol.*, **141**, 331–43.

Leon-Velarde, F. and Richalet, J. P. (2006). Respiratory control in residents at high altitude: physiology and pathophysiology. *High Alt. Med. Biol.*, **7**, 125–37.

Leon-Velarde, F., Maggiorini, M., Reeves, J. T., *et al.* (2005). Consensus statement on chronic and subacute high altitude diseases. *High Alt. Med. Biol.*, **6**, 147–57.

León-Velarde, F., De Muizon, C., Palacios, J. A., Clark, D. and Monge, C. (1996). Hemoglobin affinity and structure in high-altitude and sea-level carnivores from Peru. *Comp. Biochem. Physiol. Part A, Physiology*, **113**, 407–11.

Llanos, A. J., Riquelme, R. A., Herrera, E. A., *et al.* (2007). Evolving in thin air – lessons from the llama fetus in the altiplano. *Respir. Physiol. Neurobiol.* **158**, 298–306.

Llanos, A. J., Riquelme, R. A., Sanhueza, E. M., *et al.* (2003). The fetal llama versus the fetal sheep: different strategies to withstand hypoxia. *High Alt. Med. Biol.*, **4**, 193–202.

Loeppky, J. A., Icenogle, M., Scotto, P., *et al* (1997). Ventilation during simulated altitude, normobaric hypoxia and normoxic hypobaria. *Respir. Physiol.*, **107**, 231–9.

Longo, L. D. (1987). Respiratory gas exchange in the placenta. *Handbook of Physiology, Section 3, The Respiratory System, Volume IV. Gas Exchange*, ed. L. E. Farhi and S. M. Tenney. Bethesda, MD: American Physiological Society, pp. 351–401.

Luft, U. C. (1965). Aviation physiology – the effects of altitude. *Handbook of Physiology, Section 3: The Respiratory System, Volume II*, ed. W. O. Fenn and H. Rahn. Washington, DC: American Physiological Society, pp. 1099–145.

MacDougall, J. D., Green, H. J., Sutton, J. R., *et al.* (1991). Operation Everest II: structural adaptations in skeletal muscle in response to extreme simulated altitude. *Acta Physiol. Scand.*, **142**, 421–7.

MacMillen, R. E. and Garland, T., Jr., (1989). Adaptive physiology. In *Advances in the Study of Peromyscus (Rodentia)*, ed. G. L. Kirkland and J. N. Layne. Lubbock: Texas Tech University Press, pp. 143–68.

Maggiorini, M., Brunner-La Rocca, H. P., Peth, S., *et al.* (2006). Both tadalafil and dexamethasone may reduce the incidence of high-altitude pulmonary edema: a randomized trial. *Ann. Intern. Med.*, **145**, 497–506.

Maina, J. N. (2000). What it takes to fly: the structural and functional respiratory refinements in birds and bats. *J. Exp. Biol.*, **203**, 3045–64.

Maina, J. N., King, A. S. and Settle, G. (1989). An allometric study of pulmonary morphometric parameters in birds, with mammalian comparisons. *Phil. Trans. R. Soc. Lond. B Biol. Sci.*, **326**, 1–57.

Mairbaurl, H., Schwobel, F., Hoschele, S., *et al.* (2003). Altered ion transporter expression in bronchial epithelium in mountaineers with high-altitude pulmonary edema. *J. Appl. Physiol.*, **95**, 1843–50.

Mathieu-Costello, O. (1989). Muscle capillary tortuosity in high altitude mice depends on sarcomere length. *Respir. Physiol.*, **76**, 289–302.

Mathieu-Costello, O. (1991). Morphometric analysis of capillary geometry in pigeon pectoralis muscle. *Am. J. Anat.*, **191**, 74–84.

Mathieu-Costello, O. (2001). Muscle adaptation to altitude: tissue capillarity and capacity for aerobic metabolism. *High Alt. Med. Biol.*, **2**, 413–25.

Mathieu-Costello, O., Agey, P. J., Wu, L., Szewczak, J. M. and Macmillen, R. E. (1998). Increased fiber capillarization in flight muscle of finch at altitude. *Respir. Physiol.*, **111**, 189–99.

Matsuzawa, Y., Fujimoto, K., Kobayashi, T., *et al.* (1989). Blunted hypoxic ventilatory drive in subjects susceptible to high-altitude pulmonary edema. *J. Appl. Physiol.*, **66**, 1152–7.

Matsuzawa, Y., Kobvayashi, T., Fujimoto, K. and Schinozaki, S. (1992). Cerebral edema in acute mountain sickness. In *High-altitude Medicine*, ed. G. Ueda, J. T. and M. Sekiguchi. Matsumoto, Japan: Shinshu University, pp. 300–4.

Mejia, O., Leon-Velarde, F. and Monge, C. C. (1994). The effect of inositol hexaphosphate in the high-affinity hemoglobin of the Andean chicken (*Gallus gallus*). *Comp. Biochem. Physiol.* **109**B, 437–41.

Monge, C. (1942). Life in the Andes and chronic mountain sickness. *Science*, **95**, 79–84.

Monge, C. and Leon-Velarde, F. (1991). Physiological adaptation to high altitude: oxygen transport in mammals and birds. *Physiol. Rev.*, **71**, 1135–72.

Monge, C. and Whittembury, J. (1974). Increased hemoglobin-oxygen affinity at extremely high altitudes. *Science*, **186**, 843.

Moore, L. G., Niermeyer, S. and Vargas, E. (2007). Does chronic mountain sickness (CMS) have perinatal origins? *Respir. Physiol. Neurobiol.*, **158**, 180–9.

Mortola, J. P. (1999). How newborn mammals cope with hypoxia. *Respir. Physiol.*, **116**, 95–103.

Muza, S. R., Lyons, T. P. and Rock, P. B. (1999). Effect of altitude exposure on brain volume and development of acute mountain sickness. In *Hypoxia: into the Next Millennium*, ed. R. C. Roach, P. D. Wagner and P. H. Hackett. Advances in Experimental Medicine and Biology Series, 474. New York: Kluwer Academic/ Plenum, p. 414.

Navas, C. A. and Chaui-Berlinck, J. G. (2007). Respiratory physiology of high-altitude anurans: 55 years of research on altitude and oxygen. *Respir. Physiol. Neurobiol.* **158**, 307–13.

Neya, M., Enoki, T., Kumai, Y., Sugoh, T. and Kawahara, T. (2007). The effects of nightly normobaric hypoxia and high intensity training under intermittent normobaric hypoxia on running economy and hemoglobin mass. *J. Appl. Physiol.*, **103**, 828–34.

Oelz, O., Howald, H., Di Prampero, P. E., *et al.* (1986). Physiological profile of world-class high-altitude climbers. *J. Appl. Physiol.*, **60**, 1734–42.

Oelz, O., Maggiorini, M., Ritter, M., *et al.* (1989). Nifedipine for high altitude pulmonary oedema. *Lancet*, **2**, 1241–4.

Ostojic, H., Monge, C. C. and Cifuentes, V. (2000). Hemoglobin affinity for oxygen in three subspecies of toads (*Bufo* sp.) living at different altitudes. *Biol. Res.*, **33**, 5–10.

Packard, G. (1971). Oxygen consumption of montane and piedmont chorus frogs (*Pseudacris triseriata*). *Physiol. Zool.*, **44**, 90–7.

Podolsky, A., Eldridge, M. W., Richardson, R. S., *et al.* (1996). Exercise-induced VA/Q inequality in subjects with prior high-altitude pulmonary edema. *J. Appl. Physiol.*, **81**, 922–32.

Powell, F. L. (1993). Birds at altitude. In *Respiration in Health and Disease*, ed. P. Scheid. Stuttgart: G. Fisher, pp. 352–8.

Powell, F. L. (2003). Functional genomics and the comparative physiology of hypoxia. *Ann. Rev. Physiol.*, **65**, 203–30.

Powell, F. L. and Scheid, P. (1989). Physiology of gas exchange in the avian respiratory system. In *Form and Function in Birds*, ed. A. S. King and J. McLelland. San Diego: Academic Press, pp. 393–437.

Powell, F. L. (2000). Respiration. In *Sturkie's Avian Physiology*, ed. G. C. Whittow. San Diego: Academic Press, pp. 233–59.

Powell, F. L., Hastings, R. H. and Mazzone, R. W. (1985). Pulmonary vascular resistance during unilateral pulmonary arterial occlusion in ducks. *Am. J. Physiol.*, R39–43.

Poyart, C., Wajcman, H. and Kister, J. (1992). Molecular adaptation of hemoglobin function in mammals. *Respir. Physiol.*, **90**, 3–17.

Rose, M. S., Houston, C. S., Fulco, C. S., Coates, G., Sutton, J. R. and Cymerman, A. (1988). Operation Everest. II: nutrition and body composition. *J. Appl. Physiol.*, **65**, 2545–51.

Sartori, C., Allemann, Y., Duplain, H., *et al.* (2002). Salmeterol for the prevention of high-altitude pulmonary edema. *New Engl. J. Med.*, **346**, 1631–36.

Sartori, C., Vollenweider, L., Loffler, B. M., *et al.* (1999). Exaggerated endothelin release in high-altitude pulmonary edema. *Circulation*, **99**, 2665–8.

Saunders, D. K. and Fedde, M. R. (1994). Exercise performance of birds. In *Comparative Vertebrate Physiology: Phyletic Adaptations*, ed. J. H. Jones. San Diego: Academic Press. pp. 139–190.

Saunders, P. U., Telford, R. D., Pyne, D. B., *et al.* (2004). Improved running economy in elite runners after 20 days of simulated moderate-altitude exposure. *J. Appl. Physiol.*, **96**, 931–7.

Scherrer, U., Vollenweider, L., Delabays, A., *et al.* (1996). Inhaled nitric oxide for high-altitude pulmonary edema. *New Engl. J. Med.*, **334**, 624–9.

Schmitt, P. M., Powell, F. L. and Hopkins, S. R. (2002). Ventilation-perfusion inequality during normoxic and hypoxic exercise in the emu. *J. Appl. Physiol.*, **93**, 1980–6.

Schoene, R. B. (1982). Control of ventilation in climbers to extreme altitude. *J. Appl. Physiol.*, **53**, 886–96.

Schoene, R., Swenson, E. and Hultgren, H. (2001). High altitude pulmonary edema. In *High Altitude: an Exploration of Human Adaptation*, ed. T. Hornbein and R. Schoene. New York: Marcel Dekker, pp. 777–814.

Scott, G. R. and Milsom, W. K. (2006). Flying high: a theoretical analysis of the factors limiting exercise performance in birds at altitude. *Respir. Physiol. Neurobiol.* **154**, 284–301.

Scott, G. R. and Milsom, W. K. (2007). Control of breathing and adaptation to high altitude in the bar-headed goose. *Am. J. Physiol.*, **293**, R379–91.

Shams, H. and Scheid, P. (1989). Efficiency of parabronchial gas exchange in deep hypoxia: measurements in the resting duck. *Respir. Physiol.*, **77**, 135–46.

Shams, H. and Scheid, P. (1993). Effects of hypobaria on parabronchial gas exchange in normoxic and hypoxic ducks. *Respir. Physiol.*, **91**, 155–63.

Shams, H., Powell, F. L. and Hempleman, S. C. (1990). Effects of normobaric and hypobaric hypoxia on ventiltion and arterial blood gases in ducks. *Respir. Physiol.*, **80**, 163–70.

Shelton, G., Jones, D. R., Milsom, W. K. and Fishman, A. P. (1986). Control of breathing in ectothermic vertebrates. *Handbook of Physiology: The Respiratory System, Vol. 2*, ed. A. P. Fisherman, N. S. Cherniack, J. G. Widdicombe and S. R. Geiger. Bethesda, MD: American Physiological Society, pp. 857–909.

Sillau, A. H., Cueva, S., Valenzuela, A. and Candela, E. (1976). O_2 transport in the alpaca (*Lama pacos*) at sea level and 3,300 m. *Respir. Physiol.*, **27**, 147–55.

Singh, I., Khanna, P. K., Srivastava, M. C., Lal, M., Roy, S. B. and Subramanyam, C. S. (1969). Acute mountain sickness. *New Engl. J. Med.*, **280**, 175–84.

Smith, A. C., Abplanalp, H., Harwood, L. M. and Kelly, C. F. (1959). Poultry at high altitude. *Calif. Agricult.* **13**, 8–9.

Snyder, L. R. G., Hayes, J. P. and Chappell, M. A. (1998). Alpha-chain hemoglobin polymorphisms are correlated with altitude in the deer mouse, *Peromyscus maniculatus. Evolution*, **42**, 689–97.

Storz, J. F., Sabatino, S. J., Hoffmann, F. G., *et al.* (2007). The molecular basis of high-altitude adaptation in deer mice. *PLoS Genet.* **3**, e45.

Sun, S., Oliver-Pickett, C., Ping, Y., *et al.* (1996). Breathing and brain blood flow during sleep in patients with chronic mountain sickness. *J. Appl. Physiol.*, **81**, 611–18.

Sun, S. F., Droma, T. S., Zhang, J. G., *et al.* (1990). Greater maximal O_2 uptakes and vital capacities in Tibetan than Han residents of Lhasa. *Respir. Physiol.*, **79**, 151–61.

Swan, L. W. (1970). Goose of the Himalayas. *Nat. Hist.*, **79**, 68–75.

Swenson, E. R., Maggiorini, M., Mongovin, S., *et al.* (2002). Pathogenesis of high-altitude pulmonary edema: inflammation is not an etiologic factor. *JAMA*, **287**, 2228–35.

Tenney, S. M. (1995). Functional differences in mammalian hemoglobin affinity for oxygen. In *Hypoxia and the Brain*, ed. J. R. Sutton, C. S. Houston and M. D. Coates. Burlington: Queen City Printers.

Tschop, M., Strasburger, C. J., Hartmann, G., Biollaz, J. and Bartsch, P. (1998). Raised leptin concentrations at high altitude associated with loss of appetite. *Lancet*, **352**, 1119–20.

Tucker, V. A. (1968). Respiratory physiology of house sparrows in relation to high-altitude flight. *J. Exp. Biol.*, **48**, 55–66.

Viswanathan, R., Jain, S. K., Subramanian, S., *et al.* (1969). Pulmonary edema of high altitude. II. Clinical, aerohemodynamic, and biochemical studies in a group with history of pulmonary edema of high altitude. *Am Rev. Respir. Dis.*, **100**, 334–41.

Vitzthum, V. J. and Wiley, A. S. (2003). The proximate determinants of fertility in populations exposed to chronic hypoxia. *High Alt. Med. Biol.*, **4**, 125–39.

Wagner, P. D. (1997). Insensitivity of VO_2max to hemoglobin-P50 as sea level and altitude. *Respir. Physiol.*, **107**, 205–12.

Wagner, P. D., Gale, G. E., Moon, R. E., Torre, B. J., Stolp, B. W. and Saltzman, H. A. (1986). Pulmonary gas exchange in humans exercising at sea level and simulated altitude. *J. Appl. Physiol.*, **61**, 260–70.

Wang, T. and Hicks, J. W. (1996). The interaction of pulmonary ventilation and the right-left shunt on arterial oxygen levels. *J. Exp. Biol.* **199**, 2121–9.

Ward, M. P., Milledge, J. S. and West, J. B. (2000). *High Altitude Medicine and Physiology*. London: Arnold.

Weber, R. E. (1995). Hemoglobin adaptations to hypoxia and altitude-the phylogenetic perspective. In *Hypoxia and the Brain*, ed. J. R. Sutton, C. S. Houston and M. D. Coates. Burlington: Queen City Printers, pp. 31–44.

Weber, R. E. (2007). High-altitude adaptations in vertebrate hemoglobins. *Respir. Physiol. Neurobiol.*, **158**, 132–42.

Weil, J. V., Cherniack, N. S. and Widdicombe, J. G. (1986). Ventilatory control at high altitude. In *Handbook of Physiology The Respiratory System – Control of Breathing, Part II*, ed. N. S. Cherniack and J. G. Widdicombe. Bethesda: American Physiological Society, pp. 703–28.

Weil, J. V., Byrne-Quinn, E., Sodal, I. E., Filley, G. F. and Grover, R. F. (1971). Acquired attenuation of chemoreceptor function in chronically hypoxic man at high altitude. *J. Clin. Invest.*, **50**, 186–95.

Weinstein, Y., Bernstein, M. H., Bickler, P. E., *et al.* (1985). Blood respiratory properties in pigeons at high altitudes: effects of acclimation. *Am. J. Physiol.*, **249**, R765–75.

Weir, E. K., Tucker, A. and Reeves, J. T. (1974). The genetic factor influencing pulmonary hypertension in cattle at high altitude. *Cardiovasc. Res.*, **8**, 745–9.

West, J. B. (1983). Climbing Mt. Everest without oxygen: analysis of maximal exercise during extreme hypoxia. *Respir. Physiol.*, **52**, 265–79.

West, J. B. (1986). Highest inhabitants in the world. *Nature*, **324**, 517.

West, J. B. (1996). Prediction of barometric pressures at high altitude with the use of model atmospheres. *J. Appl. Physiol.*, **81**, 1850–4.

West, J. B., Hackett, P. H., Maret, K. H., *et al.* (1983a). Pulmonary gas exchange on the summit of Mount Everest. *J. Appl. Physiol.*, **55**, 678–87.

West, J. B., Watson, R. R. and Fu, Z. (2007). Major differences in the pulmonary circulation between birds and mammals. *Respir. Physiol. Neurobiol.* **157**, 382–90.

West, J. B., Lahiri, S., Maret, K. H., Peters, R. M., Jr. and Pizzo, C. J. (1983b). Barometric pressures at extreme altitudes on Mt. Everest: physiological significance. *J. Appl. Physiol.*, **54**, 1188–94.

Will, D. H., Hicks, J. L., Card, C. S. and Alexander, A. F. (1975). Inherited susceptibility of cattle to high-altitude pulmonary hypertension. *J. Appl. Physiol.*, **38**, 491–4.

Wiseman, C., Freer, L. and Hung, E. (2006). Physical and medical characteristics of successful and unsuccessful summiteers of Mount Everest in 2003. *Wilderness Environ. Med.*, **17**, 103–8.

Wu, T., Li, S. and Ward, M. P. (2005). Tibetans at extreme altitude. *Wilderness Environ. Med.*, **16**, 47–54.

Yagi, H., Yamada, H., Kobayashi, T. and Sekiguchi, M. (1990). Doppler assessment of pulmonary hypertension induced by hypoxic breathing in subjects susceptible to high altitude pulmonary edema. *Am. Rev. Respir. Dis.*, **142**, 796–801.

Zhuang, J., Droma, T., Sutton, J. R., *et al.* (1996). Smaller alveolar-arterial O_2 gradients in Tibetan than Han residents of Lhasa (3658 m). *Respir. Physiol.*, **103**, 75–82.

9

Surviving without any oxygen

GÖRAN E. NILSSON

9.1   Introduction

Most vertebrates cannot survive more than a few minutes without any oxygen. As pointed out in Chapter 1, the high intrinsic rate of oxygen consumption of the brain makes it one of the first organs to fail in anoxia. While medical science struggles to find ways to counteract anoxic tissue damage, unfortunately with quite limited success, evolution has solved this problem a few times, as revealed by the few vertebrates that can survive months without any oxygen. The best-studied examples of such anoxia-tolerant vertebrates are the crucian carp (*Carassius carassius*) and some North American freshwater turtles in the genera *Trachemys* and *Chrysemys*.

It is not a coincidence that these extremely anoxia-tolerant vertebrates are aquatic. The access to oxygen may be temporarily halted in many aquatic habitats, either because water oxygen content becomes severely depleted (see Chapters 1 and 5), or because lung breathers such as turtles lose their access to air for long periods, especially during overwintering. A particularly longlasting and extreme oxygen depletion occurs in many small, ice-covered lakes and ponds in the northern hemisphere. Due to a thick ice cover, which blocks oxygen diffusion as well as light needed for photosynthesis, these waters may become anoxic for several months (Holopainen and Hyvärinen, 1985; Ultsch, 1989). It is under such conditions that crucian carp and turtles have evolved their ability to survive long periods of anoxia.

As we have seen, aquatic vertebrates that live in habitats where hypoxia is common can resort to adaptations that increase their access to oxygen, such as air breathing in some fishes (Chapter 6). Some freshwater turtles can also change their route for oxygen uptake when they are forced to remain

Respiratory Physiology of Vertebrates: Life with and without Oxygen, ed. Göran E. Nilsson. Published by Cambridge University Press. © Cambridge University Press 2010.

submerged. Submerged turtles without access to air survive longer if there is oxygen in the water, because they have some capacity for extrapulmonary oxygen uptake, probably involving oxygen uptake from water entering the richly vascularized upper and lower parts of their gastrointestinal tract (Ultsch and Jackson, 1982; Ultsch, 1985).

However, in ice-covered anoxic water, anaerobic metabolism becomes the only viable option for adenosine triphospate (ATP) production. Although crucian carp and turtles have probably evolved their anoxia tolerance in response to anoxic conditions at close to 0°C, their anoxia tolerance is extended over a wide temperature range. At higher temperatures they are no longer able to survive anoxia for several months, but they do well without oxygen for at least a day or two at room temperature (Fig. 9.1). Clearly, a major reason for the temperature dependence of anoxia tolerance of crucian carp and turtles is

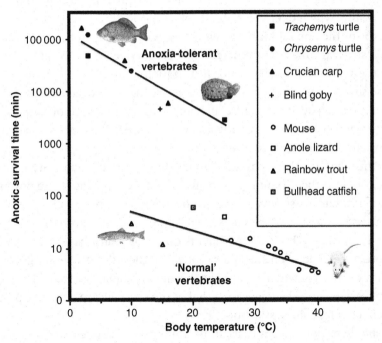

Fig. 9.1 Anoxic survival time in anoxia-tolerant vertebrates compared with 'normal' vertebrates. If temperature is taken into account, there is no major difference in survival time between trout, lizards, and mammals, but anoxia-tolerant vertebrates survive anoxia some 1000 times longer than these. Anoxia-tolerant vertebrates survive anoxia longer at low temperature because their glycogen stores last longer (i.e. they survive as long as they have fuel), whereas for anoxia-intolerant vertebrates, a low temperature slightly increases anoxic survival time because degenerative death processes initiated by a loss of ATP are slowed down. Redrawn from Lutz *et al.*, 2003.

that metabolic rate rises exponentially with temperature, resulting in a more rapid depletion of anaerobic fuel stores (glycogen). The complete exhaustion of the glycogen stores appears to be what finally limits anoxic survival in both crucian carp and turtles (Nilsson, 1990a; Warren et al., 2006). Glucose, stored as glycogen, is virtually the only fuel that can be used in anoxia. Amino acids derived from proteins cannot be utilized as fuel in the absence of oxygen because the citric acid cycle, which is closely linked to oxidative phosphorylation, is at a halt in anoxia. Similarly, fatty acid oxidation (β-oxidation) cannot take place without oxygen. Indeed, β-oxidation may even run backwards during anoxia (Van Raaij et al., 1994), and if acetyl-CoA could be produced from fat, it would still need the citric acid cycle to be running to yield ATP. Consequently, the crucian carp does not use its huge liver glycogen store when starved, but saves it for anoxia (Nilsson, 1990a).

There are some less-well-studied examples of vertebrates that can survive anoxia for a long time. These include the Californian blind goby (*Typhlogobius californiensis*), which tolerates 80 hours of anoxia at 15°C (Congleton, 1974), and embryos of the annual killifish (*Austrofundulus limnaeus*). In the latter case, the anoxic survival time is extremely long (50% survive 2 months at 25°C during diapause induced at 32 days after fertilization), but as the embryos get older anoxic survival time rapidly drops. It appears that an extreme degree of metabolic depression during diapause, including suppressed levels of Na^+/K^+ ATPase, is a prerequisite for the anoxia tolerance, and adult killifish are not particularly hypoxia tolerant (Podrabsky et al., 2007).

In this chapter, the focus will be on the crucian carp and the freshwater turtles, and on the differences and similarities seen in the strategies used by these vertebrates to survive anoxia. Also data derived from studies on the goldfish (*Carassius auratus*) will be mentioned, as the goldfish is a very close relative to the crucian carp. Whereas experiments on crucian carp have, almost without exception, been done on specimens caught in the wild, studies on goldfish are generally carried out on fish obtained through the aquarium trade. Their anoxia tolerance appears to be lower than that of the crucian carp, which is possibly a side effect of long domestication.

It should be pointed out that, unlike for most vertebrates, survival in anoxia is the normal ('control') situation in turtles and crucian carp. Therefore, it is possible to selectively block various mechanisms to evaluate their role in anoxic survival. As described in Chapter 1, anoxia is synonymous with catastrophe in mammals, particularly in the brain. Any experimental attempts to extend anoxic survival in mammals by boosting or blocking a particular mechanism are likely to be hampered by failures of other functions (Nilsson and Lutz, 2004). It can therefore be argued that mammalian models are not well suited for studying defense mechanisms against anoxia, as such mechanisms are poorly developed in these animals. Not only are the changes seen in the anoxic mammal both rapid and complex, but it is also often hard to differentiate between physiological defense mechanisms and pathophysiological events that are merely reflecting death processes.

9.2   Activity level and metabolism in anoxia

Many studies on the anoxia tolerance of crucian carp and turtles have focused on the brain, as it is likely to be the most anoxia-sensitive organ and therefore the weakest link in any strategy to survive anoxia. Unlike mammals, the crucian carp and the freshwater turtles manage to uphold brain ATP levels when exposed to anoxia (Fig. 9.2), thereby avoiding all the detrimental processes that are initiated by the failure of ATP-driven functions such as ion pumping. The key question then becomes how brain ATP levels can be maintained in anoxia, when the only viable option for ATP production is anaerobic glycolysis, which has an ATP yield less than one-tenth of that of complete glucose oxidation (Hochachka and Somero, 2002).

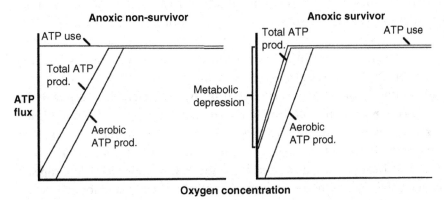

Fig. 9.2 Dying or living in anoxia depend on the ability to maintain ATP levels. The key is to match ATP use with ATP production by lowering the ATP use through metabolic depression so that anaerobic (glycolytic) ATP production is able to meet ATP use. In the anoxia-intolerant animal (left), anaerobic ATP production (glycolysis) is unable to compensate for the slowdown and stop in aerobic ATP production (oxidative phosphorylation), causing ATP levels to fall passively with falling oxygen levels. A major consequence of this is cell-membrane depolarization (as ion-pumping ATPases stop), leading to a cascade of degenerative processes. In the anoxic survivor (right), the slowdown and stop in aerobic ATP production is initially compensated for by an elevated anaerobic ATP production, and subsequently ATP use is reduced by metabolic depression, allowing long-term maintenance of ATP levels. From Nilsson and Renshaw, 2004.

There are only two ways in which ATP production and ATP use can be matched when an animal becomes anoxic. Either glycolytic production of ATP is strongly upregulated (the Pasteur effect), or ATP consumption is drastically reduced, a strategy often termed 'metabolic depression.' As pointed out by Lutz and Nilsson (1997), the turtles and the crucian carp differ by the degree to which each of these two options are utilized. This difference is clearly displayed by the level of physical activity that these animals show during anoxia: anoxic turtles are virtually comatose, whereas crucian carp still swim around in anoxia, albeit at a reduced level. In the laboratory, crucian carp exposed to 5 hours of anoxia at 9°C show a 50% reduction in spontaneous swimming activity, which probably corresponds to a 35–40% reduction in whole-body ATP use (Nilsson et al., 1993a). Also, in nature, crucian carp can be caught in traps during anoxic conditions in the winter (Vornanen and Paajanen, 2006), showing that they must retain some physical activity.

Crucian carp and turtles also show striking differences in their circulatory adjustments to anoxia. In the turtle, an 80% fall in heart rate and cardiac output during anoxia is accompanied by peripheral vasoconstriction and blunted autonomic control of the heart (Hicks and Farrell, 2000a; Hicks and Farrell, 2000b; Stecyk et al., 2004a). By contrast, heart rate, cardiac output, stroke volume, power output, autonomic control, and even ventilatory rate are maintained for several days in the anoxic crucian carp, whereas peripheral resistance falls (Fig. 9.3) (Stecyk et al., 2004b). Both the turtle and crucian carp show a doubling in brain blood flow within the first minutes of anoxia. However, while this increase in brain blood flow is sustained in the anoxic crucian carp (Nilsson et al., 1994), probably in order to allow maintained high neural activity level, cerebral circulation falls back to pre-anoxic levels within the first hours of anoxia as the turtle enters a near comatose state (Hylland et al., 1994; Stecyk et al., 2004a). Nevertheless, in both cases the increase in brain blood flow appears to be mediated by adenosine, as it can be fully blocked by aminophylline, an adenosine receptor blocker (Hylland et al., 1994; Nilsson et al., 1994).

The difference in activity between crucian carp and turtles is also reflected on the metabolic level. Whole-body metabolism (measured as heat production) is more drastically reduced in the turtle than in the crucian carp. A 90–95% reduction in body-heat production has been measured in turtles (Jackson, 1968), whereas in the goldfish, the close relative to the crucian carp, anoxia reduces heat production to one-third of the normoxic level (Van Waversveld et al., 1989). With regard to the brain, there are no direct measurements of metabolic rate in vivo in these animals. However, estimates based on lactate production in anoxia suggest at least a 70–80% fall in brain ATP turnover in the anoxic turtle brain (Lutz et al., 1984), and other studies have suggested that turtle brain-energy needs

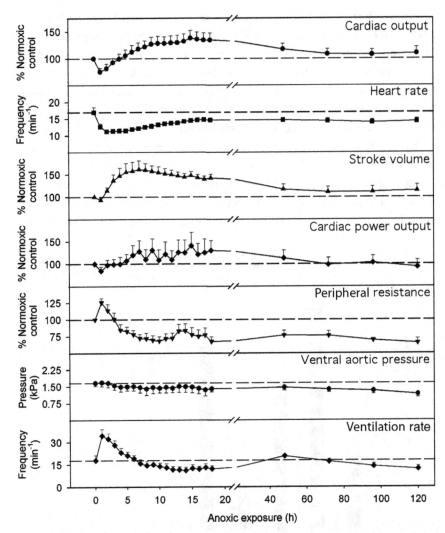

Fig. 9.3 The crucian carp has the unique ability to maintain cardiac activity during long-term anoxia (here recorded for up to 5 days of anoxia at 8°C). After some initial adjustments most variables stabilize close to the normoxic level (hatched line). The major anaerobic end product of crucian carp is ethanol. One reason why cardiac output is maintained could be that it is needed to flush ethanol out of the body though the gills. This could also explain the maintained gill ventilation rate. Otherwise, wasting energy to ventilate gills in the absence of oxygen appears pointless. Data from Stecyk *et al.*, 2004b.

can be fully met even if anaerobic glycolysis is suppressed (Storey, 1996). In the crucian carp, by contrast, measurements of lactate and heat production (using microcalorimetry) in telencephalic brain slices suggest a mere 30% reduction in ATP turnover and an upregulation of glycolysis (Johansson *et al.*, 1995).

Protein synthesis is maintained in the anoxic crucian carp brain, while it falls by about 95% in liver, and by some 50% in heart muscle and skeletal muscle (Fig. 9.4) (Smith *et al.*, 1996). In major tissues such as muscle and liver, which together constitute more than half of the fish, a suppressed protein synthesis can have significant effects on the whole-body metabolic rate. However, protein synthesis in the brain does not constitute more than one or a few percent of brain-energy use, so here it may not be worthwhile reducing protein synthesis, especially as the crucian carp brain maintains many of its functions during anoxia (Smith *et al.*, 1996). Nevertheless, in turtles, there appears to be a very strong suppression of protein synthesis in all tissues, including the brain (Fraser *et al.*, 2001). Indeed, after exposing turtles to 1–6 hours of anoxia at room temperature, protein synthesis could no longer be detected in any of the tissues

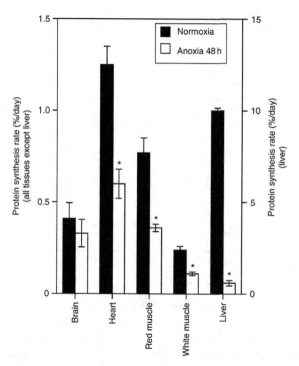

Fig. 9.4 Rates of protein synthesis in crucian carp exposed to 48 hours of anoxia at 8°C. Although brain protein synthesis is unaffected by anoxia, 50% falls are seen in muscle tissues and heart, and a 95% decrease is seen in the liver. The pattern of changes makes sense from an energy-saving perspective. Any reduction of protein synthesis in brain would be of little value because of its small size (~0.1% of body mass) and relatively low rate of protein synthesis. By contrast, reducing liver protein synthesis should lead to considerable reductions in whole-body energy use owing to the large size of the liver (~15% of body mass) and high rate of protein synthesis (25 times higher than in brain). Data from Smith *et al.*, 1996.

examined (intestine, heart, liver, brain, muscle and lungs) (Fraser *et al.*, 2001). Thus, the extreme metabolic depression of turtles compared with crucian carp is also reflected in the level of protein synthesis.

Like many animals exposed to low oxygen levels (Wood *et al.*, 1985), anoxic goldfish (and most likely anoxic crucian carp) will move to cooler water if given the opportunity (Rausch *et al.*, 2000), thereby accomplishing a general suppression of metabolic energy requirements. In nature, crucian carp and goldfish are probably exposed to anoxia only during overwintering at close to 0°C, so anoxia tolerance is normally aided by hypothermia. The ability of crucian carp and turtles to tolerate anoxia for 1–2 days at room temperature is clearly a spin-off effect of being adapted to anoxic overwintering. Still, the role of seasonality and temperature in anoxia tolerance has long been underestimated or overlooked but was a main focus in a recent review on the crucian carp (Vornanen *et al.*, 2009).

9.3   Metabolic adaptations: ethanol production or lactate buffering

In light of the clear differences in the degree of metabolic depression between turtles and crucian carp, Lutz and Nilsson (1997) suggested that there is one key feature that allows the crucian carp to remain active in anoxia: its ability to produce ethanol as the main anaerobic end product, whereas turtles accumulate lactate in anoxia. The crucian carp and the goldfish share the exotic ability of producing ethanol during anoxia (Shoubridge and Hochachka, 1980; Johnston and Bernard, 1983; Nilsson, 1988). The obvious advantage with this mechanism is that the ethanol leaves the fish and lactic acidosis is avoided. The carbonate buffering capacity of fish blood is very limited compared with that of air breathers. The reason for this is the high solubility of CO_2 in water, which facilitates CO_2 excretion to such a degree that total carbonate level in fish blood is generally only about 10% of that of air-breathing vertebrates. Indeed, because of the low buffering capacity of fish blood, it is difficult to see how any fish could tolerate anoxia as long as the crucian carp if it had to deal with high lactate and H^+ loads.

The ethanol-producing pathway appears restricted to skeletal muscle (red and white) and is carried out in three steps. Lactate is first turned into pyruvate by lactate dehydrogenase (LDH); pyruvate is subsequently turned into acetaldehyde by pyruvate dehydrogenase (PDH); and finally, acetaldehyde is turned into ethanol by alcohol dehydrogenase (ADH) (Fig. 9.5). PDH is a tightly coupled three-enzyme complex that normally turns pyruvate into acetyl-coA (which enters the citric acid cycle). In other vertebrates, acetaldehyde is merely an

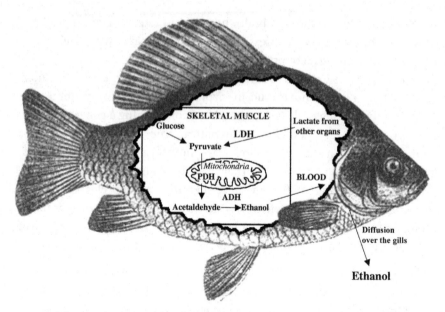

Fig. 9.5 By producing ethanol the crucian carp avoids a build-up of lactate and H^+ during anoxia. Ethanol production takes place in crucian carp skeletal muscle. Pyruvate (produced in the muscle or derived from lactate taken up from the blood) is turned into acetaldehyde (and CO_2) by a modified pyruvate dehydrogenase (PDH) in the mitochondrial inner membrane. After entering the cytosol, acetaldehyde is turned into ethanol by alcohol dehydrogenase (ADH). In crucian carp ADH is confined to skeletal muscle, suggesting that ethanol production occurs only in this tissue. The lipophilicity of ethanol allows it to freely pass cell membranes. Therefore, the ethanol diffuses out of the muscle into the blood, and when ethanol reaches the gills it diffuses out into the ambient water. LDH, lactate dehydrogenase.

intermediate that does not leave the PDH complex. Thus, the ethanol-producing pathway in crucian carp and goldfish appears to rely on a deviant form of PDH that somehow leaks acetaldehyde during anoxia (Mourik *et al.*, 1982). Molecular studies are now underway, revealing a very high expression of unique forms of PDH subunits in muscle tissue of crucian carp and goldfish. These subunits occur in addition to the normal PDH isoforms, and they are most likely responsible for the production of acetaldehyde during anoxia (Fagernes *et al.*, 2008).

Only the skeletal muscle of crucian carp and goldfish contains significant levels of ADH. Therefore, all other tissues, including the brain, have to produce lactate in anoxia. The lactate is transported in the blood to the muscle, where it is transformed to ethanol. By contrast, non-ethanol-producing vertebrates have the highest ADH activity in the liver, where it coexists with aldehyde dehydrogenase (ALDH), and in a sequential reaction ADH and ALDH function to detoxify ingested ethanol by turning it into acetate. These two enzymes do not coexist in crucian carp tissues (most of the ALDH is still found in the liver). This is fortunate, as ALDH (which has an even higher affinity for acetaldehyde than ADH) otherwise would turn the acetaldehyde into acetate, thereby circumventing the ethanol-producing pathway (Nilsson, 1988). It is worth mentioning that the unusual distribution of alcohol dehydrogenase in crucian carp is retained also during the summer, even if it is then unlikely to face anoxia (Nilsson, 1990b).

Ethanol readily penetrates cellular membranes, so probably without the need for further biochemical adaptations, ethanol merely diffuses out of the muscle into the blood, finally reaching the ambient water through diffusion over the gills.

At this point of the story, an often-asked question is how high the ethanol level gets in crucian carp blood. In other words, does the fish get drunk? The answer is that the level of ethanol in blood does probably not rise high enough to significantly suppress nervous activity; the steady-state level remains below 10 mmol l^{-1} (Van Waarde et al., 1993), which corresponds to the blood ethanol level of a human that has consumed 0.5–1.0 liters of beer. It is possible that the relatively high cardiac output of anoxic crucian carp, mentioned above, is needed to provide a sufficient rate of gill perfusion and ethanol excretion to avoid intoxication (Stecyk et al., 2004b). Thus, the crucian carp can afford to stay anoxic, but not drunk, for months during the winter. However, other roles for the maintained cardiac output of anoxic crucian carp must involve transporting glucose to anoxic organs and moving lactate to the muscle.

Although ethanol production allows the crucian carp to endure long-term anoxia without suffering lactic acidosis, it has a clear energetic drawback: ethanol, a very energy-rich hydrocarbon, is released to the water and is forever lost. Therefore, to allow for long-term survival in anoxia, autumn- and winter-acclimatized crucian carp have enormous glycogen stores, probably larger than in any other vertebrate, and the only factor that appears to limit their anoxia endurance is the total depletion of the main glycogen store in the liver (Nilsson, 1990a). During the late autumn, liver glycogen can constitute 30% of the crucian carp liver mass, and the liver may make up 15% of the body mass, whereas its glycogen store is less than one-tenth of this in the spring and summer (Hyvärinen et al., 1985). Also the muscle, brain, and heart show very high glycogen levels in the autumn (Hyvärinen et al., 1985; Vornanen, 1994; Vornanen and Paajanen, 2004; Vornanen and Paajanen, 2006).

The turtles, however, are not in possession of an ethanol-producing pathway, and even with deep metabolic depression, turtles may have to face lactate levels of up to 200 mmol l^{-1} in blood and tissues, which they have to buffer with calcium carbonate from their bones and shell (Jackson, 2002). Indeed, data suggest that the shell-buffering capacity of different turtle species correlates with their ability to withstand anoxia, clearly indicating that lactate accumulation is one limiting factor in the anoxic survival of turtles (Jackson *et al.*, 2007).

9.4   Brain activity in anoxia

The energy demand of the vertebrate brain is normally very high, as each gram of brain constantly consumes about ten times more energy than the average body tissue. Of the energy used by the brain, more than 50% appears to be devoted to ion pumping needed to maintain ionic gradients over cell membranes, a prerequisite for electrical activity as well as for transport of neurotransmitters and metabolites (Erecinska and Silver, 1994). Therefore, a strategy that involves suppressing electrical activity should lead to significant energy savings. Indeed, freshwater turtles appear to rely strongly on this strategy. During anoxia, the turtle brain electroencephalogram (EEG) is virtually a flat line with a few bursts of minor periodic activity (Fernandes *et al.*, 1997). Also electrical responses (evoked potentials), experimentally induced by electrical stimulation in the brain, are strongly suppressed in anoxic turtles (Feng *et al.*, 1988a; Feng *et al.*, 1988b).

An EEG has not been recorded in crucian carp (a difficult enterprise in fish due to electrical disturbance from water movements). Nevertheless, it can hardly be as reduced as in the anoxic turtle, as the crucian carp remains active in anoxia and obviously need its brain to be 'turned on' (Lutz and Nilsson, 1997; Nilsson, 2001). However, vision and hearing appear to be suppressed in anoxic crucian carp and goldfish. Thus, light-evoked potentials virtually disappear in the anoxic crucian carp retina and in the optic tectum of its brain (Johansson *et al.*, 1997). Similarly, the activity of the auditory nerve of goldfish (Suzue *et al.*, 1987) is strongly suppressed during anoxia. Vision and hearing are senses that are likely to be of minor importance during the long anoxic winter, and can therefore be temporarily sacrificed. Interestingly, turtles exposed to anoxia do not show such a profound suppression of light-evoked potentials in their retina, suggesting that there is some advantage for turtles to retain vision during anoxia (Stensløkken *et al.*, 2008a). One speculative possibility is that turtles need vision to detect the disappearance of the ice cover in the spring, triggering them to move to the surface.

Unlike anoxia-sensitive vertebrates, turtles and crucian carp maintain brain-ion homeostasis during anoxia, showing no, or only minor, rises in extracellular [K$^+$] (Fig. 9.6) and intracellular [Ca^{2+}] (Sick *et al.*, 1982; Nilsson *et al.*, 1993b;

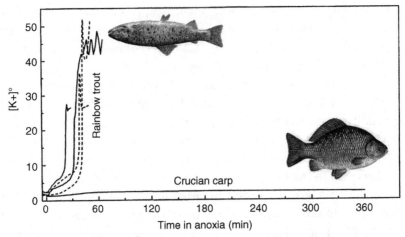

Fig. 9.6 Extracellular levels of K⁺ in the brain of rainbow trout and crucian carp exposed to anoxia at 10°C . Note that the crucian carp maintains a steady level of extracellular K⁺ characteristic of an anoxia-tolerant vertebrate brain. The fast and massive rise in extracellular K⁺ in the rainbow trout (traces represent four different individuals) reveals a general anoxic depolarization of the brain that is similar to that seen in anoxic mammalian brains. Indeed, the time to depolarization is similar to that of mammals if the difference in metabolic rate related to temperature (10°C versus 37°C) is taken into account. Data from Nilsson *et al.*, 1993b.

Johansson and Nilsson, 1995; Nilsson, 2001). There is a slow and relatively small increase in intracellular [Ca²⁺] in turtles during anoxia, which may be caused by a rise in extracellular [Ca²⁺] derived from the breakdown of bone and shell needed to buffer the lactate load (Bickler, 1998). This moderate and probably controlled elevation of intracellular [Ca²⁺] (<300 nmol l⁻¹) may actually initiate neuroprotective mechanisms in turtles (Bickler 1998; Bickler and Hansen, 1998; Bickler and Donohoe, 2002; Bickler, 2004). Although very high levels of intracellular [Ca²⁺] are clearly deadly, mammalian studies have suggested that too little intracellular [Ca²⁺] may have apoptotic effects, and neurons may have a [Ca²⁺] set-point or window for survival (Johnson *et al.*, 1992; Zipfel *et al.*, 2000).

9.5   Mechanisms of anoxic metabolic depression

9.5.1   *Direct sensing of energy deficiency*

In energetically compromised tissues, a net breakdown of phosphory-lated adenylates (ATP, adenosine diphosphate [ADP], and adenosine monopho-sphate [AMP]) can lead to the activation of protective mechanisms, either initiated by changes in the levels of the phosphorylated adenylates, or by the main breakdown product of these, adenosine. Several studies have suggested a role for adenosine in anoxia tolerance. Adenosine has been termed a 'retaliatory metabolite,' as it has its own set of receptors that initiate a variety of mechanisms aimed at reducing metabolic demand and increasing energy supply (Lipton, 1999). Even if the turtle brain does not show any drastic changes in ATP levels, there are small but significant falls in brain ATP, ADP, and AMP initially during anoxia (Lutz *et al.*, 1984). As a result of this net breakdown of phosphorylated adenylates, there is a tenfold increase in extracellular levels of adenosine in the brain of turtles exposed to anoxia (Fig. 9.7) (Nilsson and Lutz, 1992). The rise in adenosine is only temporary, and at the same time, there is an adenosine-induced increase in brain blood flow (Hylland *et al.*, 1994). In addition, adenosine has been suggested to cause a reduction in cellular K^+ outflow (Pek and Lutz., 1997), a downregulation in neuronal conductance and glutamate NMDA receptor activity (Buck and Bickler, 1998; Ghai and Buck, 1999; but see Pamenter *et al.*, 2008b), and a reduction in glutamate efflux (Milton *et al.*, 2002). (The neurotransmitter glutamate will be further discussed below.) Moreover, if adenosine A1 receptors are inhibited pharmacologically, there is a rapid depolarization of anoxic turtle brain slice preparations (Pérez-Pinzón *et al.*, 1993). Recent results also suggest the possibility that adenosine may inhibit anoxia-induced apoptosis in turtle brain by modulating the activity of kinases involved in the downstream regulation of apoptosis

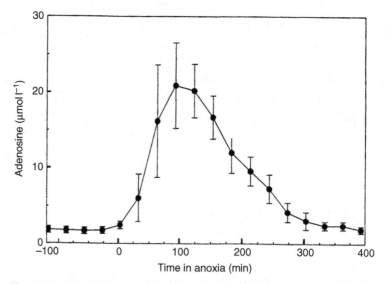

Fig. 9.7 Increase in the extracellular level of the inhibitory neuromodulator adenosine in the brain of turtles exposed to anoxia at room temperature. The adenosine level was monitored using microdialysis. From Nilsson and Lutz, 1992.

(Milton *et al.*, 2008). One may expect adenosine to be particularly effective in the turtle brain, as it is acting on a system that is capable of making ATP production meet energy needs. By contrast, in a mammal (or any anoxia-sensitive vertebrate), the mechanisms activated by adenosine may very well be present, but they are apparently not efficient enough to maintain energy charge during such a drastic event as anoxia.

The evidence for a role of adenosine in anoxic survival of crucian carp is not as firm as for the turtle. Microdialysis measurements in the brain of crucian carp have so far failed to detect any increase in extracellular adenosine (P. Hylland and G. E. Nilsson, unpublished). One reason for this could be that adenosine release is very local and not large enough to be picked up by a relatively large microdialysis probe. It is, for example, very likely that the anoxia-induced increase in brain blood flow seen in anoxic crucian carp is mediated by adenosine, as it can be blocked by superfusing the brain with the adenosine receptor blocker aminophylline (Nilsson *et al.*, 1994). Moreover, adding adenosine to goldfish hepatocytes causes a fall in both protein synthesis and Na^+/K^+ ATPase activity, indicative of an induction of metabolic depression (Krumschnabel *et al.*, 2000). Furthermore, glutamate release can be suppressed by adenosine in goldfish cerebellar slices (Rosati *et al.*, 1995). Finally, treating crucian carp with the adenosine receptor blocker aminophylline causes a threefold increase in the rate of ethanol release to the water during anoxia (whereas it has no effect on oxygen consumption in normoxia), indicating a significant role of adenosine in anoxic metabolic depression (Nilsson, 1991).

Adenosine monophosphate-activated protein kinase (AMPK) has been termed a metabolic master switch, particularly in mammalian heart and muscle. It senses cellular energy charge by being activated (i.e. phosphorylated) during conditions of increased AMP/ATP ratios. After being phosphorylated it acts by inhibiting anabolic energy-consuming pathways and stimulating energy-producing catabolic pathways. Recent studies have shown increased levels of phosphorylated AMPK in anoxic crucian carp brain, heart, and liver (Stensløkken *et al.*, 2008b), and in goldfish liver (Jibb and Richards, 2008), as well as in anoxic zebrafish embryos (Mendelsohn *et al.*, 2008). Inhibiting AMPK activity in anoxic crucian carp pharmacologically (with Compound C) causes significant increases in ethanol production (i.e. in metabolic rate) and falls in cellular energy charge, but the changes are relatively modest and do not lead to mortality during anoxia. Thus, at least in crucian carp, AMPK appears to play a significant but limited role in activating key anoxic survival mechanisms.

Potassium ion channels regulated by ATP (K_{ATP} channels), which occur in many (probably all) vertebrates, open in response to low ATP levels. Opened K_{ATP} channels and activated adenosine A1 receptors may be linked, and they appear to be involved in mediating a reduced K^+ flux out of anoxic turtle neurons (Pek and Lutz, 1998). That an opening of K_{ATP} channels will lead to a reduction in K^+ flux may seem paradoxical, but if the K_{ATP} channels are hyperpolarizing structures involved in excitatory neurotransmission, the net result may be suppressed electrical activity and, therefore, an overall reduction in K^+ fluxes in the brain. There is also K_{ATP} channel-like activity in mitochondrial membranes, and a recent study has suggested that activation of such mitochondrial currents in the anoxic turtle cortex uncouples mitochondria and reduces mitochondrial Ca^{2+} uptake, thereby increasing intracellular Ca^{2+}, which in turn acts to reduce the activity of excitatory glutamate receptors of the NMDA type (Pamenter *et al.*, 2008a; see section 5.2). With regard to fish, one study has suggested a protective role of both mitochondrial and cell-membrane K_{ATP} channels in hypoxic goldfish heart at room temperature (Chen *et al.*, 2005), but these channels do not appear to play any significant role in the anoxia tolerance of the crucian carp heart, at least not at low temperatures (Paajanen and Vornanen, 2004). Moreover, the K_{ATP} channel blocker glibenclamide has no significant effect on the K^+ homeostasis in crucian carp brain (Johansson and Nilsson, 1995). Thus, at present, there is good evidence for a significant, but possibly limited, role of K_{ATP} channels in anoxic survival of freshwater turtles; but such a role in crucian carp is less certain.

9.5.2   Ion channels

As ion pumping is the major energy-consuming process in neurons, directly reducing ion permeability could lead to considerable energy savings. Such a strategy, often termed 'channel arrest,' was suggested more than 20 years ago (Hochachka, 1986; Lutz *et al.*, 1985), before any proof of such a process existed. The possible role of 'channel arrest' in diving animals is discussed in Chapter 7. Reduced ion-channel activity does not seem to play a major role in the anoxia defense of the crucian carp, but it is clear that anoxia tolerance in turtles at least partly relies on a substantial downregulation in K^+, Na^+, and Ca^{2+} fluxes over neuronal membranes (Bickler, 1992; Pérez-Pinzón *et al.*, 1992; Pek and Lutz, 1998), although terming it 'arrest' seems too drastic. There is, for example, a 40% reduction in the density of voltage-gated Na^+ channels in the anoxic turtle cerebellum (Pérez-Pinzón *et al.*, 1992). A reduced density of voltage-gated Na^+ channels is possibly responsible for the increased action potential threshold seen in the anoxic turtle brain (Sick *et al.*, 1993).

A significant proportion of the ion fluxes in brain occurs through ligand-gated ion channels, for which the ligands are usually various neurotransmitters. Thus, these include many of the neurotransmitter receptors. Downregulation of excitatory ligand-gated channels that allow the entrance of Na^+ and Ca^{2+} into

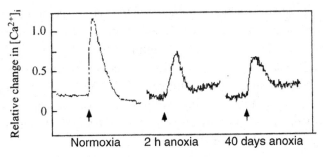

Fig. 9.8 Increases in intracellular Ca^{2+} induced by stimulation the NMDA glutamate receptor in brain slices from turtles kept in normoxic conditions or exposed to 2 hours or 40 days of anoxia. Arrows mark application of the receptor agonist NMDA to the preparations. Data from Bickler, 1998.

neurons would provide reduced neural activity and therefore reduced ATP use. With regard to anoxic turtles, the best-studied ligand-gated channel is the NMDA glutamate receptor. This is a high-flux cation channel with a high permeability for Ca^{2+}. In the anoxic/ischemic mammalian brain, excessive stimulation of this receptor from uncontrolled glutamate release results in a massive inflow of Ca^{2+} that activates an array of death processes (see Lipton, 1999 for a review). In the turtle brain, the NMDA receptor activity is reduced during anoxia (Fig. 9.8) (Bickler, 1998; Bickler *et al.*, 2000). Suggested mediators of this downregulation include phosphatase 1 or 2A (Bickler and Donohoe, 2002), adenosine receptors (Buck and Bickler, 1998), and, most recently, K_{ATP} channels (Pamenter *et al.*, 2008a). Application of adenosine to turtle brain slices results in a reduction in NMDA receptor open probability and whole-cell conductance (Buck and Bickler, 1995; Buck and Bickler, 1998; Ghai and Buck, 1999). However, more recent data have played down the role of adenosine in suppressing NMDA-receptor activity during anoxia (Pamenter *et al.*, 2008b). In addition to NMDA receptors, also the conductivity of glutamate receptors of the AMPA type (another major excitatory cation channel) has been found to be reduced in the anoxic turtle brain (Pamenter *et al.*, 2008c).

In the crucian carp, neural K^+ or Ca^{2+} permeability or fluxes appear to be maintained during anoxia (Fig. 9.9) (Johansson and Nilsson, 1995; Nilsson, 2001). A recent survey of the expression of genes involved in excitatory neurotransmission in crucian carp exposed to 1 week of anoxia (12°C) has shown maintained expression of voltage-gated Ca^{2+} channels and AMPA receptors, a 50% upregulation of voltage-gated Na^+ channels, and a 50% fall in some subunits of the NMDA receptor (Ellefsen *et al.*, 2008). Thus, this study gave no indication of a broad reduction of excitatory neurotransmission at the level of gene expression. However, the indicated fall in NMDA receptor function is corroborated by a recent study on goldfish in which whole-cell patch-clamp recordings of slices from the telencephalon indicated a 40–50% reduction in NMDA receptor activity during 40 minutes of acute anoxia at room temperature (Wilkie *et al.*, 2008). It should be mentioned that the amino acid sequence of the crucian carp NR-1 subunit, a key element in the function of all NMDA receptors, is very similar to that of other vertebrates, suggesting that anoxia tolerance in crucian carp does not rely on any major structural changes altering the function of this major excitatory receptor (Ellefsen *et al.*, 2008).

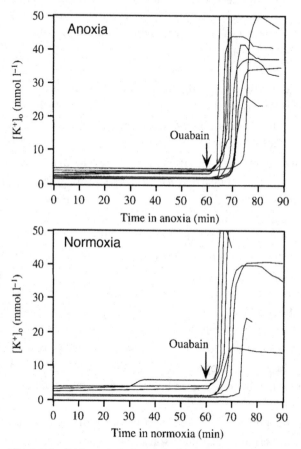

Fig. 9.9 Lack of anoxia-induced changes in K^+ permeability in crucian carp brain. The graphs show ouabain-induced increases in the extracellular level of K^+ seen in the brain of several crucian carp exposed to normoxia and anoxia (each trace representing one fish). Ouabain is a selective inhibitor of the Na^+/K^+ pump, and when this pump is blocked the outflow of K^+ should correlate with the K^+ permeability of the brain cells. The lack of significant differences between the normoxic and anoxic fish suggests that K^+ permeability is maintained in anoxia. From Johansson and Nilsson, 1995.

Although 'channel arrest' may form an important component of the mechanisms that send the anoxic turtle into a nearly comatose state during anoxia, it is likely that a considerable reduction in ion-channel activity is not compatible with the comparatively high level of physical and neural activity displayed by anoxic crucian carp. This leads us to faster and more dynamic methods by which neural activity and energy consumption can be altered in the brain: changes in the release of neurotransmitters.

9.5.3 Neurotransmitters

A major disastrous event in the anoxia-intolerant brain involves the release of excitatory neurotransmitters such as glutamate and dopamine to the extracellular space (see Lipton 1999; Lutz et al., 2003 for reviews), an event that also takes place in the brain of hypoxia-sensitive fish (Hylland et al., 1995). Glutamate is the most abundant excitatory neurotransmitter in the vertebrate brain, and any increase in extracellular glutamate is likely to stimulate neural activity and, therefore, increase energy use, which is precisely what the brain needs to avoid in anoxia. The receptors activated by glutamate include the NMDA and AMPA receptors mentioned above, and for the mammalian brain the result is an uncontrolled flow of Ca^{2+} into neurons, which activates various deadly processes (Lipton, 1999). By contrast, the brains of crucian carp and freshwater turtles have been found to maintain normal extracellular levels of glutamate during anoxia (Nilsson and Lutz, 1991; Hylland and Nilsson, 1999). Similarly, most dopamine receptors are excitatory, and studies of turtle brain show that the level of dopamine is also maintained during anoxia (Milton and Lutz, 1998).

Gamma-amino butyric acid (GABA), by contrast, is the major inhibitory neurotransmitter in the brain of vertebrates. It activates ion channels that either increase the membrane conductance to Cl^- (through $GABA_A$ receptors) or K^+ (through $GABA_B$ receptors). In both cases, the result is normally a hyperpolarization or clamping of the membrane potential. Thus, GABA plays a completely opposite role to glutamate, as it inhibits membrane depolarization and the formation of action potentials. Not surprisingly, the $GABA_A$ receptor is the target for most general anesthetic drugs (Franks, 2008). Extracellular levels of GABA rise during anoxia in anoxia-tolerant vertebrates. In the turtle, the rise in GABA is massive, reaching 80 times the normoxic level within 6 hours (Fig. 9.10) (Nilsson and Lutz, 1991). At such high levels, GABA can be expected to function as an endogenous anesthetic, mediating the near comatose state that characterizes anoxic turtles. The crucian carp brain shows a more moderate and more variable increase in extracellular GABA, on average showing a 50% increase after 6 hours of anoxia at 10 °C (Hylland and Nilsson, 1999). Thus, for the anoxic

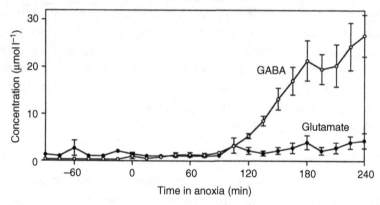

Fig. 9.10 Changes in the extracellular levels of the inhibitory neurotransmitter GABA and the excitatory neurotransmitter glutamate in the brain of turtles exposed to anoxia. Whereas there is an 80-fold rise in the GABA level, no significant change is seen in that of glutamate. The neurotransmitter levels were monitored with microdialysis. Data from Nilsson and Lutz, 1991.

crucian carp, GABA may play a sedative, rather than an anesthetic, role. It is interesting to note that anesthesia has long been used to counteract the deleterious effects of brain hypoxia or brain trauma in humans. The rise in GABA levels in anoxic turtle and crucian carp brains indicates that there is an evolutionary precedent for such treatments.

In turtles exposed to 24 h anoxia at room temperature, the rise in GABA is accompanied by an increase in the number of $GABA_A$ receptors, which may further increase the inhibitory action of GABA (Lutz and Leone-Kabler, 1995). By contrast, a study of the gene expression of numerous components of GABAergic neurotransmission in crucian carp (Ellefsen *et al.*, 2009) revealed a slight fall in mRNA levels of $GABA_A$ receptor subunits after 1–7 days of anoxia at 8°C, again suggesting a more modest GABAergic inhibition in the anoxic crucian carp brain than in the turtle brain. Still, blocking either GABA receptors or GABA synthesis in crucian carp makes the fish release up to three times more ethanol to the water during anoxia, suggesting a significant role for GABA in initiating whole-body metabolic depression in anoxic crucian carp (Nilsson, 1992).

When measuring extracellular GABA levels in crucian carp brain using microdialysis, Hylland and Nilsson (1999) found that running a high-[K^+] saline solution through the microdialysis probe (which forces cells surrounding the probe to depolarize) causes the extracellular GABA level to rise 14 times, while that of glutamate is only doubled (Hylland and Nilsson, 1999). Thus, the potential for GABA release in the crucian carp brain appears to be much higher than for glutamate. Moreover, when the crucian carp brain is forced into energy deficiency by blocking glycolysis with iodoacetate (causing neural ATP levels to plummet), it releases GABA faster and more massively (a tenfold increase after 30 minutes) than glutamate (a threefold increase after 2 hours) (Hylland and Nilsson, 1999). Thus, the crucian carp brain may have a second line of defense when faced with energy deficiency: an 'emergency brake' in the form of a major GABA release that strongly suppresses neural activity and allows ATP levels to be restored.

The mechanisms responsible for the elevated extracellular GABA levels seen in anoxia-tolerant vertebrates during anoxia are not well understood, which is not surprising when we deal with such a complex organ as the brain. Two contributing mechanisms have been suggested. One involves a suppression of GABA reuptake from the extracellular space: a study of anoxia-induced changes in gene expression recently found that the mRNA levels of transporter proteins in the GAT2 family, which are responsible for a major part of GABA reuptake, fall by ~75% during anoxia (Ellefsen et al., 2009). A second factor that may play a role in elevating extracellular GABA levels, at least in the long term, is the intimate metabolic interrelation between GABA and glutamate. GABA is synthesized directly from glutamate by glutamate decarboxylase, a reaction that is independent of oxygen. By contrast, both the synthesis of glutamate and the breakdown of GABA are linked to oxygen-dependent metabolic processes. As a result of this metabolic arrangement, which is common to all animals, the concentration of GABA rises in anoxic tissues, whereas that of glutamate falls, and the rate of change is dependent on the size and turnover of the GABA and glutamate pools (Siesjö, 1978; Nilsson and Lutz, 1993). For example, in crucian carp exposed to 17 days of anoxia at 8°C, there is a fivefold increase in the whole brain content of GABA and a corresponding fall in that of glutamate (Nilsson, 1990a). It is possible that these long-term changes in tissue levels will be reflected in similar changes in extracellular levels.

When considering the metabolic interrelation of GABA and glutamate, it is interesting to note that GABA is the major inhibitory neurotransmitter and glutamate the major excitatory neurotransmitter, not only in all vertebrates, but also in many invertebrates, including primitive groups such as the flatworms (Gerschenfeld, 1973; Usherwood, 1978; Koopowitz and Keenan, 1982; McGeer and McGeer, 1989; Restifo and White, 1990). Thus, the opposing roles of GABA and glutamate appear to have been fixed early in evolution and then subsequently maintained. Nilsson and Lutz (1993) suggested that the underlying selection pressure has been hypoxia, and the advantage of the arrangement is that the inhibitory neurotransmitter level automatically rises, and the excitatory one falls, in hypoxia, providing a mechanism for initiating and maintaining hypoxic metabolic depression.

9.6   Conclusions

In this chapter we have focused on the best-studied examples of anoxia tolerance among vertebrates: North American freshwater turtles in the genera *Trachemys* and *Chrysemys*, and Eurasian cyprinid fishes in the genus *Carassius* (the crucian carp and the goldfish). The anoxic survival strategies utilized by the turtles on the one hand, and the fishes on the other, show both similarities and differences (Fig. 9.11). They have all evolved anoxia tolerance in response to overwintering in ice-covered anoxic freshwater habitats. By matching ATP use with glycolytic ATP production, they both defend their brain ATP levels during anoxia. This allows them to maintain ion homeostasis, thereby avoiding nerve-cell depolarization, which in anoxia-sensitive animals such as mammals triggers a cascade of disastrous events.

However, the turtles and *Carassius* differ considerably in the way by which energy status is defended. The turtles strongly depress both brain and heart activity: the heart of cold, anoxic turtles may beat only once per minute. The neural depression in turtles is initially achieved through adenosine release and then maintained through a massive release of the inhibitory neurotransmitter GABA, combined with a downregulation of ion-channel conductances, including the NMDA receptor that in mammals is responsible for much of the unwanted ion fluxes during anoxia. As a result, the turtle essentially anesthetizes itself and becomes virtually comatose during anoxia.

By contrast, the crucian carp remains active in anoxia, although at a reduced level. With exception of the NMDA receptor, crucian carp and goldfish do not appear to suppress ion conductances or release massive amounts of GABA. Instead, they upregulate glycolysis and downregulate selected neural functions such as vision and hearing, senses that are probably of little importance during the dark, anoxic winter. A moderate, regulated increase in GABAergic activity appears to be responsible for depressing energy use in the crucian carp brain to a level that is still compatible with some physical activity.

The ability of the genus *Carassius* to produce ethanol as the major anaerobic end product, thereby avoiding the enormous lactic acid load that turtles have to deal with, is probably the single most important factor allowing these fishes to maintain activity in anoxia. Because turtles lack the ability to produce alternative anaerobic end products during anoxia, they have to reduce their metabolism to a minimum and boost their blood-buffering capacity to accommodate the rising lactic acid levels by releasing carbonate from bones and shell.

As pointed out at the beginning of this chapter, the turtles and the crucian carp prove that evolution has repeatedly solved the problem of long-term anoxic survival, whereas the attempts of biomedical science to do the same have been

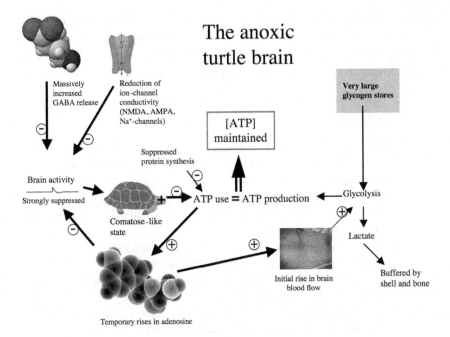

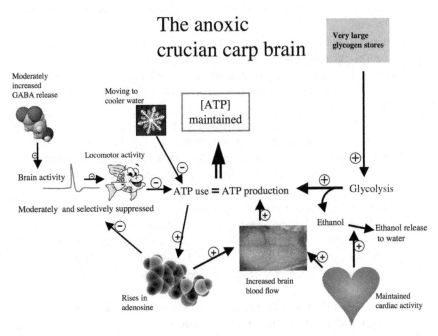

Fig. 9.11 Summary of the major mechanisms believed to allow the brains of freshwater turtles and crucian carp to survive without oxygen.

hampered by disappointments, and progress has been slow. Clearly, studying anoxia tolerance in these animals provides unique opportunities for finding adaptive physiological, biochemical, and molecular mechanisms that allow survival without oxygen for extended periods.

References

Bickler, P. E. (1992). Cerebral anoxia tolerance in turtles, regulation of intracellular calcium and pH. *Am. J. Physiol. Regul. Integr. Comp. Physiol.*, **263**, R1298–302.

Bickler, P. E. (1998). Reduction in NMDA receptor activity in cerebrocortex of turtles (*Chrysemys picta*) during 6 wk of anoxia. *Am. J. Physiol. Regul. Integr. Comp. Physiol.*, **275**, R86–91.

Bickler, P. E. (2004). Clinical perspectives, neuroprotection lessons from hypoxia-tolerant organisms. *J. Exp. Biol.*, **207**, 3243–9.

Bickler, P. E. and Donohoe, P. H. (2002). Adaptive responses of vertebrate neurons to hypoxia. *J. Exp. Biol.*, **205**, 3579–86.

Bickler, P. E. and Hansen, B. M. (1998). Hypoxia-intorerant neonatal CA1 neurons: relationship of survial to evoked glutamate release and glutamate receptor-mediated calcium changes in hippocampal slices. *Dev. Brain. Res.*, **106**, 57–69.

Bickler, P. E., Donohoe, P. H. and Buck, L. T. (2000). Hypoxia-induced silencing of NMDA receptors in turtle neurons. *J. Neurosci.*, **20**, 3522–8.

Buck, L. T. and Bickler, P. E. (1995). Role of adenosine in NMDA receptor modulation in the cerebral-cortex of an anoxia-tolerant tutle (*Chrysemys picta belli*). *J. Exp. Biol.*, **198**, 1621–8.

Buck, L. T. and Bickler, P. E. (1998). Adenosine and anoxia reduce N-methyl-D-aspartate receptor open probability in turtle cerebrocortex. *J. Exp. Biol.*, **201**, 289–97.

Chen, J., Zhu J. X., Wilson, I. and Cameron, J. S. (2005). Cardioprotective effects of KATP channel activation during hypoxia in goldfish *Carassius auratus*. *J. Exp. Biol.* **208**, 2765–72.

Congleton, J. L. (1974). The respiratory response of asphyxia of *Typhlogobius californiensis* (Teleostei, Gobiidae) and some related gobies. *Biol. Bull.*, **146**, 186–205.

Ellefsen, S., Sandvik, G. K., Larsen, H. K., Stensløkken, K. O., Hov, D. A. S., Kristensen, T. A. and Nilsson, G. E. (2008). Expression of genes involved in excitatory neurotransmission in anoxic crucian carp (*Carassius carassius*) brain. *Physiol. Genomics*, **35**, 5–17.

Ellefsen, S., Stenslokken, K. O., Fagernes, C. E., Kristensen, T. A. and Nilsson, G. E. (2009). Expression of genes involved in GABAergic neurotransmission in anoxic crucian carp brain (*Carassius carassius*). *Physiol. Genomics*, **36**, 61–8.

Erecinska, M. and Silver, I. A. (1994). Ions and energy in mammalian brain. *Prog. Neurobiol.*, **43**, 37–71.

Fagernes, C., Ellefsen, S., Stenslokken, K. O., Berenbrink, M. and Nilsson, G. (2008). Molecular background to ethanol production in crucian carp (*Carassius carassius*). *Comp. Biochem. Physiol.*, **150**A (Suppl. 1), S112.

Feng, Z.-C., Rosenthal, M. and Sick, T. J (1988a). Suppression of evoked potentials with continued ion transport during anoxia in turtle brain. *Am. J. Physiol. Regul. Integr. Comp. Physiol.*, **255**, R478–84.

Feng, Z.-C., Sick, T. J. and Rosenthal, M. (1988b). Orthodromic field potentials and recurrent inhibition during anoxia in turtle brain. *Am. J. Physiol. Regul. Integr. Comp. Physiol.*, **255**, R485–91.

Fernandes, J. A., Lutz, P. L., Tannenbaum, A., Todorov, A. T., Liebovitch, L. and Vertes, R. (1997). Electroencephalogram activity in the anoxic turtle brain. *Am. J. Physiol. Regul. Integr. Comp. Physiol.*, **273**, R911–19.

Franks, N. P. (2008). General anaesthesia, from molecular targets to neuronal pathways of sleep and arousal. *Nature Rev. Neurosci.*, **9**, 370–86.

Fraser, K. P., Houlihan, D. F., Lutz, P. L., Leone-Kabler, S., Manuel, L. and Brechin, J. G (2001). Complete suppression of protein synthesis during anoxia with no post-anoxia protein synthesis debt in the red-eared slider turtle *Trachemys scripta elegans. J. Exp. Biol.*, **204**, 4353–60.

Gerschenfeld H. M. (1973). Chemical transmission in invertebrate central nervous systems and neuromuscular junctions. *Physiol. Rev.*, **53**, 1–119.

Ghai, H. S. and Buck, L. T. (1999). Acute reduction in whole cell conductance in anoxic turtle brain. *Am. J. Physiol. Regul. Integr. Comp. Physiol.*, **277**, R887–93.

Hicks, J. M. T. and Farrell, A. P. (2000a). The cardiovascular responses of the red eared slider (*Trachemys scripta*) acclimated to either 22 or 5°C. I. Effects of anoxia exposure on in vivo cardiac performance. *J. Exp. Biol.*, **203**, 3765–74.

Hicks, J. M. T. and Farrell, A. P. (2000b). The cardiovascular responses of the red eared slider (*Trachemys scripta*). acclimated to either 22 or 5°C. II. Effects of anoxia on adrenergic and cholinergic control. *J. Exp. Biol.*, **203**, 3775–84.

Hochachka, P. W. (1986). Defense strategies against hypoxia and hypothermia. *Science*, **231**, 234–41.

Hochachka, P. W. and Somero, G. N. (2002). *Biochemical Adaptation, Mechanism and Process in Physiological Evolution.* Oxford: Oxford University Press.

Holopainen, I. J. and Hyvärinen, H. (1985). Ecology and physiology of crucian carp (*Carassius carassius* (L.)) in small Finnish ponds with anoxic conditions in winter. *Verh. Internat. Verein. Limnol.*, **22**, 2566–70.

Hylland, P. and Nilsson, G. E. (1999). Extracellular levels of amino acid neurotransmitters during anoxia and forced energy deficiency in crucian carp brain. *Brain Res.*, **823**, 49–58.

Hylland, P., Nilsson, G. E. and Lutz, P. L (1994). Time course of anoxia induced increase in cerebral blood flow rate in turtles: evidence for a role of adenosine. *J. Cereb. Blood Flow Metab.*, **14**, 877–81.

Hylland, P., Nilsson, G. E. and Johansson, D. (1995). Anoxic brain failure in an ectothermic vertebrate, release of amino acids and K^+ in rainbow trout thalamus. *Am. J. Physiol. Regul. Integr. Comp. Physiol.*, **269**, R1077–84.

Hyvärinen, H., Holopainen, I. J. and Piironen, J. (1985). Anaerobic wintering of crucian carp (*Carassius carassius* L.). I. Annual dynamics of glycogen reserves in nature. *Comp. Biochem. Physiol.*, **82A**, 797–803.

Jackson, D. C. (1968). Metabolic depression and oxygen depletion in the diving turtle. *J. Appl. Physiol.*, **24**, 503–9.

Jackson, D. C. (2002). Hibernation without oxygen, physiological adaptations in the painted turtle. *J. Physiol.*, **543**, 731–7.

Jackson, D. C., Taylor, S. E., Asare, V. S., Villarnovo, D., Gall, J. M. and Reese, S. A. (2007.). Comparative shell buffering properties correlate with anoxia tolerance in freshwater turtles. *Am. J. Physiol. Regul. Integr. Comp. Physiol.*, **292**, R1008–15.

Jibb, L. A. and Richards, J. G. (2008). AMP-activated protein kinase activity during metabolic rate depression in the hypoxic goldfish, *Carassius auratus. J. Exp. Biol.*, **211**, 3111–22.

Johansson, D. and Nilsson, G. E. (1995). Roles of energy status, K_{ATP} channels, and channel arrest in fish brain K^+ gradient dissipation during anoxia. *J. Exp. Biol.*, **198**, 2575–80.

Johansson, D., Nilsson, G. E. and Døving, K. B. (1997). Anoxic depression of light-evoked potentials in retina and optic tectum of crucian carp. *Neurosci. Lett.*, **237**, 73–6.

Johansson, D., Nilsson, G. E. and Törnblom, E. (1995). Effects of anoxia on energy metabolism in crucian carp brain slices studied with microcalorimetry. *J. Exp. Biol.*, **198**, 853–9.

Johnson, E. M. Jr, Koike, T. and Franklin, J. (1992). A 'calcium set-point hypothesis' of neuronal dependence on neurotrophic factor. *Exp. Neurol.*, **115**, 163–6.

Johnston, I. A. and Bernard L. M. (1983). Utilization of the ethanol pathway in carp following exposure to anoxia. *J. Exp. Biol.*, **104**, 73–8.

Koopowitz, H. and Keenan, L. (1982). The primitive brain of platyhelminthes. *Trends Neurosci.*, **5**, 77–9.

Krumschnabel, G., Biasi, C. and Wieser, W. (2000). Action of adenosine on energetics, protein synthesis and K^+ homeostasis in teleost hepatocytes. *J. Exp. Biol.*, **203**, 2657–65.

Lipton, P. (1999). Ischemic cell death in brain neurons. *Physiol. Rev.*, **79**, 1431–568.

Lutz, P. L. and Leone-Kabler, S. A. (1995). Upregulation of GABAA receptor during anoxia in the turtle brain. *Am. J. Physiol. Regul. Integr. Comp. Physiol.*, **268**, R1332–5.

Lutz, P. L. and Nilsson, G. E. (1997). Contrasting strategies for anoxic brain survival: glycolysis up or down. *J. Exp. Biol.*, **200**, 411–9.

Lutz, P. L., McMahon, P., Rosenthal, M. and Sick, T. J (1984). Relationships between aerobic and anaerobic energy production in turtle brain in situ. *Am. J. Physiol. Regul. Integr. Comp. Physiol.*, **247**, R740–4.

Lutz, P. L., Nilsson, G. E. and Prentice, H. (2003). *The Brain Without Oxygen*, 3rd edn. Dordrecht: Kluwer Academic Publishers.

Lutz, P. L., Rosenthal, M. and Sick, T. (1985). Living without oxygen, turtle brain as a model of anaerobic metabolism. *Mol. Physiol.*, **8**, 411–25.

McGeer, P. L. and McGeer, E. G. (1989). Amino acid neurotransmitters. In *Basic Neurochemistry*, ed. G. J. Siegel, B. Agranoff and R. W. Alberts. New York: Raven Press, pp. 311–32.

Mendelsohn, B. A., Kassebaum B. L. and Gitlin J. D. (2008). The zebrafish embryo as a dynamic model of anoxia tolerance. *Dev. Dyn.*, **237**, 1780–8.

Milton, S. L, and Lutz, P. L. (1998). Low extracellular dopamine levels are maintained in the anoxic turtle brain. *J. Cereb. Blood Flow Metab.* **18**, 803–7.

Milton, S. L., Dirk, L. J., Kara, L. F. and Prentice, H. M (2008). Adenosine modulates ERK1/2, PI3K/Akt, and p38MAPK activation in the brain of the anoxia-tolerant turtle *Trachemys scripta. J. Cereb. Blood Flow. Metab.* **28**, 1469–77.

Milton, S. L., Thompson, J. W. and Lutz, P. L. (2002). Mechanisms for maintaining extracellular glutamate in the anoxic turtle striatum. *Am. J. Physiol. Regul. Integr. Comp. Physiol.*, **282**, R1317–23.

Mourik, J., Raeven, P., Steur, K. and Addink, A. D. F. (1982). Anaerobic metabolism of red skeletal muscle of goldfish, *Carassius auratus* (L.). *FEBS Lett* **137**, 111–14.

Nilsson, G. E. (1988). A comparative study of aldehyde dehydrogenase and alcohol dehydrogenase activity in crucian carp and three other vertebrates, apparent adaptations to ethanol production. *J. Comp. Physiol. B* **158**, 479–85.

Nilsson, G. E. (1990a). Long term anoxia in crucian carp, changes in the levels of amino acid and monoamine neurotransmitters in the brain, catecholamines in chromaffin tissue, and liver glycogen. *J. Exp. Biol.*, **150**, 295–320.

Nilsson, G. E. (1990b). Distribution of aldehyde dehydrogenase and alcohol dehydrogenase in summer acclimatized crucian carp (*Carassius carassius* L.). *J. Fish Biol.* **36**, 175–9.

Nilsson, G. E. (1991). The adenosine receptor blocker aminophylline increases anoxic ethanol production in crucian carp. *Am. J. Physiol. Regul. Integr. Comp. Physiol.*, **261**, R1057–60.

Nilsson, G. E. (1992). Evidence for a role of GABA in metabolic depression during anoxia in crucian carp (*Carassius carassius* L.). *J. Exp. Biol.*, **164**, 243–59.

Nilsson, G. E. (2001). Surviving anoxia with the brain turned on. *News Physiol. Sci.* **16**, 218–21.

Nilsson, G. E. and Lutz, P. L. (1991). Release of inhibitory neurotransmitters in response to anoxia in turtle brain. *Am. J. Physiol. Regul. Integr. Comp. Physiol.*, **261**, R32–7.

Nilsson, G. E. and Lutz, P. L. (1992). Adenosine release in the anoxic turtle brain, a possible mechanism for anoxic survival. *J. Exp. Biol.*, **162**, 345–51.

Nilsson, G. E. and Lutz, P. L. (1993). Role of GABA in hypoxia tolerance, metabolic depression and hibernation – possible links to neurotransmitter evolution. *Comp. Biochem. Physiol.*, **105C**, 329–36.

Nilsson, G. E. and Lutz. P. L. (2004). Anoxia tolerant brains. *J. Cereb. Blood Flow Metab.*, **24**, 475–86.

Nilsson, G. E. and Renshaw, G. M. C. (2004). Hypoxic survival strategies in two fishes: extreme anoxia tolerance in the North European crucian carp and natural hypoxic preconditioning in a coral-reef shark. *J. Exp. Biol.*, **207**, 3131–9.

Nilsson, G. E., Hylland, P. and Löfman, C. O. (1994). Anoxia and adenosine induce increased cerebral blood flow in crucian carp. *Am. J. Physiol. Regul. Integr. Comp. Physiol.*, **267**, R590–5.

Nilsson, G. E., Pérez-Pinzón, M., Dimberg, K. and Winberg, S. (1993b). Brain sensitivity to anoxia in fish as reflected by changes in extracellular potassium-ion activity. *Am. J. Physiol. Regul. Integr. Comp. Physiol.*, **264**, R250–3.

Nilsson, G. E., Rosén, P. and Johansson, D. (1993a). Anoxic depression of spontaneous locomotor activity in crucian carp quantified by a computerized imaging technique. *J. Exp. Biol.*, **180**, 153–63.

Paajanen, V. and Vornanen, M. (2004). Regulation of action potential duration under acute heat stress by by I_{KATP} and I_{K1} in fish cardiac myocytes. *Am. J. Physiol. Regul. Integr. Comp. Physiol.*, **286**, R405–15.

Pamenter, M. E., Shin, D. S. and Buck, L. T. (2008b). Adenosine A1 receptor activation mediates NMDA receptor activity in a pertussis toxin-sensitive manner during normoxia but not anoxia in turtle cortical neurons. *Brain Res.*, **1213**, 27–34.

Pamenter, M. E., Shin, D. S. and Buck, L. T. (2008c). AMPA receptors undergo channel arrest in the anoxic turtle cortex. *Am. J. Physiol. Regul. Integr. Comp. Physiol.*, **294**, R606–13.

Pamenter, M. E., Shin, D. S., Cooray, M. and Buck, L. T. (2008a). Mitochondrial ATP-sensitive K^+ channels regulate NMDAR activity in the cortex of the anoxic western painted turtle. *J. Physiol.*, **586**, 1043–58.

Pek, M, and Lutz, P. L. (1997). Role for adenosine in 'channel arrest' in the anoxic turtle brain. *J. Exp. Biol.*, **200**, 1913–17.

Pek, M. and Lutz, P. L. (1998). K^+_{ATP} channel activation provides transient protection in anoxic turtle brain. *Am. J. Physiol. Regul. Integr. Comp. Physiol.*, **275**, R2023–7.

Pérez-Pinzón, M. A., Lutz, P. L., Sick, T., Rosenthal, M. (1993). Adenosine, a 'retaliatory' metabolite, promotes anoxia tolerance in turtle brain. *J. Cereb. Blood Flow Metab.* **13**, 728–32.

Peréz-Pinzón, M. A., Rosenthal, M., Sick, T. J., Lutz, P. L., Pablo, J. and Mash, D. (1992). Down-regulation of sodium channels during anoxia, a putative survival strategy of turtle brain. *Am. J. Physiol. Regul. Integr. Comp. Physiol.*, **262**, R712–15.

Podrabsky, J. E., Lopez, J. P., Fan, T. W. M. Higashi, R. and Somero, G. N (2007). Extreme anoxia tolerance in embryos of the annual killifish *Austrofundulus limnaeus*, insights from a metabolomics analysis. *J. Exp. Biol.*, **210**, 2253–66.

Rausch, R. N., Crawshaw, L. I. and Wallace, H. L (2000). Effects of hypoxia, anoxia, and endogenous ethanol on thermoregulation in goldfish, *Carassius auratus*. *Am. J. Physiol. Regul. Integr. Comp. Physiol.*, **278**, R545–55.

Restifo, L. L. and White, K. (1990). Molecular and genetic approaches to neurotransmitter and neuromodulator systems in *Drosophila*. In *Advances in Insect Physiology, Vol. 22*, ed. P. D. Evans and V. B. Wigglesworth. London: Academic Press, pp. 115–219.

Rosati, A. M., Traversa, U., Lucchi, R. and Poli, A. (1995). Biochemical and pharmacological evidence for the presence of A1 but not A2a adenosine receptors in the brain of the low vertebrate teleost *Carassius auratus* (goldfish). *Neurochem. Int.*, **26**, 411–23.

Shoubridge, E. A. and Hochachka, P. W. (1980). Ethanol, novel endproduct in vertebrate anaerobic metabolism. *Science* **209**, 308–9.

Sick, T. J., Pérez-Pinzón, M., Lutz, P. L. and Rosenthal, M. (1993). Maintaining coupled metabolism and membrane function in anoxic brain, a comparison between the turtle and rat. In *Surviving Hypoxia, Mechanisms of Control and Adaptation*, ed. P. W. Hochachka, P. L. Lutz, T. Sick, M. Rosenthal and G. van den Thillart. Boca Raton: CRC Press, pp. 351–63.

Sick, T. J., Rosenthal, M., LaManna, J. C. and Lutz, P. L. (1982). Brain potassium homeostasis, anoxia, and metabolic inhibition in turtles and rats. *Am. J. Physiol. Regul. Integr. Comp. Physiol.*, **243**, R281–8.

Siesjö, B. K. (1978). *Brain Energy Metabolism*, Chichester: Wiley.

Smith, R. W., Houlihan, D. F., Nilsson, G. E. and Brechin, J. G. (1996). Tissue specific changes in protein synthesis rates in vivo during anoxia in crucian carp. *Am. J. Physiol. Regul. Integr. Comp. Physiol.*, **271**, R897–904.

Stecyk, J. A. W., Overgaard, J., Farrell, A. P. and Wang, T. (2004a). Alpha-adrenergic regulation of systemic peripheral resistance and blood flow distribution in the turtle *Trachemys scripta* during anoxic submergence at 5°C and 21°C. *J. Exp. Biol.*, **207**, 269–83.

Stecyk, J. A. W., Stensløkken, K.-O., Farrell, A. P. and Nilsson, G. E. (2004b). Maintained cardiac pumping in anoxic crucian carp. *Science*, **306**, 77.

Stensløkken, K.-O., Ellefsen, S., Stecyk, J. A. W., Dahl, M. B. Nilsson, G. E. and Vaage, J. (2008b). Differential regulation of AMP-activated kinase and AKT kinase in response to oxygen availability in crucian carp (*Carassius carassius*). *Am. J. Physiol. Regul. Integr. Comp. Physiol.*, **295**, R1803–14.

Stensløkken K.-O., Milton, S. L., Lutz, P. L., *et al.* (2008a). Effect of anoxia on the electroretinogram of three anoxia-tolerant vertebrates. *Comp. Biochem. Physiol.* **150A**, 395–403.

Storey, K. B. (1996). Metabolic adaptations supporting anoxia tolerance in reptiles, recent advances. *Comp. Biochem. Physiol.*, **113B**, 23–35.

Suzue, T, Wu, G.-B. and Furukawa, T. (1987). High susceptibility to hypoxia of afferent synaptic transmission in the goldfish sacculus. *J. Neurophysiol.*, **58**, 1066–79.

Ultsch, G. R. (1985). The viability of nearctic freshwater turtles submerged in anoxia and normoxia at 3 and 10°C. *Comp. Biochem. Physiol.*, **81A**, 607–11.

Ultsch, G. R. (1989). Ecology and physiology of hibernation and overwintering among freshwater fishes, turtles and snakes. *Biol. Rev.*, **64**, 435–516.

Ultsch, G. R. and Jackson, D. C. (1982). Long-term submergence at 3°C of the turtle *Chrysemys picta belli* in normoxic and severely hypoxic water I. Survival, gas exchange and acid-base status. *J. Exp. Biol.*, **96**, 11–28.

Usherwood, P. N. R. (1978). Amino acids as neurotransmitters. *Adv. Comp. Physiol. Biochem.*, **7**, 227–309.

Van Raaij, M. T. M., Breukel B.-J., van den Thillart, G. E. E. J. M. and Addink, A. D. F. (1994). Lipid metabolism of goldfish, *Carassius auratus* (L.) during normoxia and anoxia. Indications for fatty acid chain elongation. *Comp. Biochem. Physiol.*, **107**B, 75–84.

Van Waarde, A., van den Thillart, G. and Verhagen, M. (1993). Ethanol formation and pH regulation in fish. In *Surviving Hypoxia, Mechanisms of Control and Adaptation*, ed. P. W. Hochachka, P. L. Lutz, T. Sick, M. Rosenthal and G. van den Thillart. Boca Raton: CRC Press, pp. 157–70.

Van Waversveld, J., Addink, A. D. F. and van den Thillart, G. (1989). Simultaneous direct and indirect calorimetry on normoxic and anoxic goldfish. *J. Exp. Biol.*, **142**, 325–35.

Vornanen, M. (1994). Seasonal adaptation of crucian carp (*Carassius carassius* L.). heart, glycogen stores and lactate dehydrogenase activity. *Can. J. Zool.*, **72**, 433–42.

Vornanen, M. and Paajanen, V. (2004). Seasonality of dihydropyridine receptor binding in the heart of an anoxia-tolerant vertebrate, the crucian carp (*Carassius carassius* L.). *Am. J. Physiol. Regul. Integr. Comp. Physiol.*, **287**, R1263–9.

Vornanen, M. and Paajanen V. (2006). Seasonal changes in glycogen content and Na^+-K^+-ATPase activity in the brain of crucian carp. *Am. J. Physiol. Regul. Integr. Comp. Physiol.*, **291**, R1482–9.

Vornanen, M., Stecyk, J. A. W. and Nilsson, G. E. (2009). The anoxia-tolerant crucian carp (*Carassius carassius* L.). In *Fish Physiology, Vol. 27, Hypoxia*. ed. J. G. Richards, A. P. Farrell and C. Brauner. Amsterdam: Elesevier/Academic Press.

Warren, D. E., Reese, S. A. and Jackson, D. C (2006). Tissue glycogen and extracellular buffering limit the survival of red-eared slider turtles during anoxic submergence at 3°C. *Physiol. Biochem. Zool.* **79**, 736–44.

Wilkie, M. P., Pamenter, M. E., Alkabie, S., Carapic, D., Shin, D. S. and Buck, L. T. (2008). Evidence of anoxia-induced channel arrest in the brain of the goldfish (*Carassius auratus*). *Comp. Biochem. Physiol.*, **148C**, 355–62.

Wood, S. C., Dupre, R. K. and Hicks, J. W. (1985). Voluntary hypothermia in hypoxic animals. *Acta. Physiol. Scand.* **124** (Suppl. 542), 46.

Zipfel, G. J., Babcock, D. J., Lee, J. M. and Choi, D. W. (2000). Neuronal apoptosis after CNS injury, the roles of glutamate and calcium. *J. Neurotrauma*, **17**, 857–69.

Index

Printed in the United States
by Baker & Taylor Publisher Services